Methods in Molecular Biology

Series Editor
John M. Walker
School of Life and Medical Sciences
University of Hertfordshire
Hatfield, Hertfordshire, UK

For further volumes:
http://www.springer.com/series/7651

For over 35 years, biological scientists have come to rely on the research protocols and methodologies in the critically acclaimed *Methods in Molecular Biology* series. The series was the first to introduce the step-by-step protocols approach that has become the standard in all biomedical protocol publishing. Each protocol is provided in readily-reproducible step-by-step fashion, opening with an introductory overview, a list of the materials and reagents needed to complete the experiment, and followed by a detailed procedure that is supported with a helpful notes section offering tips and tricks of the trade as well as troubleshooting advice. These hallmark features were introduced by series editor Dr. John Walker and constitute the key ingredient in each and every volume of the *Methods in Molecular Biology* series. Tested and trusted, comprehensive and reliable, all protocols from the series are indexed in PubMed.

Docking Screens for Drug Discovery

Edited by

Walter Filgueira de Azevedo Jr.

Escola de Ciências da Saúde, Pontifícia Universidade Católica do Rio Grande do Sul—PUCRS, Porto Alegre, Ria Grande do Sul, Brazil

Humana Press

Editor
Walter Filgueira de Azevedo Jr.
Escola de Ciências da Saúde
Pontifícia Universidade Católica do Rio Grande do Sul—PUCRS
Porto Alegre, Ria Grande do Sul, Brazil

ISSN 1064-3745 ISSN 1940-6029 (electronic)
Methods in Molecular Biology
ISBN 978-1-4939-9751-0 ISBN 978-1-4939-9752-7 (eBook)
https://doi.org/10.1007/978-1-4939-9752-7

This Humana imprint is published by the registered company Springer Science+Business Media, LLC, part of Springer Nature.
The registered company address is: 233 Spring Street, New York, NY 10013, U.S.A.

Dedication

This book is dedicated to my beloved mother Marion de Fátima Pereira de Azevedo and my darling wife Maria do Carmo Dantas de Santana Azevedo.

Preface

The data explosion in the number of biological macromolecules deposited in the Protein Data Bank (PDB) [1–3] opened the possibility to investigate the correlation of these experimentally determined structures with biological information, which is a favorable scenario for the application of computational systems biology approaches to develop a mathematical model to predict ligand-binding affinity for this target protein. It is also possible to use these three-dimensional structures to study target proteins employed in the development and design of drugs [4–10]. The use of structural information for a target protein makes it possible to apply virtual screening methodology to identify new hits and guide the future development of new medicines. The primary approach to investigate potential new hits for a target protein is the methodology of protein-ligand docking simulation [11].

Docking is a simulation method that predicts the structure of a receptor-ligand complex, in which the receptor is a protein and the ligand is a small molecule [12–16]. This simulation is equivalent to the key-lock theory of enzyme specificity [17, 18], in which the lock is the receptor and the key is the ligand. The goal in any protein-ligand docking simulation is to adjust the position of the key (ligand) in the lock (ligand-binding pocket in a protein). From the computational view, we see the protein-ligand docking as an optimization problem, where our goal is to find the best solution (right position for the ligand) from a set of possible locations. Protein-ligand docking often makes use of one or more of the following computational methodologies: genetic algorithm, differential evolution, Lamarckian genetic algorithm, fast shape matching, incremental construction, distance geometry, simulated annealing, and others [19]. Protein-ligand docking methodology can produce several positions for the key in the lock. Therefore, we need a scoring function that will allow evaluations of all possible positions of the key, and then a selection can be carried out for the best location. For general reviews of the principles underlying molecular docking programs, see references [12–16].

Also, to evaluate the ligand-binding affinity for a specific target protein, we can employ a scoring function to compute scores that resemble ligand-binding energy functions. For both approaches, experimental information is vital to validate protein-ligand docking simulations and the ability of scoring functions to estimate ligand-binding affinity [20].

For protein-ligand docking simulations, it is common to start investigating if the computational approach is capable of reproducing an experimental 3D structure for a complex involving a protein and at least one ligand. If such structure is available, we employ it to check whether a specific molecular docking protocol is capable of predicting the crystallographic position for the ligand in the protein structure, a procedure called redocking. The most used criteria to evaluate redocking success are the root-mean-square deviation (RMSD) between the crystallographic position for the ligand and the pose (generated by the computer simulation). In docking simulations, we expect that the best results generate RMSD values less than 2.0 Å compared with crystallographic structures [12–16].

Furthermore, if we have more than one structure complexed with a ligand, we can take the validation process further, applying the molecular docking protocol to an ensemble of complexes structures. In this ensemble, we could have the same protein structure in complex with different ligands. For instance, a search in the PDB for structures containing the name cyclin-dependent kinases (CDKs) and for which there is inhibition constant (K_i) information returned 31 structures. These structures have water molecules close to the active ligand and without repeated ligands (search carried out on March 20, 2019). This data set is an ensemble of CDK structures, where each entry is a structure complexed with a different ligand. This ensemble of structures can be employed to validate a docking strategy for a specific protein target. Moreover, it could also be used to test scoring functions.

For validation of scoring functions, it is common to investigate the correlation between the experimental binding affinity with scoring functions. Here we evaluated the predictive performance using squared Pearson's (R^2) or Spearman's (ρ) correlation coefficients [21]. Application of machine learning methods can improve the predictive performance of scoring functions trained against data sets composed of experimentally determined structures for which ligand-binding data is available [22–32].

The focus of the present book is on recent developments in docking simulations for target proteins. We have chapters dealing with specific techniques or applications for docking simulations. For instance, we describe the major docking programs. Also, we explain the scoring functions developed for the analysis of docking results and to predict ligand-binding affinity. Due to the importance of docking simulations for the initial stages of drug discovery, we believe that the present volume will appeal to those interested in molecular docking simulation and also in the application of these methodologies for drug discovery.

Finally, I would like to express my gratitude to all authors who accepted the challenge of bringing to a book their scientific knowledge. I want to thank Prof. John M. Walker (series editor for the *Methods in Molecular Biology* series) for his patience and assistance during the editorial process. This book wouldn't be possible without the aid of Anna Rakovsky (Assistant Editor at Springer Science + Business Media, LLC). Many others contributed directly or indirectly to this book. I want to thank all my students who tested the tutorials and protocols described here. They did a great job of helping to improve the quality of the material described in this work. This book is a dream coming true, and it wouldn't be possible without the comprehension and love of my wife Carminha (Maria do Carmo Dantas de Santana Azevedo) who understood my absence and helped me during the months of preparation of this book. To her: "Obrigado minha linda. Este livro é para você. Te amo muito."

Porto Alegre, RS, Brazil ***Walter Filgueira de Azevedo Jr.***

References

1. Berman HM, Westbrook J, Feng Z, Gilliland G, Bhat TN, Weissig H et al (2000) The protein data bank. Nucleic Acids Res 28(1):235–242
2. Berman HM, Battistuz T, Bhat TN, Bluhm WF, Bourne PE, Burkhardt K et al (2002) The protein data bank. Acta Crystallogr D Biol Crystallogr 58(Pt 6 No 1):899–907
3. Westbrook J, Feng Z, Chen L, Yang H, Berman HM (2003) The protein data bank and structural genomics. Nucleic Acids Res 31(1):489–491
4. de Azevedo WF Jr (2010) Structure-based virtual screening. Curr Drug Targets 11:261–263
5. de Ávila MB, Bitencourt-Ferreira G, de Azevedo Jr. WF (2019) Structural basis for inhibition of enoyl-[acyl carrier protein] reductase (InhA) from mycobacterium tuberculosis. Curr Med Chem doi: 10.2174/0929867326666181203125229

6. Volkart PA, Bitencourt-Ferreira G, Souto AA, de Azevedo WF (2019) Cyclin-dependent kinase 2 in cellular senescence and cancer. A structural and functional review. Curr Drug Targets doi: 10.2174/1389450120666181204165344
7. Canduri F, Fadel V, Basso LA, Palma MS, Santos DS, de Azevedo WF Jr (2005) New catalytic mechanism for human purine nucleoside phosphorylase. Biochem Biophys Res Commun. 327 (3):646–649
8. Canduri F, Teodoro LG, Fadel V, Lorenzi CC, Hial V, Gomes RA et al (2001) Structure of human uropepsin at 2.45 A resolution. Acta Crystallogr D Biol Crystallogr 57(Pt 11): 1560–1570
9. de Azevedo WF Jr, Dias R (2008) Experimental approaches to evaluate the thermodynamics of protein-drug interactions. Curr Drug Targets 9(12):1071–1076
10. Delatorre P, Rocha BA, Souza EP, Oliveira TM, Bezerra GA, Moreno FB et al (2007) Structure of a lectin from Canavalia gladiata seeds: new structural insights for old molecules. BMC Struct Biol 7:52
11. Gschwend DA, Good AC, Kuntz ID (1996) Molecular docking towards drug discovery. J Mol Recognit 9:175–186
12. Azevedo LS, Moraes FP, Xavier MM, Pantoja EO, Villavicencio B, Finck JA et al (2012) Recent progress of molecular docking simulations applied to development of drugs. Curr Bioinform 7:352–365
13. DesJarlais RL, Dixon JS (1994) A shape- and chemistry-based docking method and its use in the design of HIV-1 protease inhibitors. J Comput Aided Mol Des 8:231–242
14. de Azevedo WF Jr (2010) Structure-based virtual screening. Curr Drug Targets 11:261–263
15. de Azevedo WF Jr (2010) MolDock applied to structure-based virtual screening. Curr Drug Targets 11:327–334
16. Dias R, de Azevedo WF Jr (2008) Molecular docking algorithms. Curr Drug Targets 9:1040–1047
17. Fischer E (1890) Ueber die optischen Isomeren des Traubezuckers, der Gluconsäure und der Zuckersäure. Ber Dtsch Chem Ges 23:2611–2624
18. Fischer E (1894) Einfluss der Configuration auf die Wirkung der Enzyme. Ber Dtsch Chem Ges 27:2985–2993
19. Heberlé G, de Azevedo WF Jr (2011) Bio-inspired algorithms applied to molecular docking simulations. Curr Med Chem 18:1339–1352
20. de Azevedo WF Jr, Dias R (2008) Computational methods for calculation of ligand-binding affinity. Curr Drug Targets 9:1031–1039
21. Zar JH (1972) Significance testing of the spearman rank correlation coefficient. J Am Stat Assoc 67:578–580
22. Bitencourt-Ferreira G, de Azevedo Jr WF (2018) Development of a machine-learning model to predict Gibbs free energy of binding for protein-ligand complexes. Biophys Chem 240:63–69
23. de Ávila MB, de Azevedo WF Jr (2018) Development of machine learning models to predict inhibition of 3-dehydroquinate dehydratase. Chem Biol Drug Des 92:1468–1474
24. Russo S, de Azevedo WF (2019) Advances in the understanding of the Cannabinoid Receptor 1—focusing on the inverse agonists interactions. Curr Med Chem doi: 10.2174/0929867325666180417165247
25. Amaral MEA, Nery LR, Leite CE, de Azevedo Junior WF, Campos MM (2018) Pre-clinical effects of metformin and aspirin on the cell lines of different breast cancer subtypes. Invest New Drugs 36:782–796
26. Levin NMB, Pintro VO, Bitencourt-Ferreira G, Mattos BB, Silvério AC, de Azevedo Jr WF (2018) Development of CDK-targeted scoring functions for prediction of binding affinity. Biophys Chem 235:1–8
27. Freitas PG, Elias TC, Pinto IA, Costa LT, de Carvalho PVSD, Omote DQ et al (2018) Computational approach to the discovery of phytochemical molecules with therapeutic potential targets to the PKCZ protein. Lett Drug Des Discov 15:488–499
28. Pintro VO, Azevedo WF (2017) Optimized virtual screening workflow. Towards target-based polynomial scoring functions for HIV-1 protease. Comb Chem High Throughput Screen 20:820–827
29. de Ávila MB, Xavier MM, Pintro VO, de Azevedo WF (2017) Supervised machine learning techniques to predict binding affinity. A study for cyclin-dependent kinase 2. Biochem Biophys Res Commun 494:305–310
30. Heck GS, Pintro VO, Pereira RR, de Ávila MB, Levin NMB, de Azevedo WF (2017) Supervised machine learning methods applied to predict ligand-binding affinity. Curr Med Chem 24:2459–2470

31. Levin NM, Pintro VO, de Ávila MB, de Mattos BB, De Azevedo WF Jr (2017) Understanding the structural basis for inhibition of cyclin-dependent kinases. New pieces in the molecular puzzle. Curr Drug Targets 18:1104–1111
32. Xavier MM, Heck GS, de Avila MB, Levin NM, Pintro VO, Carvalho NL et al (2016) SAnDReS a computational tool for statistical analysis of docking results and development of scoring functions. Comb Chem High Throughput Screen 19:801–812

Acknowledgments

This work was supported by grants from CNPq (Brazil) (308883/2014-4). This study was financed in part by the Coordenação de Aperfeiçoamento de Pessoal de Nivel Superior—Brasil (CAPES)—Finance Code 001. WFA is a researcher for CNPq (Brazil) (Process Numbers: 308883/2014-4 and 309029/2018-0).

Contents

Contributors

PEDRO J. BALLESTER • *Cancer Research Center of Marseille, INSERM U1068, Marseille, France; Institut Paoli-Calmettes, Marseille, France; Aix-Marseille Université, Marseille, France; CNRS UMR7258, Marseille, France*

GABRIELA BITENCOURT-FERREIRA • *Escola de Ciências da Saúde, Pontifícia Universidade Católica do Rio Grande do Sul—PUCRS, Porto Alegre, RS, Brazil*

ERNESTO R. CAFFARENA • *Programa de Computação Científica, Fundação Oswaldo Cruz, Rio de Janeiro, RJ, Brazil*

NELSON J. F. DA SILVEIRA • *Laboratory of Molecular Modeling and Computer Simulation/MolMod-CS, Institut of Exact Science/ICEx, Federal University of Alfenas/UNIFAL-MG, Alfenas, MG, Brazil*

WALTER FILGUEIRA DE AZEVEDO JR. • *Escola de Ciências da Saúde, Pontifícia Universidade Católica do Rio Grande do Sul—PUCRS, Porto Alegre, RS, Brazil*

THIAGO C. ELIAS • *Laboratory of Molecular Modeling and Computer Simulation/MolMod-CS, Institut of Exact Science/ICEx, Federal University of Alfenas/UNIFAL-MG, Alfenas, MG, Brazil*

RAFAELA S. FERREIRA • *Laboratório de Modelagem Molecular e Planejamento de Fármacos, Departamento de Bioquímica e Imunologia, Universidade Federal de Minas Gerais, Belo Horizonte, MG, Brazil*

TIAGO HENRIQUE • *Departament of Molecular Biology, Medical School of São José do Rio Preto/FAMERP, São José do Rio Preto, SP, Brazil*

FELIPE SICONHA S. PEREIRA • *Laboratory of Computacional Modeling, National Laboratory of Scientific Computing (LNCC), Petrópolis, RJ, Brazil*

VAL OLIVEIRA PINTRO • *Escola de Ciências da Saúde, Pontifícia Universidade Católica do Rio Grande do Sul—PUCRS, Porto Alegre, RS, Brazil*

LUCIANNA H. S. SANTOS • *Laboratório de Modelagem Molecular e Planejamento de Fármacos, Departamento de Bioquímica e Imunologia, Universidade Federal de Minas Gerais, Belo Horizonte, MG, Brazil*

PAWEL SIEDLECKI • *Institute of Biochemistry and Biophysics PAS, Warsaw, Poland; Department of Systems Biology, Institute of Experimental Plant Biology and Biotechnology, University of Warsaw, Warsaw, Poland*

MARTINA VEIT-ACOSTA • *Escola de Ciências da Saúde, Pontifícia Universidade Católica do Rio Grande do Sul—PUCRS, Porto Alegre, RS, Brazil*

MACIEJ WÓJCIKOWSKI • *Institute of Biochemistry and Biophysics PAS, Warsaw, Poland*

About the Editor

Walter Filgueira de Azevedo Jr. is Frontiers Section Editor (Bioinformatics and Biophysics) for the *Current Drug Targets*, member of the editorial board of *Current Bioinformatics*, and section editor (Bioinformatics in Drug Design and Discovery) for the *Current Medicinal Chemistry*. Prof. Azevedo graduated in physics (BSc in physics) from the University of São Paulo (USP) in 1990. He completed a Master's Degree in Applied Physics also from the USP (1992), working under the supervision of Prof. Yvonne P. Mascarenhas, the founder of crystallography in Brazil. His dissertation was about X-ray crystallography applied to organometallic compounds. During his Ph.D., he worked under the supervision of Prof. Sung-Hou Kim (University of California, Berkeley), on a split Ph.D. program with a fellowship from Brazilian Research Council (CNPq) (1993–1996). His Ph.D. was about the crystallographic structure of CDK2. At present, he is the coordinator of the Structural Biochemistry Laboratory at Pontifical Catholic University of Rio Grande do Sul (PUCRS). His research interests are interdisciplinary with two major emphases: molecular simulations and protein-ligand interactions. He published over 160 scientific papers about protein structures and computer simulation methods applied to the study of biological systems (H-index: 40, RG Index > 41.0). These publications have over 5000 citations.

Chapter 1

Building Machine-Learning Scoring Functions for Structure-Based Prediction of Intermolecular Binding Affinity

Maciej Wójcikowski, Pawel Siedlecki, and Pedro J. Ballester

Abstract

Molecular docking enables large-scale prediction of whether and how small molecules bind to a macromolecular target. Machine-learning scoring functions are particularly well suited to predict the strength of this interaction. Here we describe how to build RF-Score, a scoring function utilizing the machine-learning technique known as Random Forest (RF). We also point out how to use different data, features, and regression models using either R or Python programming languages.

Key words Machine learning, Scoring function, Docking, Binding affinity

1 Introduction

Molecular docking is the most widely used high-throughput structure-based tool. Docking enables large-scale prediction of whether and how small molecules bind to a macromolecular target. Although there are many relatively accurate scoring functions for pose generation, the inaccuracies of scoring functions to predict binding affinity are known to be a major limiting factor for the reliability of docking [1]. Therefore, studies have focused on improving the prediction of binding affinity by using benchmarks based on X-ray crystal structures rather than docking poses [2–9].

This is also our focus here, and hence, we explain how to generate machine-learning scoring functions for binding affinity prediction using free resources. These scoring functions permit investigating which are the optimal description of complexes, data set partition steps, regression models, and best modeling practices for the prediction of binding affinities from X-ray crystal structures of protein–ligand complexes [10]. This is of great theoretical value, as confounding factors can be eliminated and one can get an assessment of exactly how well a given approach or theory works in practice. By contrast, these scoring functions are less suited for

Walter Filgueira de Azevedo Jr. (ed.), *Docking Screens for Drug Discovery*, Methods in Molecular Biology, vol. 2053,
https://doi.org/10.1007/978-1-4939-9752-7_1,

docking applications such as Virtual Screening or Potency Optimization. However, machine-learning scoring functions can also be built to excel at these related applications [11–17] (this requires another way of building them, which is out of the scope of this chapter). An analysis of the different types of machine-learning scoring functions is made in a recent review [18].

2 Components

The following are the three main components of a machine-learning scoring function:

(a) The data to train and test the scoring function.

(b) The procedure to generate the features describing each protein–ligand complex.

(c) The regression model used to link the features or descriptors of a protein–ligand complex with its binding affinity (a classification model can also be used if a binary score, for example, binder/nonbinder, is convenient in a given case).

Here we explain how to build the original RF-Score [2] (RF-Score v1) using the R programming language. Those readers with experience in using R will be able to substitute Random Forest (RF) [19] with other machine-learning techniques or use alternative features to describe complexes. In addition, we will use the notes to indicate how to expand its functionality using the Open Drug Discovery Toolkit (ODDT) [8]. ODDT employs the Python programming language, and hence, it provides an easier route to build custom machine-learning scoring functions for those with more experience in using Python.

2.1 Prerequisites

The R software environment must be installed, which can be freely downloaded from http://www.r-project.org/. Another requisite is to have a C compiler installed, for instance, the gcc compiler in Dev-C++, which is also free and can be downloaded from http://www.bloodshed.net/devcpp.html.

In addition, the RF-Score code is available at http://ballester.marseille.inserm.fr/RF-Score-v1.zip. Uncompress this file and save the following files to the same directory:

(a) PDBbind_refined07-core07.txt

(b) PDBbind_core07.txt

(c) RF-Score_desc.c

(d) RF-Score_desc.h

(e) RF-Score_pred.r

(a) and (b) specify the training and test sets, respectively. (c) and (d) calculate RF-Score v1 descriptors or features (*see* **Note 1**) while preparing training and test sets. (e) builds the model using the prepared training and test sets.

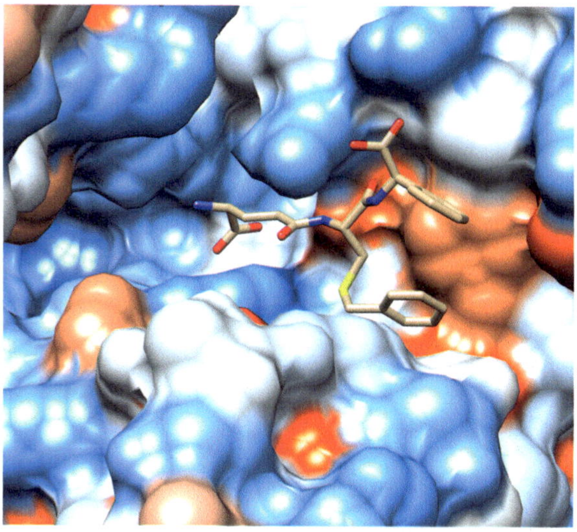

Fig. 1 An illustrative example of a high-quality protein–ligand complex (PDB-code: 10gs), which was included in the refined set of the 2016 release of the PDBbind database (http:/www.pdbbind.org.cn). The protein surface is colored by hydrophobicity scale of Kyte and Doolittle [27] using UCSF Chimera version 1.10

2.2 Data Acquisition

1. Scoring function has been primarily calibrated or trained on high-quality X-ray crystal structures (*see* **Note 2**). Figure 1 shows an example of such complexes, with the corresponding ligand bound to its protein pocket.
2. Therefore, the first step is to acquire such data from databases such as PDBbind [20] or Binding MOAD [21]. Here we will use the PDBbind database. Start by downloading the 2007 version of PDBbind database from http://www.pdbbind.org.cn (*see* **Note 3**). This will require registering a free account (follow the website instructions).
3. Once logged into http://www.pdbbind.org.cn, click on the DOWNLOAD tab and see the list of available files. From there, download "pdbbind_v2007.tar.gz," which contains the entire database.
4. Untar and uncompress "pdbbind_v2007.tar.gz". Save the resulting directory "v2007" within the same directory where the RF-Score files are located.
5. Alternatively, **Note 4** explains how to install ODDT and **Note 5** explains how ODDT pre-processes the 2016 PDBbind data for further use in scoring function training. Additionally, **Notes 5–11** describe all the subsequent steps to build a machine-learning scoring function using Python via ODDT.

2.3 Feature Generation

1. Note that "PDBbind_refined07-core07.txt" and "PDBbind_core07.txt" specify training complexes and test complexes, respectively. Further details about this and other data partitions

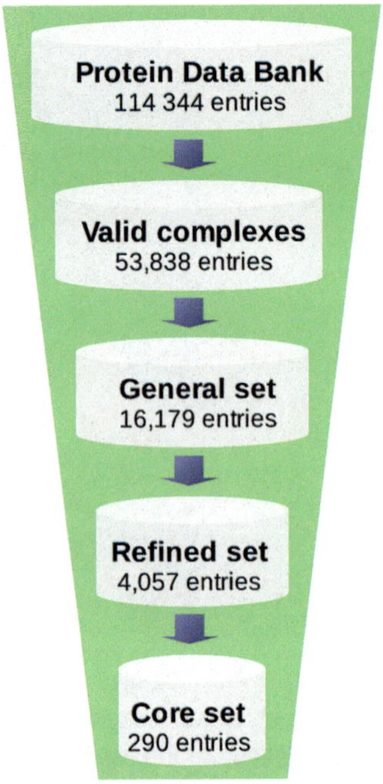

Fig. 2 Steps describing the preparation of PDBbind v2016 data sets. Increasingly stringent filters result in smaller sets of increasing structural and interaction data quality. More details can be found in the PDBbind website: http:/www.pdbbind-cn.org/

can be found in RF-Score publications [4, 8–10]. Figure 2 sketches the contents of the latest release of the PDBbind database.

2. Calculate 36 intermolecular features for each test set complex with "RF-Score_desc.c" by (a) opening "RF-Score_desc.c" from Dev-C++ (File ⇒ Open Project or File), (b) making sure that txt input and csv output files are called "PDBbind_-core07.{csv,txt}" (at lines 77 and 81), and (c) compiling and running it (Execute ⇒ Compile & Run). Output file "PDBbind_core07.csv" should have 195 entries, one per protein–ligand complex and will be the first input file in RF-Score_pred.r (see the next section).
3. Calculate 36 intermolecular features for each training set complex with "RF-Score_desc.c" by: (a) opening "RF-Score_desc.c" from Dev-C++ (File ⇒ Open Project or File), (b) making sure that txt input and csv output files are called "PDBbind_refined07-core07.{txt,csv}" (at lines 77 and 81), and (c) compiling and running it (Execute ⇒ Compile & Run). Output file "PDBbind_refined07-core07.csv" should have 1105 entries, one per protein–ligand complex. "PDBbind_refined07-core07.csv" will be the second input file in "RF-Score_pred.r" (see the next section).
4. These are RF-Score v1 features, which were designed to be simple and hence serve as a performance baseline for more comprehensive sets of intermolecular features (*see* **Note 1**).

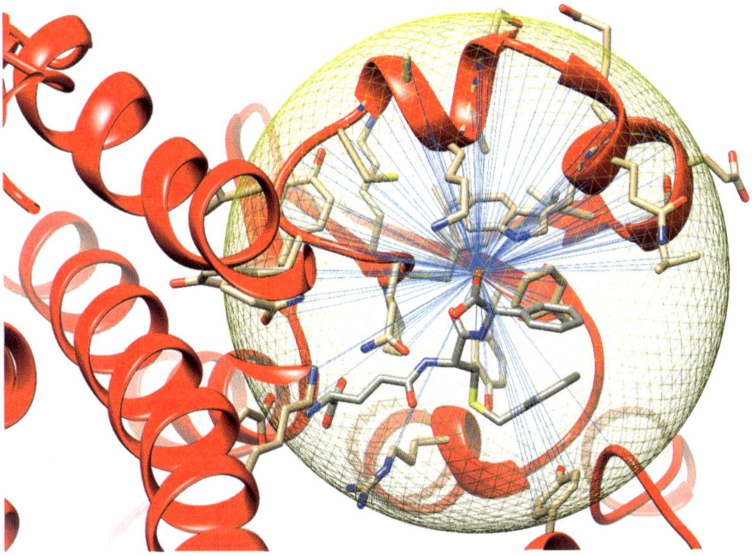

Fig. 3 RF-Score features describing protein–ligand complexes are generated by tallying atoms in close contact (<12 Å for v1 [2] and v3 [22]). Atoms are additionally grouped by their atomic number on the ligand and protein sides (the plot shows a particular oxygen atom in the ligand with protein atoms within a 12 Å neighborhood). The plot shows human glutathione S-transferase protein (PDB code: 10GS) interacting with its ligand (HET code: VWW) using UCSF Chimera [27] version 1.10. Additionally, v2 [10] introduces distance grouping with 2 Å bins and v3 supplements v1 features by including intermolecular Autodock Vina [28] terms

Note that we are directly using the complexes as provided by the PDBbind database. Instead, the user is invited to follow standard protocols to prepare the structures and investigate whether any performance improvement is achieved in this way. Ligand protocols include generating tautomers and assigning bond orders. Protein protocols typically append missing side chains, add hydrogen atoms, and assign charges according to the physiological pH.

5. For partitioning these data sets in ODDT, *see* **Note 6**.
6. For preparing these data sets for feature generation using ODDT, *see* **Note 7**.
7. For generating RF-Score v1 features in ODDT, *see* **Note 8**. Figure 3 illustrates how the inter-atomic distances of each ligand atom to close protein atoms are calculated as a first step toward generating the features for the complex.

2.4 Model Building

1. Build RF-Score and use it to predict the test set by (a) opening "RF-Score_pred.r" from the R Graphical User Interface (version 2.8 is suggested), (b) setting the working directory to the directory containing all the files mentioned in previous steps (File ⇒ Change dir), (c) making sure that the package randomForest is installed (Packages ⇒ Install Packages, then select

closest mirror server and randomForest package) and (d) running "RF-Score_pred.r" (File ⇒ Source R code). The three figures in the paper [2] will be generated. Another output is "RF-Score_pred.csv," which contains the predicted binding affinities (pK or log K units) for the 195 test complexes.

2. Alternatively, other machine-learning techniques can be used instead of RF to build the underlying regression model (*see* **Note 9**).

3 Methods

Figure 4 shows a typical training and testing (evaluation) workflow of a machine-learning scoring function for binding affinity prediction. We continue our example building and testing the original RF-Score.

3.1 Training the Model

1. Training is carried out for model building using two control parameters: number of trees ("ntree"=500) and maximum number of features considered at the tree node split ("mtry,"

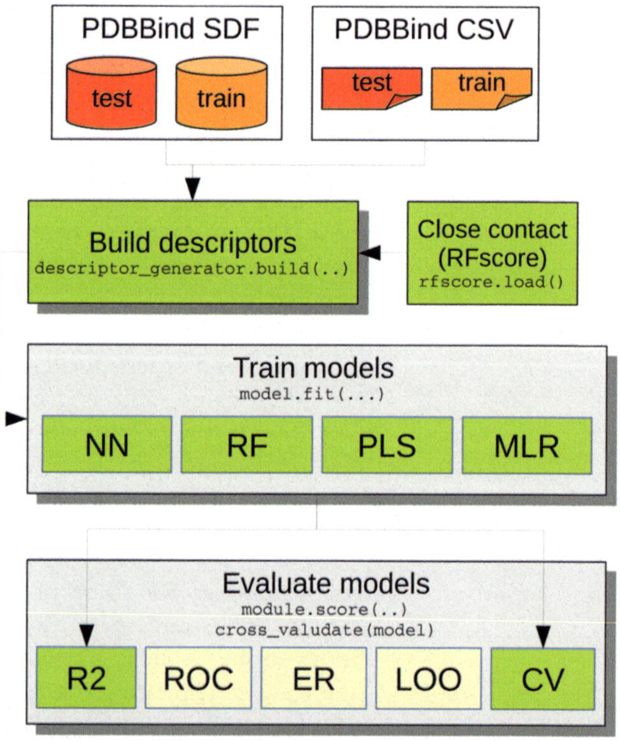

Fig. 4 Training and testing workflow showing main options in ODDT. The blocks are interchangeable. For example, pre-existing descriptor generator function may be loaded from a different ODDT scoring function (NNScore, Vina, etc.), or any of the four currently supported machine learning models. At the end, a suite of metrics and cross-validation techniques can be chosen to assess the performance of the resulting scoring function

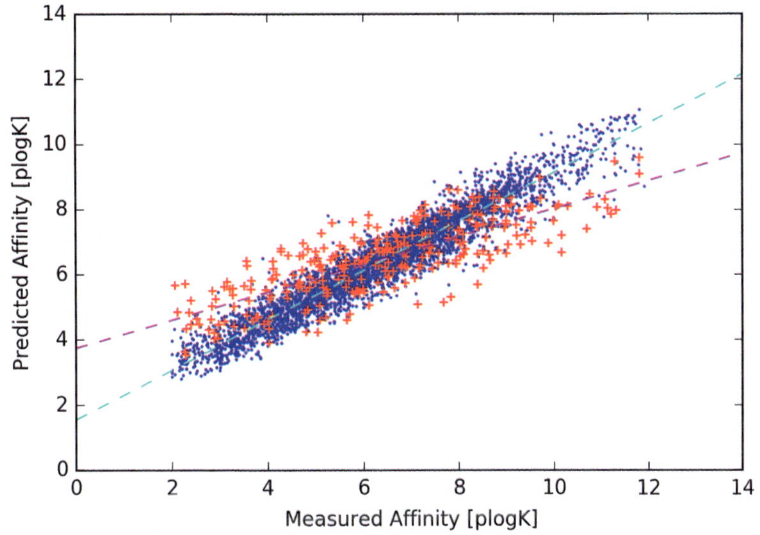

Fig. 5 Correlation plots from RF-Score v1 trained on PDBbind 2016 using ODDT. Training set (blue dots) and test set (red crosses). Horizontal axis represents the measured activity each complex, whereas vertical axis shows its predicted value by model (RF-Score v1)

selected by internal validation). This process is fully explained in this paper [2]. Further details can be found as comments in "RF-Score_pred.r".

2. Other control parameters of the algorithm or other values of these parameters can be used, which will result in a slightly different RF model.

3.2 Testing the Model

1. The trained model, now RF-Score v1, can be used on any test set, in particular, the provided test set with 195 complexes.
2. There are several metrics to measure test set performance (*see* **Note 10**).
3. Figure 5 shows the high correlation achieved in the test set (correlation in the training set is even higher but irrelevant for the quality of RF-Score v1 [2]).
4. *See* **Note 11** for instructions of how to apply RF-Score v1 to a different test set.

4 Notes

1. There are very many ways to describe a complex from its 3D structural model, each giving rise to a particular set of features [9, 10, 12, 18, 22–24].

2. However, in some scenarios, there is advantage in training with lower quality structures [25] or even docked poses [26] of the protein–ligand complex.
3. Alternatively, the latest version can be downloaded (this is described in Fig. 2), which contains more data and thus will lead to more accurate and widely applicable machine-learning scoring functions. Note that the ODDT workflow below is employing data from the 2016 version of PDBbind.
4. The easiest way to get ODDT on any operating system is by using Conda package manager. Go to https://conda.io/miniconda.html for the latest Miniconda installer and install it to your system. Next you need to install molecular toolkits (either openbabel or RDKit—or both). Here we will use openbabel toolkit:

   conda install -c openbabel openbabel

   Now you are ready to install ODDT including all needed dependencies:

```
conda install -c mwojcikowski oddt
```

   After introducing these two commands, ODDT should be available both in python ("*import oddt*") and CLI ("*oddt_cli --help*").
5. PDBbind is already pre-processed for use in ODDT in the form of CSV files. To use the prepared CSV files, follow the python code below. Note that the CSV file contains many versions of PDBbind, in this example, we will use the latest 2016_refined version.

```
import oddt
import pandas as pd
data = pd.read_csv(oddt.__path__[0] + "/scoring/functions/RFScore/rfscore_descs_v1.csv")
```

6. With ODDT, the user has to partition the PDBbind data set into training set ("refined set" with excluded "core set") and testing set ("core set"). It is important to make sure that these sets do not overlap in order to avoid over optimistic results. With this purpose, execute the following python code:

```
# Exclude test set from training set
training_data = data[data['2016_refined'] & ~data['2016_core']]
# select last 36 columns of the CSV containing features
```

```
features = training_data.iloc[:, -36:].values
# select activity values
activity = training_data['act'].values
```

7. For every complex, we will need the protein and ligand objects and the measured affinity (activity). Different databases have their own way of storing data. Here we show an example where protein and ligand files are separate files stored in a single directory named with PDBID string. Affinity measures are stored in a csv file for all complexes.

```
for pdbid in ['10gs', '4da4']:
protein = next(oddt.toolkit.readfile('pdb',
    'directory/%s/%s_protein.pdb' % (pdbid, pdbid)))
ligand = next(oddt.toolkit.readfile('mol2',
    'directory/%s/%s_ligand.mol2' % (pdbid, pdbid)))
activities = pd.read_csv('activitity.csv')
```

If you want to use PDBbind dataset, ODDT implements a convenient wrapper for automating this task.

```
from oddt.datasets import pdbbind
dataset = pdbbind('/home/directory/pdbbind/v2016/',
version=2016,
    default_set='refined')
for pid in dataset:
protein = pid.protein
ligand = pid.ligand
activity = dataset.activities
```

8. Now that all data points are available in ODDT, we are ready to generate features. ODDT allows easy generation of RF-Score v1 features using the following lines of python code:

```
from oddt.scoring.functions import rfscore
desc_gen = rfscore(version=1).descriptor_generator
features = desc_gen.build([ligand], protein)
```

For other versions of RF-Score, set the "version" parameter to "2" or "3." Note that if you wish to generate features for multiple ligands targeting the same protein, then the last line of the script above must be substituted by

```
ligands = list(oddt.toolkit.readfile('mol2',
```

```
'ligands.mol2'))
        features = desc_gen.build(ligands, protein)
```

9. ODDT adopts the models and API from scikit-learn (http://scikit-learn.org/), which makes it trivial to use just call the ".fit ()" method of the model. Moreover, ODDT provides a variety of ML models such as SVM, feed forward neural network, and Random Forest. Algorithms such as SVM and neural networks are bundled with a preprocessing step, which normalize input data. In this example, code we will train the random forest regressor using 500 trees, with the aim of correlating the RFScore v1 features with activity data.

```
            from oddt.scoring.models.regressors import
randomforest,
        neuralnetwork, svm
        model = randomforest(n_estimators=500)
        model.fit(features, activity)
        You can also train neural network model substituting
the model line with
        model = neuralnetwork()
```

10. Evaluating the model can be done by estimating how well the predicted values correlate with the measured ones. The most common metrics are Pearson's R, Spearman's R, and Kendall's Tau. Here we show how Pearson's (R_p) correlation coefficient and its square (R_p^2) can be computed with ODDT/scikit-learn

```
        testing_data = data[data['2016_core']]
        testing_features = testing_data.iloc[:, -36:].values
        testing_activity = testing_data['act'].values
        model.score(testing_features, testing_activity)
```

11. Now that the machine-learning model is trained and tested, we are ready to apply it to prospective data. In order to score a new series of protein–ligand complexes we need to assemble a custom object in ODDT, which will act as a scoring function:

```
        from oddt.scoring import scorer
            scoring_function = scorer(model, desc_gen,
score_title='my_custom_score')
            protein = next(oddt.toolkit.readfile('mol2',
'protein.mol2'))
        docked_poses = list(oddt.toolkit.readfile('mol2','-
docked.mol2'))
```

```
scoring_function.set_protein(protein)
scores = scoring_function.predict(docked_poses)
```

In the above example, we use our own machine-learning scoring function with a single protein ("protein.mol2") and a series of docked molecules ("docked.mol2"). What is more, custom scoring object can be saved to a file by "*scoring_function.save('my_sf.pkl')*" method and used directly in the command line:

```
oddt_cli -score_file = my_sf.pkl docked.mol2 -protein
protein.mol2 -O scores.csv
```

Such scoring functions can be shared between users, as they depend only on ODDT being installed. Also, you can change the output of the scoring process by substituting the "csv" file extension with "sdf" or other supported by formats.

Acknowledgments

This work was supported by INSERM and the Polish Ministry of Science and Higher Education POIG.02.02.00-14-024/08-00 and POIG.02.03.00-00-003/09-00.

References

1. Huang S-Y, Grinter SZ, Zou X (2010) Scoring functions and their evaluation methods for protein-ligand docking: recent advances and future directions. Phys Chem Chem Phys 12:12899–12908
2. Ballester PJ, Mitchell JBO (2010) A machine learning approach to predicting protein-ligand binding affinity with applications to molecular docking. Bioinformatics 26:1169–1175
3. Kramer C, Gedeck P (2010) Leave-cluster-out cross-validation is appropriate for scoring functions derived from diverse protein data sets. J Chem Inf Model 50:1961–1969
4. Ballester PJ, Mitchell JBO (2011) Comments on "leave-cluster-out cross-validation is appropriate for scoring functions derived from diverse protein data sets": significance for the validation of scoring functions. J Chem Inf Model 51:1739–1741
5. Kinnings SL, Liu N, Tonge PJ, Jackson RM, Xie L, Bourne PE (2011) A machine learning-based method to improve docking scoring functions and its application to drug repurposing. J Chem Inf Model 51:408–419
6. Zilian D, Sotriffer CA (2013) SFCscore(RF): a random forest-based scoring function for improved affinity prediction of protein-ligand complexes. J Chem Inf Model 53:1923–1933
7. Ashtawy HM, Mahapatra NR (2015) A comparative assessment of predictive accuracies of conventional and machine learning scoring functions for protein-ligand binding affinity prediction. IEEE/ACM Trans Comput Biol Bioinform 12:335–347
8. Wójcikowski M, Zielenkiewicz P, Siedlecki P (2015) Open drug discovery toolkit (ODDT): a new open-source player in the drug discovery field. J Cheminform 7:26
9. Pires DEV, Ascher DB (2016) CSM-lig: a web server for assessing and comparing protein-small molecule affinities. Nucleic Acids Res 44:W557–W561
10. Ballester PJ, Schreyer A, Blundell TL (2014) Does a more precise chemical description of protein-ligand complexes lead to more accurate prediction of binding affinity? J Chem Inf Model 54:944–955

11. Li L, Wang B, Meroueh SO (2011) Support vector regression scoring of receptor-ligand complexes for rank-ordering and virtual screening of chemical libraries. J Chem Inf Model 51:2132–2138
12. Ding B, Wang J, Li N, Wang W (2013) Characterization of small molecule binding. I. Accurate identification of strong inhibitors in virtual screening. J Chem Inf Model 53:114–122
13. Zhan W, Li D, Che J, Zhang L, Yang B, Hu Y et al (2014) Integrating docking scores, interaction profiles and molecular descriptors to improve the accuracy of molecular docking: toward the discovery of novel Akt1 inhibitors. Eur J Med Chem 75:11–20
14. Sun H, Pan P, Tian S, Xu L, Kong X, Li Y, Li D, Hou T (2016) Constructing and validating high-performance MIEC-SVM models in virtual screening for kinases: a better way for actives discovery. Sci Rep 6:24817
15. Pereira JC, Caffarena ER, dos Santos CN (2016) Boosting docking-based virtual screening with deep learning. J Chem Inf Model 56:2495–2506
16. Wójcikowski M, Ballester PJ, Siedlecki P (2017) Performance of machine-learning scoring functions in structure-based virtual screening. Sci Rep 7:46710
17. Ragoza M, Hochuli J, Idrobo E, Sunseri J, Koes DR (2017) Protein–ligand scoring with convolutional neural networks. J Chem Inf Model 57:942–957
18. Ain QU, Aleksandrova A, Roessler FD, Ballester PJ (2015) Machine-learning scoring functions to improve structure-based binding affinity prediction and virtual screening. Wiley Interdiscip Rev Comput Mol Sci 5:405–424
19. Breiman L (2001) Random forests. Mach Learn 45:5–32
20. Cheng T, Li X, Li Y, Liu Z, Wang R (2009) Comparative assessment of scoring functions on a diverse test set. J Chem Inf Model 49:1079–1093
21. Ahmed A, Smith RD, Clark JJ, Dunbar JB, Carlson HA (2015) Recent improvements to binding MOAD: a resource for protein-ligand binding affinities and structures. Nucleic Acids Res 43:465–469
22. Li H, Leung K-S, Wong M-H, Ballester PJ (2015) Improving AutoDock Vina using random Forest: the growing accuracy of binding affinity prediction by the effective exploitation of larger data sets. Mol Inform 34:115–126
23. Li H, Leung K-S, Wong M-H, Ballester PJ (2014) Substituting random forest for multiple linear regression improves binding affinity prediction of scoring functions: Cyscore as a case study. BMC Bioinformatics 15:291
24. Durrant JD, McCammon JA (2011) BINANA: a novel algorithm for ligand-binding characterization. J Mol Graph Model 29:888–893
25. Li H, Leung K-S, Wong M-H, Ballester P (2015) Low-quality structural and interaction data improves binding affinity prediction via random Forest. Molecules 20:10947–10962
26. Li H, Leung K-S, Wong M-H, Ballester PJ (2016) Correcting the impact of docking pose generation error on binding affinity prediction. BMC Bioinformatics 17:308
27. Pettersen EF, Goddard TD, Huang CC, Couch GS, Greenblatt DM, Meng EC, Ferrin TE (2004) UCSF chimera--a visualization system for exploratory research and analysis. J Comput Chem 25:1605–1612
28. Trott O, Olson AJ (2010) AutoDock Vina: improving the speed and accuracy of docking with a new scoring function, efficient optimization, and multithreading. J Comput Chem 31:455–461

Chapter 2

Integrating Molecular Docking and Molecular Dynamics Simulations

Lucianna H. S. Santos, Rafaela S. Ferreira, and Ernesto R. Caffarena

Abstract

Computational methods, applied at the early stages of the drug design process, use current technology to provide valuable insights into the understanding of chemical systems in a virtual manner, complementing experimental analysis. Molecular docking is an in silico method employed to foresee binding modes of small compounds or macromolecules in contact with a receptor and to predict their molecular interactions. Moreover, the methodology opens up the possibility of ranking these compounds according to a hierarchy determined using particular scoring functions. Docking protocols assign many approximations, and most of them lack receptor flexibility. Therefore, the reliability of the resulting protein–ligand complexes is uncertain. The association with the costly but more accurate MD techniques provides significant complementary with docking. MD simulations can be used before docking since a series of "new" and broader protein conformations can be extracted from the processing of the resulting trajectory and employed as targets for docking. They also can be utilized a posteriori to optimize the structures of the final complexes from docking, calculate more detailed interaction energies, and provide information about the ligand binding mechanism. Here, we focus on protocols that offer the docking–MD combination as a logical approach to improving the drug discovery process.

Key words Molecular docking, Molecular dynamics, Virtual screening, Flexible docking, Enhanced sampling methods

1 Introduction

Over the past few decades, technological and scientific advances have fueled genomic, proteomics, and related fields. One of the most profitable areas, and also one of the most challenging fields, is drug discovery and development. Today, techniques such as X-ray crystallography, nuclear magnetic resonance (NMR) spectroscopy, high-throughput screening, combinatorial chemistry, and computational approaches are well-established and affordable methods often employed toward the search and characterization of targets and development of drugs of interest. Although there are no stiff guidelines to the drug design process, a combination of experimental techniques and computational methods may be the most cost-

Walter Filgueira de Azevedo Jr. (ed.), *Docking Screens for Drug Discovery*, Methods in Molecular Biology, vol. 2053, https://doi.org/10.1007/978-1-4939-9752-7_2,

efficient among the drug design approaches. For example, some commercial drugs arose as the product of this type of strategy [1].

Currently, drug development involves biological targets, genetic studies, molecular biology, gene technology, and protein knowledge [2]. Therefore, the availability of the three-dimensional structure of the biomolecule is a prominent component in the discovery of a new drug. The use of this piece of information from macromolecular targets is what comprises structure-based drug design (SBDD) methods [3]. Therefore, SBDD methods are enabled by the ever-expanding collection of high-resolution protein structures, usually from X-ray crystallography or NMR spectroscopy, or by comparative computational modeling and other protein structure prediction techniques. With the use of computational tools in SBDD, it is possible not only to visualize compounds bound to their biological targets, providing details regarding molecular interactions (hydrogen bonds, salt bridges, van der Waals repulsive and attractive forces) driving the binding process, but also to score them in a proper and reliable way [2].

The most popular method in computational drug design is molecular docking, which is based on the "lock and key" concept [4] created by Emil Fischer (in 1894). In this framework, molecular recognition occurs when the binding site of a receptor protein is exactly complementary to ligand shape, just like a key to a lock. Nowadays, this theoretical idea has been updated, and it is well known that numerous entropic and enthalpic aspects contribute to the binding of a ligand to a receptor. Currently, the most up-to-date docking algorithms are capable of predicting possible binding modes of a ligand, a small molecule, a peptide, or a protein, by sampling its orientation, conformation, and interactions when bound to an enzyme or another protein receptor of a different kind. Although the algorithms permit the flexibility of the ligand with minimum cost, efficiently accounting for protein flexibility is still challenging.

Operationally speaking, setting up a docking experiment is relatively simple, and it usually does not require much computational power [5]. However, despite docking programs being fast, one of the major methodological issues is obtaining accurate results [5–7]. Hence, it is imperative that molecular docking be combined with other computational techniques to provide more reliable results. A widely used practice to optimize outcomes is pairing docking with Molecular Dynamics (MD) simulations. By performing MD simulations, the dynamic behavior of molecular arrangements can be monitored and probed at different timescales, allowing studies from fast internal motions and slow conformational changes to complex processes such as ligand binding to an active site or protein folding [8–10].

MD is also a very popular and well-established method with a high number of published studies reporting its use. The number of applications of MD to drug design is ever increasing, and it would be almost impossible to name them all. For further readings, *see* [11, 12]. Consequently, joining both approaches is a practical habit to improving computational drug design projects. A combined approach unites the ligand binding mode prediction provided by docking, alongside the induced fit effect of the receptor around the ligand and the more accurate description of the energies involved explored by MD simulations [13].

It is worth mentioning that in silico approaches do not substitute or provide the same information as experimental methods do. Therefore, a set of prioritized compounds still needs to be synthesized, and their biological properties are determined by using several experimental platforms [14]. When employed jointly, in silico and experimental procedures offer knowledge of the elaborated characteristics of intermolecular recognition, making such procedure usually a good practice in drug discovery [15].

2 Materials

For docking experiments, the availability of coordinate files for receptor and ligand structures, which can be obtained in a variety of formats such as pdb, mol2, cif, and sdf, is necessary. Moreover, libraries of small molecules with expected drug-like properties can be downloaded from specialized databases such as the ZINC database [16]. A large number of docking programs and web servers [5] can be used, including AutoDock [17], GOLD [18], GLIDE [19, 20], and FlexX [21]. Analysis of docking results can be done by programs for visualization, such as Pymol [22], UCSF Chimera [23], and VMD [24].

For MD simulations, an initial coordinate file containing the atomic coordinates of the ligand–receptor complex is required. Programs for MD simulations of biomolecules include AMBER [25], CHARMM [26], GROMACS [27], and NAMD [28], among others. Trajectory analysis can be done with tools such as the ones found in GROMACS packages, AmberTools [29], and VMD [24] plug-ins.

3 Methods

3.1 Molecular Docking

The aims of molecular docking techniques are twofold: to predict the conformation of the guest molecule (also known as a ligand) within its target (also referred to as a receptor) binding site [13] and to provide an estimation of the affinity of this particular interaction [30]. Although the first goal is achieved more often to a great extent, the second is still an inherent computational difficulty associated with simplified approximations. These two components are linked, where the first element is the docking per se carried out with the docking algorithm, while the second is referred to as scoring and it is also calculated with the docking program using a predetermined scoring function. In general, most scoring functions consider the ligand size, flexibility, internal conformation energy, and atomic positions [31].

Alternatively, molecular docking can also be used in integrated ways to achieve goals beyond protein–ligand binding mode prediction. For instance, ligand docking can help in the computational design or redesign of binding pockets by altering ligand–protein interactions. This method uses a binding pocket of an already known target as a scaffold and mutations are introduced in the pocket by several molecular modeling tools in order to enhance the affinity between the interacting molecules [32]. In one of the steps, ligands are docked in the binding pocket of a predefined protein of interest and a combined energy score is used in the identification of promising pockets to be created by a protein design program [32–34]. Enriquez et al. [35] presented another example of integrating molecular docking. In their method, the conformational space of the peptides is searched by MD simulations to obtain relaxed structures of each conformer, while the docking of the peptide is performed using the given ligand as a target, and the sequence space is searched by the Monte Carlo method. This method was used to design a decapeptide able to bind the potent HIV-1 inhibitor efavirenz, and most of the predicted contacts between peptide and efavirenz were confirmed by NMR experiments [35].

Although all docking programs perform conformational sampling of the ligand and some even include receptor flexibility, issues such as the explicit consideration of desolvation and entropic effects, and inaccurate scoring functions, remain to most of them [36]. Hence, the choice of a program will depend on the kind of docking experiment to be performed. For instance, the level of receptor flexibility and the type of hardware to be used are critical at the moment of deciding which program fits better to the biological problem to be solved (*see* **Note 1**).

3.2 Assembling Molecular Docking Experiments

1. Obtain or generate the ligand coordinate file. The three-dimensional structure of a ligand can either be obtained by experimental coordinates from the Protein Data Bank (PDB) [37], from compound databases like ZINC, or be built using one of the many molecular editors (Avogadro [38], MarvinSketch [39], ACD/ChemSketch [40], etc.).
2. Obtain the three-dimensional structure to be used as a target. Structures can be downloaded from the PDB when available. NMR structures can also be found at the PDB (*see* **Note 2**). Additionally, in the absence of an experimentally obtained

structure of a biological target, comparative modeling (also known as template-based homology modeling) and ab initio modeling can be used to build a receptor model [41] (*see* **Note 3**).

3. Prepare ligand and receptor structures. Such preparation entails removing alternative residue conformations, co-factors, and unwanted water molecules, adding hydrogen atoms and atomic partial charges when required (*see* **Note 4**). The last two steps also hold for ligand preparation, the details of which may depend on the source of the ligand structure (*see* **Note 5**).
4. Set up other specific predocking preparatory steps such as definition and calculation of a grid (*see* **Note 6**). Usually, a user-defined rectangular box is chosen as the search space, encompassing entirely (*see* **Note 7**) or partially the receptor (including the binding site), where the ligand conformations will be sampled by the docking algorithms [42] (*see* **Note 8**).
5. After all preparation steps, docking simulations can be performed (*see* **Note 9**) for a single ligand or for a library of compounds in a structure-based virtual screening (VS) approach (*see* **Note 10**) using one or multiple receptor structures (*see* **Note 11**). A step-by-step flowchart is found in Fig. 1. Although most docking protocols and algorithms account for ligand flexibility, the size and complexity of macromolecules turn difficult a comprehensive incorporation of receptor

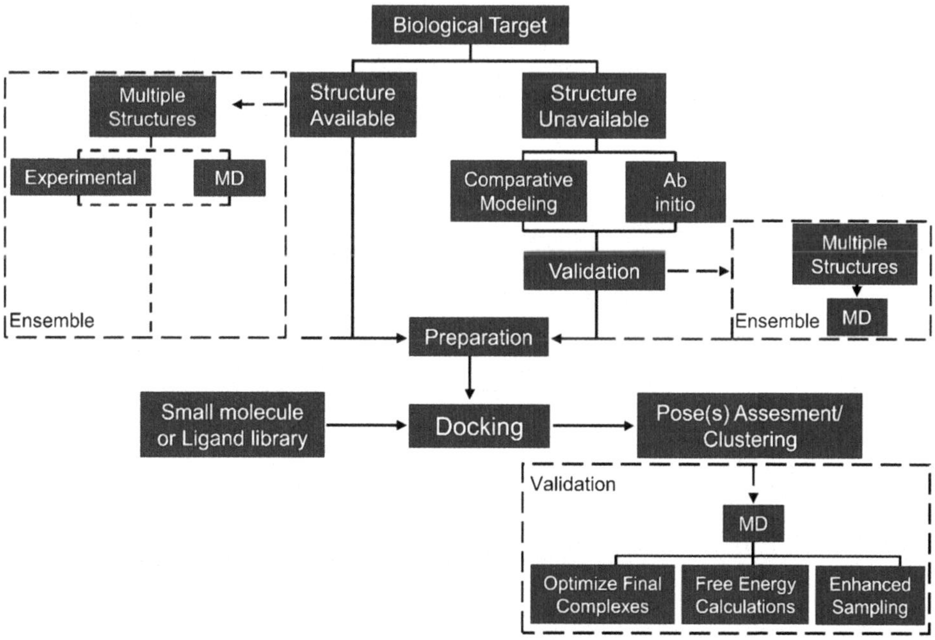

Fig. 1 Flowchart of molecular docking steps

flexibility during docking [13]. However, a few established methods incorporate partial receptor flexibility during different stages of the docking process (*see* **Note 12**).

6. Visualize docking outcomes (known as poses) with molecular visualization software (*see* **Note 13**). It is expected that the best pose is scored higher than any other sampled conformation by the scoring function. However, this assumption may not be true. Therefore, a wide number of resulting poses need to be examined or reevaluated (*see* **Note 14**).

3.3 Molecular Docking Combined with MD

Docking protocols are usually fast and demand little computational power due to their many approximations and lack of protein flexibility. However, these approximations may interfere with the reliability of the resulting protein–ligand complexes. Therefore, it is the combination of the expensive but more accurate MD techniques that might provide better complementary with docking.

MD simulations are a useful and broadly applied computational method for understanding biological macromolecule behavior [13]. Since MD is based on classical mechanics, Newton's equations of motion are applied to calculate the position and speed of each atom of the studied system. Therefore, MD simulations carry out a more intensive conformational search than molecular docking methods do and provide a more accurate representation of protein motions [43].

Target flexibility is taken into account in a more realistic way since enzymes and receptors can experience conformational changes during the molecular recognition process [44]. The acting forces on each particle of the system are given by the calculation of the spatial gradient of an effective molecular interaction potential function, usually parameterized using quantum chemical calculations or experimental data (*see* **Note 15**). Currently, simulated systems often include an explicit model for water molecules, counterions, and even entire membrane environments, and they can be recorded into a trajectory over a period of tens to thousands of nanoseconds (ns) from an initial conformation [45] (*see* **Note 16**).

Despite all its usability and progress, setting up an MD simulation is not overall simple, especially when the choice of software is concerned, since it will depend on an adequate force field to better represent the biological system. Most modern force field parameters can describe proteins and their interactions adequately (*see* **Note 17**).

Another limitation is the high computational cost required to simulate large systems, comprising thousands of atoms. Although computational processing has evolved, some of the conformational changes undertaken by receptors occur on time scales exceeding the available computational capacity [46], and specific approaches are needed to solve this problem [47] (*see* **Note 18**).

3.3.1 Assembling MD Simulations

1. Choose the system to reproduce. Before starting any MD simulation, it is mandatory to know thoroughly the system (or similar ones) to be simulated and to consider if a simulation would provide the properties of interest and answer the question that prompted its application.
2. Determine which MD software and force field will be used to perform the simulations. This step is not trivial since the choice of software depends on the force field compatible with the program that might provide the appropriate representation of the system (*see* **Note 15**).
3. Obtain a file with the atomic coordinates of all molecules in the system. The file can either be retrieved from the PDB or be generated by comparative modeling or even consist of a protein–ligand complex originated from molecular docking (*see* **Note 19**).
4. Produce a topology file, inferred from the original file (*see* **Note 20**). A topology file specifies relevant information about the system, such as the atoms that are connected to one another through chemical bonds, the angles formed by three connected atoms, and the dihedral angles formed by four atoms linearly connected.
5. Choose the method to represent solvent in the system, in either an explicit or an implicit form (*see* **Note 21**).
6. Define a simulation box large enough to contain the molecular system (*see* **Note 22**). Counterions to neutralize the system may also be considered in the solvated system.
7. Create new coordinate and topology files for the solvated and neutralized system.
8. Perform energy minimization (*see* **Note 23**). Configuration files with specific MD software parameters are needed (*see* **Note 24**).
9. Perform temperature and density equilibration of the system (*see* **Note 25**). Equilibration simulations need to run for an adequate time to permit the system to relax before initiating MD production (*see* **Note 26**).
10. Execute the production stage of MD. This stage of the MD simulation also requires sufficient time so that the property of interest can be observed (*see* **Note 27**).
11. Analyze the output data from an MD simulation, the so-called "trajectory," to obtain information on the system (*see* **Note 28**).

The information provided by an MD simulation can be used before docking, to achieve a series of "new" and broader conformations of the protein to be used as targets for docking. Alternatively, it can be employed to optimize the structures of the final complexes from docking, calculate more detailed interaction energies, and provide information about the binding mechanism of the ligand.

3.3.2 MD Simulations to Generate Receptor Conformations

A way to take the receptor flexibility into account is to apply molecular docking against multiple conformations of the receptor, experimentally solved, bound to a diverse range of ligands [48]. However, only for a few targets, we are fortuned enough to have structural ensembles with such conformational variation available [44]. Therefore, the throughout conformational sampling employed by MD can provide alternative conformations for the studied target not experimentally observed before. Conformations of the system can be extracted from the MD trajectory at regular intervals or by clustering methods, thus reducing conformational redundancy.

Another method that employs both crystallographic ensemble of structures and multiple computer-generated conformations from MD simulation is the Relaxed Complex Scheme (RCS) [44, 49, 50]. First, in the RCS, an ensemble of high-resolution crystallographic structures is selected, and VS of a compound library is performed in all the structures. The top-ranked compounds are then chosen to compose a new and reduced screening library. After MD, simulations of receptor–ligand crystallographic complexes are done on a time scale of ten to hundreds of nanoseconds to allow the receptor to explore new regions in its conformational space. The simulations are followed by RMSD-based clustering of the MD trajectories to select a diverse ensemble of conformations, and the new compound library is then screened against all MD resulting structures. RCS was successfully applied to the identification of two compounds that inhibit HIV-1 reverse transcriptase activity at concentrations of 60 nM [51].

Nevertheless, all ensemble-based approaches are limited by the demanding docking phase, which must be repeated for each receptor conformation, and by the nontrivial selection of the best conformations generated by MD. Selection of conformations, for both crystal- and MD-generated structures, may be done through retrospective VS experiments aimed at measuring the discrimination abilities of each conformation to distinguish known inhibitors from noninhibitors [52] (Fig. 2). Therefore, a hierarchical approach is necessary to test each conformation to identify the best ones. Although comparative studies showed that the discrimination abilities for some MD originated structures are better than (or comparable to) their respective crystal structures [53, 54], the enrichment enhancement seemed to depend on a reduced number of MD structures rather than the whole generated ensemble [53].

Another issue that must be borne in mind is to regard the induced fit effect of ligands when performing an exhaustive search of ligand poses within the binding site. Usually, the observation of multiple X-ray structures can point out the residues that suffer

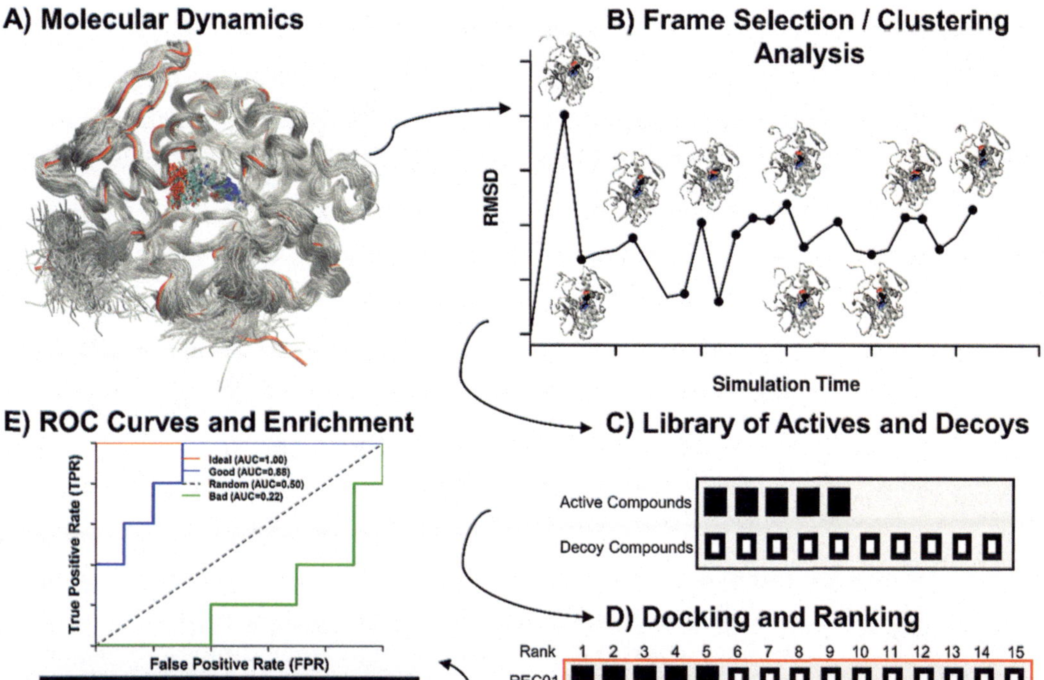

REC	AUC	EF1%	EF5%	EF10%
01	1.00	10	10	5
02	0.88	4	4	3
03	0.22	0	0	0

Fig. 2 Basic steps for assessing receptor discrimination abilities to distinguish known actives from nonactives (decoys) in a VS-based approach. (**a**) MD simulations can generate receptor conformations from a target bound to a ligand. (**b**) The conformations can be extracted by selecting specific frames or by clustering analysis. (**c**) A compound library containing known active compounds of the target and nonactive compounds can be created to evaluate the MD generated structures. (**d**) Docking and ranking of compounds are performed by a docking program. (**e**) VS-based metrics such as ROC curves and enrichment factor can be employed to measure the discrimination abilities of the conformations. The metrics can be used to point out the conformations to use in prospective VS runs

conformation changes during ligand recognition. However, in a recent work, Gao et al. [55] inspected the ability of MD simulations to prospectively predict regions of ligand-binding sites capable of undergoing induced fit effects without the need for inspecting multiple structures. The authors raised some caveats on the use of *apo* and *holo* simulation frames obtained straightforwardly from MD simulations for molecular docking, due to unfavorable residue deviations from the initial binding site arrangements in the structures. Their results showed that the choice of force field could influence the ability of the MD simulation to sample-induced changes in the active site.

3.3.3 Pose Validation Using MD, Free Energy Calculations, and Enhanced Sampling Methods

A good practice for validating poses obtained by molecular docking is to complement computational experiments with MD, enhanced sampling methods, and free energy of binding calculations. The incorporated flexibility of both ligand and receptor, granted by the MD-based methods, can better capture interactions and complementarity. Since the dynamic behavior of the ligand–receptor complex is monitored along the simulation, its stability and consistency can be measured. Therefore, an incorrectly docked ligand is expected to generate unstable trajectories, while an exact pose will display a more stable behavior [13].

Yadav et al. [56] performed an ensemble-based molecular docking and molecular dynamics study to discover inhibitors of the epidermal growth factor receptor tyrosine kinase (EGRF-TK), an attractive target for cancer therapy. After docking a library of 134 curcumin (diferuloylmethane[1,7-bis(4-hydroxy-3-methoxyphenyl)-1,6-hepatidiene-3,5dione]) analogs against five EGFR wild-type crystal structures, five top-ranked compounds were selected. MD simulations of these analogs confirmed the stability of the complexes, making them promising scaffolds for developing effective leads capable of inhibiting EGFR. A similar combination of molecular docking and MD simulations was used by Watanabe et al. [57] to investigate the role of water molecules in inhibitor (α-naphthoflavone) and substrate (7-ethoxyresorufin) recognition in the active site of cytochrome 1A2 (CYP1A2). CYP1A2 is a drug-metabolizing enzyme that affects the pharmacokinetics of drugs used in asthma, antipsychotics, and antiarrhythmic therapies. Docking was performed in an ensemble of conformations extracted from a 100 ns ligand-free MD simulation, and the complexes with the highest docking score were selected. During MD simulations of these complexes, they found that water molecules were necessary for CYP1A2 substrate recognition, while for ligand recognition, no water molecules seemed to be required.

While the stability of a receptor–ligand complex is important, it may be necessary to apply a more rigorous approach capable of discriminating between ligand poses by offering accurate estimations of their binding free energy. Methods such as the thermodynamic integration (TI) and free energy perturbation (FEP) are among the MD-based methodologies available for the calculation of free energies. Both free energy methods involve a set of long MD simulations on a pathway connecting nonphysical states to determine the relative free energy of binding between two states. Although free energy methods provide a useful approach for obtaining accurate predictions of protein–ligand binding free energies and increment of a degree of certainty about the correct docked poses, they are computationally expensive, limiting the application to only a small number of ligands.

In Carlevaro et al. [58], the authors applied FEP calculations to estimate the relative free energy of binding between two isomers of a particular ligand within the binding site of a α4β1 integrin head-piece. In their work, all docking solutions generated by the docking program Vina [59] were clustered, and three different plausible binding modes were observed. The best scored pose ranked by Vina and was the binding mode that presented larger deviation from the experimental results, leading to a significant positive ΔG value. On the other hand, an intermediate ranked solution, in which charged groups of the ligand and the protein established a salt bridge, matched well with experiments, with $\Delta G < 0$ despite not being the best-scored solution [60]. Based on these studies, it is clear that when calculations were performed using wrong initial coordinates of the ligand in the binding site, experimental results could not be reproduced. The same conclusion was found by Wang et al. [61]. In their work, three hypothetical binding poses of the same eEF2K ligand were generated by docking and reproduced for seven analogs of this ligand. FEP calculations showed that only one of the supposed binding poses was in good correlation (r^2 coefficient = 0.96) with experimental IC_{50} values.

Currently, MD simulations can run up to a few milliseconds. However, the unbinding kinetics of some drug-like molecules may take up to several minutes [46]. Therefore, classical MD simulations, even if running on dedicated hardware, may never describe such rare events. Consequently, enhanced sampling methods are interesting techniques to validate docking poses. Clark et al. [62] proposed an approach to increase the accuracy of protein–ligand binding poses, by combining the induced fit docking procedure with metadynamics (MetaD). In general, enhanced sampling methods allow crossing free energy barriers in the free energy surface by introducing artificial bias into the simulated complex, speeding the sampling of the relative stability in less computational time than classical MD [63]. Their results showed that with the use of MetaD, it was possible to discriminate the lowest free energy binding mode for a protein–ligand complex from possible alternatives originated from an induced fit docking protocol. The relevance of identifying the right pose to obtain good experimental results may also be illustrated by a MetaD published by Brandt et al. [64]. In their work, the authors performed a set of calculations describing the unbinding of immunogenic peptides from the cleft of the alpha subunit of MHC class I. They found that simulations that run starting from complexes with inaccurate initial peptide docking configurations provided differences over 2.0 kcal/mol in free energy of dissociation ($\Delta\Delta G_d$) against experimental data. Therefore, a properly calibrated MetaD can be used to discriminate a correct binding pose from wrong ones.

However, an obstacle when employing many enhanced sampling simulations is that a reaction coordinate has to be set up a priori [63]. The reaction coordinate is not simple to determine since previous knowledge of the system arrangement is required. Consequently, when studying a new configuration of a ligand–receptor system, enhanced sampling methods might not be as useful since an accurate calibration to find the best reaction coordinate would be necessary.

4 Notes

1. Approaches to validate docking programs can help pointing out an efficient protocol to a specific target (Fig. 3). Approaches such as docking of a ligand into the receptor from which it was extracted (a re-docking procedure) or using another receptor structure from the same protein complexed with a different ligand (a cross-docking procedure) provide an assessment of the docking program accuracy to reproduce crystallographic binding modes [65]. The use of existing metrics of VS success, such as Enrichment Factors (EF), Receiver operating characteristic (ROC), and Area Under the Curve (AUC), using sets of compounds (DUD [66], DUDE [67], PDBBinding [68], and so on) with measured binding affinities, are also very helpful [69].
2. If the target structure comes from X-ray crystallography, particular attention is required to the X-ray resolution, B-factor and occupancy, and other structure details, since they might contain incomplete chains, missing amino acids, or undesirable mutations [5]. Choosing the structure to be used for docking experiments is a crucial step, especially since these methods usually employ rigid docking protocols, in which the receptor structure is kept fixed [70]. This simplified approximation increases the speed of computations, although at the cost of a more realistic representation.
3. It is worth mentioning that predicted three-dimensional models are computationally derived approximations of structure and need to be submitted to validation processes and quality estimation before any SBDD approach can be carried out [71].
4. Knowledge about the biological target binding sites, pockets, cavities, and interaction interfaces is essential for biomolecular modeling and simulation experiments [70]. These details can be inferred from experimental or in silico studies of the biological target. Significant attention has to be paid for the correct assignment of protonation and tautomeric states of both receptor and the ligand. Moreover, receptor minimization including only hydrogen atoms (fixed heavy atoms) or all receptor atoms can be performed to achieve a low-energy conformation with appropriate bond length and angles.

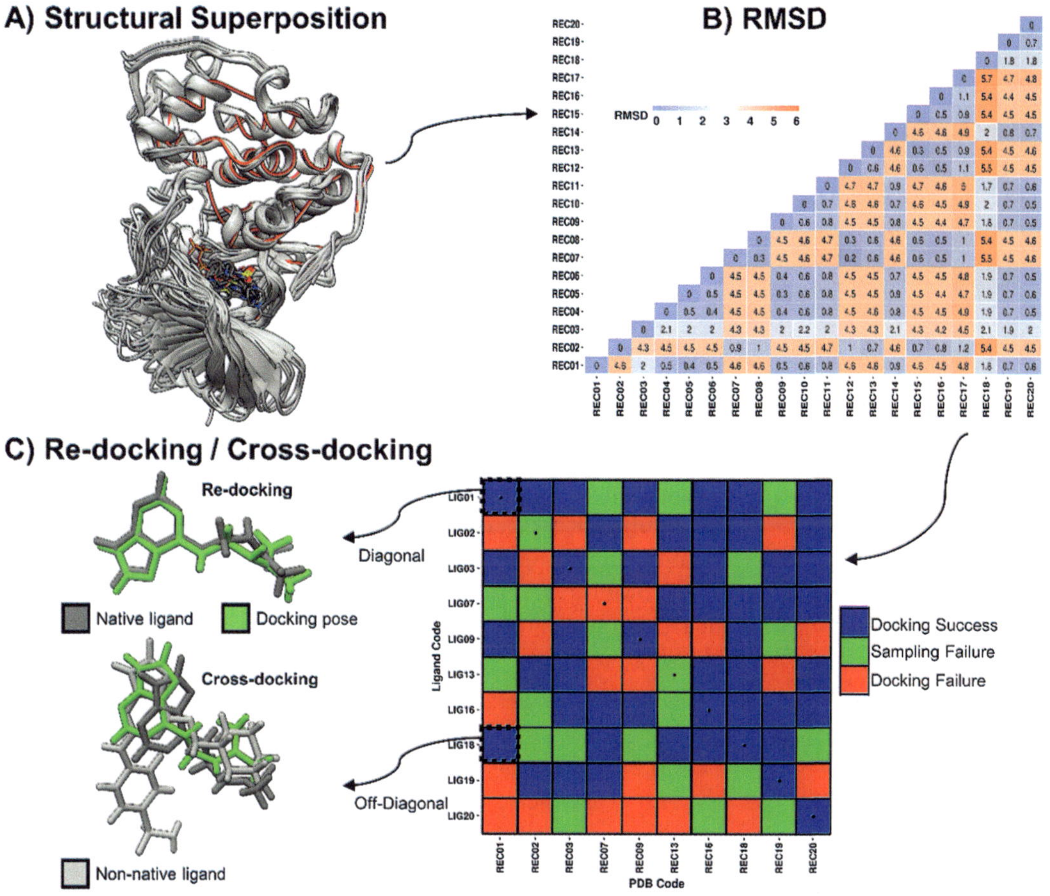

Fig. 3 Basic steps for validating docking programs using a multiple structure approach. (**a**) Superposition of different ligand bound structures of the same target. (**b**) RMSD calculation between all the structures to determine which structures would provide variability in the ensemble. Low RMSD values show targets with close structural arrangements, while high RMSD values display structural deviation between the structures. (**c**) After preparing the chosen group of structures, re-docking (diagonal squares marked with a dot symbol in the heatmap) and cross-docking (off-diagonal squares in the heatmap) are performed. Docking success (blue squares) shows that software can reproduce the native position of the ligand in a 2.0 Å cut-off as the best scoring pose. Sampling failure (red squares) illustrates the incapability of the software to reproduce the native position, and scoring failure (green squares) displays that the best scoring pose is not the closest to the native position

5. For instance, when working with novel compounds or databases with only two-dimensional ligand information, obtaining the three-dimensional structure of the ligand must precede any additional pre-docking preparation.
6. Some docking programs employ grids composed of a set of points, for which potentials are pre-calculated and used during rigid receptor docking to determine interaction energies. In such cases, specific parameters such as the space between grid points, the so-called "grid spacing" (usually low values of

0.3 Å), the center and size of the search space enclosed by the grid are established. Determination of the grid is typically done rapidly and with low computational cost [72]. The pre-calculated grids are subsequently utilized in the scoring stage of docking.

7. Another molecular docking strategy, the so-called "blind docking" involves detecting possible binding sites through exploration of the entire protein surface using a particular compound as a probe [73, 74]. Although blind docking might predict known binding sites retrospectively [74, 75], the computational cost of applying docking considering the whole target is often exorbitant, and the resulting conformations might not be reliable [76]. Ghersi and Sanchez [76] provided a protocol to minimize this concern starting with binding site prediction using blind docking, followed by focused docking of small molecules into the predicted sites of a set of 77 known protein–ligand complexes and 19 non-ligand-bound structures. This combined approach improved the sampling and accuracy in the predicted regions when compared with blind docking alone.

8. Structures with no known binding sites can be submitted to pocket prediction methods capable of indicating and ranking possible binding regions in the receptor [77, 78]. For example, the AutoDock suite has a module called AutoLigand [79] that identifies likely binding sites on a receptor surface using the free energy force field of AutoDock [80]. Therefore, an all-in-one docking protocol with AutoDock can include binding site prediction and ligand binding mode search [81].

9. A configuration file is usually needed for running molecular docking. In this file, specific algorithm parameters are discriminated, such as the number of runs performed by the program, time spent sampling ligand conformations in the search, and the amount of returned docking poses for analysis. Extensive conformation sampling might improve the quality of the poses. However, computational time increases linearly with increasing the depth of the algorithm's search parameters.

10. In the context of drug discovery, an important application of docking and scoring is in structure-based virtual screening (VS) experiments. Molecular docking-based VS can be seen as a complementary computational approach to the more time- and resource-consuming high-throughput screening (HTS) technique [82]. In VS, chemical databases are screened applying computational methods, and compounds are sorted out according to their predicted binding strength to a chosen protein site [83] and some other filters such as Lipinski rule of five [84], toxicity, partition coefficient (log *P*), and so on.

In the end, only a small fraction of the screened compounds are further examined as possible hits for biological trials. Throughout the years, VS has become a broadly useful and highly employed approach, and numerous libraries of small molecules with expected drug-like properties are accessible, such as the ZINC database [16].

11. A popular docking approach is to perform docking against several slightly different global conformations of the same receptor to increase the chances of accommodating a ligand in an appropriate conformation [85]. This approach is based on the fact that, during binding, a protein undergoes conformational changes to accommodate the ligand. Therefore, different receptor conformations in complex with a broad range of ligands and not just a single structure can provide a broader vision of the macromolecule binding site [48]. However, it is important to keep in mind that different receptor conformations differ on stability, and ideally, this should be accounted for in the scoring function.

12. One of the simplest approaches to account for receptor flexibility is called soft docking, where the repulsive terms of the Lennard-Jones potential are reduced to allow for a closer approximation between ligand and receptor [86]. In this method, no major changes are made in receptor conformation, since it is maintained rigid during docking and the scoring function handles for the difference. A more comprehensive method provides the option of selecting multiple conformations for side chains of chosen residues, usually in the binding site, during or after ligand docking. This method uses rotamer libraries, a set of commonly observed amino acid side chain conformations, turning its computational cost higher but improving ligand fit somehow [87]. Some approaches of induced fit docking (IFD) provide limited backbone variations alongside side chain flexibility during docking [88]. However, considering major backbone conformational changes, such as the opening and closing of subdomains, during the docking process remains challenging [85].

13. Available receptor–ligand structures can be useful to analyze a binding site and suggest important interactions between the ligand and its receptor that may be reproduced by a novel compound after docking.

14. Commonly used scoring functions use approximations to define both intramolecular and intermolecular interactions in the formed complex and also to determine the strength of interactions between receptor and ligand [70]. Therefore, individual scoring functions are not ideal, and one should carefully test or combine alternative functions to enhance the quality of docking results.

15. A force field is responsible for describing the interactions between atoms (or particles) of the system regarding parameters of covalently bound atoms (bonds, angles, and torsions) and nonbonded parameters (van der Waal and electrostatic interactions) [12]. These established parameters and functional form of the function constitute the force-field, which is indispensable to determine the contribution of each type of interaction to the general function [8]. A force field is responsible for describing the interactions between atoms (or particles) of the system regarding parameters of covalently bound atoms (bonds, angles, and torsions) and nonbonded parameters (van der Waal and electrostatic interactions) [12]. AMBER [25], OPLS [89], CHARMM [90], and GROMOS [91] are conventionally used force-fields. If ligands are in the simulation, particular parametrization will have to be performed by a generalized force-field such as GAFF [92], OPLS-AA [93], and CGenFF [94, 95].
16. Biomolecular simulations have advanced significantly since the first protein MD simulation of the bovine pancreatic trypsin inhibitor (BPTI), which was performed in 1977 for almost 10 ps in vacuo by McCammon, Gelin, and Karplus [96].
17. Parameterization of nonstandard molecules such as ligands can be problematic [13]. Some missing parameters can be easily determined, while others demand a more time-consuming process to parameterize. Ligand parameterization can be a bottleneck, especially when one is interested in a significant number of ligands. In this case, MD might not be the most practical methodology to apply.
18. The so-called "enhanced sampling methods" can accelerate the rare events not sampled by conventional MD methods and can be used to study ligand binding, estimate free energies, and kinetics [97]. These methods involve free-energy perturbation [98], metadynamics (MetaD) [99], steered MD [100], accelerated MD [101], umbrella sampling [102], replica exchange [103], and possible combinations of these approaches.
19. The initial coordinate file, containing the system to be simulated, has to be properly checked and cleaned up of undesired particles.
20. The topology file to be generated will once again depend on the MD program, and particular modules/programs can be used to produce it. For instance, pdb2gmx [104, 105] is used to generate topology files in GROMACS format, PSFgen [106] constructs topology files for NAMD format, and Leap [107] creates topology files for AMBER format.
21. The use of an explicit solvent model can be done by using specific models according to the force field to be utilized

(TIP3P [108], TIP4P [108], and SPC [109] water models) to resemble the cellular atmosphere closely. When an implicit representation of solvent is used, the model treats the solvent as a continuous medium. The explicit representation demands the explicit addition of a particular number of water molecules (or any other kind of solvent) taken into account physicochemical parameters. The calculation of molarity, molality, or concentration to be used may help determine the right number of water molecules to be added to the simulation box to mimic physiological medium. Although explicit solvent models are more computationally expensive than implicit ones, it is the most broadly used method for carrying out MD simulations.

22. The concept of periodic boundary conditions (PBC) might be applicable here. Use of PBC involves surrounding the simulated system with the same virtual unit cells that can interact with the atoms in the real system [28]. This concept recreates a more faithful representation of the in vivo environment and helps avoid boundary effects.
23. Energy minimization comprises systematically changing the positions of atoms in a predetermined number of iterations and calculating the energy up until the stress in the molecule is relaxed. Minimization is also required to fix any structural clashes caused during the system preparation.
24. Typically, thousands of minimization iterations, a number provided in the configuration file, are necessary to reach energy convergence, where the energy gradient approaches zero. In general, three minimization protocols can be chosen: steepest descent, conjugate gradient, and Newton-Raphson. More than one minimization optimization in tandem might be needed.
25. Equilibration comprises simulations in the canonical ensemble (NVT—substance (*N*), volume (*V*), and temperature (*T*) kept constant), and isothermal-isobaric ensemble (NPT—substance (*N*), pressure (*P*), and temperature (*T*) conserved). Equilibration in the NVT ensemble, where the energy of endothermic and exothermic processes is exchanged with a thermostat, should be done to bring up the temperature from zero Kelvin to the temperature of interest. Equilibration in the NPT ensemble is needed to stabilize the density by using a thermostat and a barostat.
26. In general MD protocols, minimization and equilibration are systematically performed in several steps, often imposing and releasing position restraints on the solvent and solute in the system. After every minimization-equilibration cycle, it is a good idea to search for error messages in the output files. Moreover, it is important to visualize the last equilibration cycle using visualization software such as VMD, to ensure a

consistent structure has been achieved before starting the production MD run.

27. In the production stage of MD, thermodynamic averages and new configurations of the system are sampled by solving Newton's equation of motion.

28. The standard trajectory analysis consists of root-mean-square deviation (RMSD), measuring distances, radii of gyration, clustering of conformations, time correlations, among other investigations. Trajectory analysis can be done with tools such as the ones found in GROMACS packages, AmberTools [29], and VMD [24] plug-ins.

5 Final Considerations

Accurate identification of the right pose of a ligand within a receptor binding, through computational methods, is difficult to achieve. Effects such as induced fit, which involves the adaptation of the neighboring residues to the presence of the ligand, polarizability effects, and the presence of water molecules, cofactors, or ions, may hinder binding mode predictions using static approaches such as docking.

The combination of molecular docking and MD simulations help out in the correction of these issues offering a more realistic picture, although it does not eliminate the ambiguity completely. MD simulations are expensive to carry out but with the advances in hardware and the application of high-throughput molecular dynamics new aspects of the binding nature of ligands can be achieved.

Although this issue is still to be solved, the joint use of both techniques can be very useful and insightful during drug development.

References

1. Sliwoski G, Kothiwale S, Meiler J, Lowe EW (2014) Computational methods in drug discovery. Pharmacol Rev 66:334–395
2. Lounnas V, Ritschel T, Kelder J, McGuire R, Bywater RP, Foloppe N (2013) Current progress in structure-based rational drug design marks a new mindset in drug discovery. Comput Struct Biotechnol J 5:1–14
3. Salum LB, Polikarpov I, Andricopulo AD (2008) Structure-based approach for the study of estrogen receptor binding affinity and subtype selectivity. J Chem Inf Model 48:2243–2253
4. Fischer E (1894) Influence of configuration on the action of enzymes. Ber Dtsch Chem Ges 27:2985–2993
5. Chen Y-C (2015) Beware of docking! Trends Pharmacol Sci 36:78–95
6. Hou X, Du J, Zhang J, Du L, Fang H, Li M (2013) How to improve docking accuracy of AutoDock4. 2: a case study using different electrostatic potentials. J Chem Inf Model 53:188–200
7. Lee MR, Sun Y (2007) Improving docking accuracy through molecular mechanics generalized born optimization and scoring. J Chem Theory Comput 3:1106–1119

8. Karplus M, McCammon JA (2002) Molecular dynamics simulations of biomolecules. Nat Struct Biol 9:646–652
9. Doerr S, Harvey M, Noé F, De Fabritiis G (2016) HTMD: High-throughput molecular dynamics for molecular discovery. J Chem Theory Comput 12:1845–1852
10. Buch I, Giorgino T, De Fabritiis G (2011) Complete reconstruction of an enzyme-inhibitor binding process by molecular dynamics simulations. Proc Nati Acad Sci U S A 108:10184–10189
11. Durrant JD, McCammon JA (2011) Molecular dynamics simulations and drug discovery. BMC Biol 9:71
12. Mortier J, Rakers C, Bermudez M, Murgueitio MS, Riniker S, Wolber G (2015) The impact of molecular dynamics on drug design: applications for the characterization of ligand–macromolecule complexes. Drug Discov Today 20:686–702
13. Alonso H, Bliznyuk AA, Gready JE (2006) Combining docking and molecular dynamic simulations in drug design. Med Res Rev 26:531–568
14. Fang Y (2012) Ligand–receptor interaction platforms and their applications for drug discovery. Expert Opin Drug Discovery 7:969–988
15. Weigelt J (2010) Structural genomics—impact on biomedicine and drug discovery. Exp Cell Res 316:1332–1338
16. Irwin JJ, Shoichet BK (2005) ZINC–a free database of commercially available compounds for virtual screening. J Chem Inf Model 45:177
17. Morris GM, Goodsell DS, Halliday RS, Huey R, Hart WE, Belew RK et al (1998) Automated docking using a Lamarckian genetic algorithm and an empirical binding free energy function. J Comput Chem 19:1639–1662
18. Jones G, Willett P, Glen RC, Leach AR, Taylor R (1997) Development and validation of a genetic algorithm for flexible docking. J Mol Biol 267:727–748
19. Friesner RA, Banks JL, Murphy RB, Halgren TA, Klicic JJ, Mainz DT et al (2004) Glide: a new approach for rapid, accurate docking and scoring. 1. Method and assessment of docking accuracy. J Med Chem 47:1739–1749
20. Halgren TA, Murphy RB, Friesner RA, Beard HS, Frye LL, Pollard WT et al (2004) Glide: a new approach for rapid, accurate docking and scoring. 2. Enrichment factors in database screening. J Med Chem 47:1750–1759
21. Rarey M, Kramer B, Lengauer T, Klebe G (1996) A fast flexible docking method using an incremental construction algorithm. J Mol Biol 261:470–489
22. DeLano WL (2002) The PyMOL Molecular Graphics System. De-Lano Scientific, San Carlos, CA. http://www.pymol.org
23. Pettersen EF, Goddard TD, Huang CC, Couch GS, Greenblatt DM, Meng EC, Ferrin TE (2004) UCSF Chimera—a visualization system for exploratory research and analysis. J Comput Chem 25:1605–1612
24. Humphrey W, Dalke A, Schulten K (1996) VMD: visual molecular dynamics. J Mol Graph 14:33–38
25. Merz KM Jr, Ferguson DM, Spellmeyer DC, Fox T, Caldwell JW, Kollman PA (1995) A second generation force field for the simulation of proteins, nucleic acids, and organic molecules. J Am Chem Soc 117:5179–5197
26. Brooks BR, Bruccoleri RE, Olafson BD, States DJ, Swaminathan S, Karplus M (1983) CHARMM: A program for macromolecular energy, minimization, and dynamics calculations. J Comput Chem 4:187–217
27. van der Spoel D, van Maaren PJ, Caleman C (2012) GROMACS molecule & liquid database. Bioinformatics 28:752–753
28. Nelson MT, Humphrey W, Gursoy A, Dalke A, Kalé LV, Skeel RD et al (1996) NAMD: a parallel, object-oriented molecular dynamics program. Int J High Perform Comput Appl 10:251–268
29. Case D, Berryman J, Betz R, Cerutti D, Cheatham T III, Darden T et al (2015) AMBER. University of California, San Francisco
30. Yuriev E, Agostino M, Ramsland PA (2011) Challenges and advances in computational docking: 2009 in review. J Mol Recognit 24:149–164
31. Jain AN (2006) Scoring functions for protein-ligand docking. Curr Protein Pept Sci 7:407–420
32. Malisi C, Schumann M, Toussaint NC, Kageyama J, Kohlbacher O, Höcker B (2012) Binding pocket optimization by computational protein design. PLoS One 7: e52505
33. Leaver-Fay A, Tyka M, Lewis SM, Lange OF, Thompson J, Jacak R et al (2011) ROSETTA3: an object-oriented software suite for the simulation and design of macromolecules. Methods Enzymol 487:545
34. Gainza P, Roberts KE, Georgiev I, Lilien RH, Keedy DA, Chen C-Y et al (2013) OSPREY: protein design with ensembles, flexibility, and

provable algorithms. Methods Enzymol 523:87

35. Hong Enriquez RP, Pavan S, Benedetti F, Tossi A, Savoini A, Berti F et al (2012) Designing short peptides with high affinity for organic molecules: a combined docking, molecular dynamics, and Monte Carlo approach. J Chem Theory Comput 8:1121–1128
36. Jorgensen WL (2004) The many roles of computation in drug discovery. Science 303:1813–1818
37. Berman HM, Westbrook J, Feng Z, Gilliland G, Bhat TN, Weissig H et al (2000) The protein data bank. Nucleic Acids 28:235–242
38. Hanwell MD, Curtis DE, Lonie DC, Vandermeersch T, Zurek E, Hutchison GR (2012) Avogadro: an advanced semantic chemical editor, visualization, and analysis platform. J Chem 4:17
39. Csizmadia P (2000) MarvinSketch and MarvinView: molecule applets for the World Wide Web. In: Proceedings of ECSOC-3 The Third International Electronic Conference on Synthetic Organic Chemistry, September 1–30, 1999, pp 367–369
40. Ultra C (2001) CambridgeSoft. Cambridge, MA, USA
41. Mullins JG (2012) 5 structural modelling pipelines in next generation sequencing projects. Adv Protein Chem Struct Biol 89:117
42. Feinstein WP, Brylinski M (2015) Calculating an optimal box size for ligand docking and virtual screening against experimental and predicted binding pockets. J Cheminform 7:18
43. Morra G, Genoni A, Neves M, Merz J, Colombo G (2010) Molecular recognition and drug-lead identification: what can molecular simulations tell us? Curr Med Chem 17:25–41
44. Ivetac A, Andrew McCammon J (2011) Molecular recognition in the case of flexible targets. Curr Pharm Des 17:1663–1671
45. Klepeis JL, Lindorff-Larsen K, Dror RO, Shaw DE (2009) Long-timescale molecular dynamics simulations of protein structure and function. Curr Opin Struct Biol 19:120–127
46. Lu H, Tonge PJ (2010) Drug–target residence time: critical information for lead optimization. Curr Opin Chem Biol 14:467–474
47. De Vivo M, Masetti M, Bottegoni G, Cavalli A (2016) Role of molecular dynamics and related methods in drug discovery. J Med Chem 59:4035–4061
48. Barril X, Morley SD (2005) Unveiling the full potential of flexible receptor docking using multiple crystallographic structures. J Med Chem 48:4432–4443
49. Amaro RE, Baron R, McCammon JA (2008) An improved relaxed complex scheme for receptor flexibility in computer-aided drug design. J Comput Aided Mol Des 22:693–705
50. Lin J-H, Perryman AL, Schames JR, McCammon JA (2002) Computational drug design accommodating receptor flexibility: the relaxed complex scheme. J Am Chem Soc 124:5632–5633
51. Ivetac A, Swift SE, Boyer PL, Diaz A, Naughton J, Young JA et al (2014) Discovery of novel inhibitors of HIV-1 reverse transcriptase through virtual screening of experimental and theoretical ensembles. Chem Biol Drug Des 83:521–531
52. Rueda M, Bottegoni G, Abagyan R (2010) Recipes for the selection of experimental protein conformations for virtual screening. J Chem Inf Model 50:186
53. Nichols SE, Baron R, Ivetac A, McCammon JA (2011) Predictive power of molecular dynamics receptor structures in virtual screening. J Chem Inf Model 51:1439–1446
54. Tian S, Sun H, Pan P, Li D, Zhen X, Li Y et al (2014) Assessing an ensemble docking-based virtual screening strategy for kinase targets by considering protein flexibility. J Chem Inf Model 54:2664–2679
55. Gao C, Desaphy J, Vieth M (2017) Are induced fit protein conformational changes caused by ligand-binding predictable? A molecular dynamics investigation. J Comput Chem 38:1229–1237
56. Yadav IS, Nandekar PP, Shrivastava S, Sangamwar A, Chaudhury A, Agarwal SM (2014) Ensemble docking and molecular dynamics identify knoevenagel curcumin derivatives with potent anti-EGFR activity. Gene 539:82–90
57. Watanabe Y, Fukuyoshi S, Kato K, Hiratsuka M, Yamaotsu N, Hirono S et al (2017) Investigation of substrate recognition for cytochrome P450 1A2 mediated by water molecules using docking and molecular dynamics simulations. J Mol Graph Model 74:326–336
58. Carlevaro CM, Martins-Da-Silva JH, Savino W, Caffarena ER (2013) Plausible binding mode of the active α4β1 antagonist, Mk-0617, determined by docking and free energy calculations. J Theor Comput Chem 12:1250108

59. Trott O, Olson AJ (2010) AutoDock Vina: improving the speed and accuracy of docking with a new scoring function, efficient optimization, and multithreading. J Comput Chem 31:455–461
60. Silva JHM, Dardenne LE, Savino W, Caffarena ER (2010) Analysis of α4 β1integrin specific antagonists binding modes: structural insights by molecular docking, molecular dynamics and linear interaction energy method for free energy calculations. J Braz Chem Soc 21:546–555
61. Wang Q, Edupuganti R, Tavares CD, Dalby KN, Ren P (2015) Using docking and alchemical free energy approach to determine the binding mechanism of eEF2K inhibitors and prioritizing the compound synthesis. Front Mol Biosci 2:9
62. Clark AJ, Tiwary P, Borrelli K, Feng S, Miller EB, Abel R, Friesner RA, Berne BJ (2016) Prediction of protein–ligand binding poses via a combination of induced fit docking and metadynamics simulations. J Chem Theory Comput 12:2990–2998
63. Sinko W, Lindert S, McCammon JA (2013) Accounting for receptor flexibility and enhanced sampling methods in computer-aided drug design. Chem Biol Drug Des 81:41–49
64. Brandt AM, Batista PR, Souza-Silva F, Alves CR, Caffarena ER (2016) Exploring the unbinding of *Leishmania* (L.) *amazonensis* CPB derived-epitopes from H2 MHC class I proteins. Proteins 84:473–487
65. Sutherland JJ, Nandigam RK, Erickson JA, Vieth M (2007) Lessons in molecular recognition. 2. Assessing and improving cross-docking accuracy. J Chem Inf Model 47:2293–2302
66. Huang N, Shoichet BK, Irwin JJ (2006) Benchmarking sets for molecular docking. J Med Chem 49:6789–6801
67. Mysinger MM, Carchia M, Irwin JJ, Shoichet BK (2012) Directory of useful decoys, enhanced (DUD-E): better ligands and decoys for better benchmarking. J Med Chem 55:6582–6594
68. Wang R, Fang X, Lu Y, Yang C-Y, Wang S (2005) The PDBbind database: methodologies and updates. J Med Chem 48:4111–4119
69. Cross JB, Thompson DC, Rai BK, Baber JC, Fan KY, Hu Y et al (2009) Comparison of several molecular docking programs: pose prediction and virtual screening accuracy. J Chem Inf Model 49:1455–1474
70. Biesiada J, Porollo A, Meller J (2012) On setting up and assessing docking simulations for virtual screening. Methods Mol Biol 928:1–16
71. Schmidt T, Bergner A, Schwede T (2014) Modelling three-dimensional protein structures for applications in drug design. Drug Discov Today 19:890–897
72. Wu G, Robertson DH, Brooks CL, Vieth M (2003) Detailed analysis of grid-based molecular docking: a case study of CDOCKER—A CHARMm-based MD docking algorithm. J Comput Chem 24:1549–1562
73. Zhou M, Luo H, Li R, Ding Z (2013) Exploring the binding mode of HIV-1 Vif inhibitors by blind docking, molecular dynamics and MM/GBSA. RSC Adv 3:22532–22543
74. Hetényi C, van der Spoel D (2002) Efficient docking of peptides to proteins without prior knowledge of the binding site. Protein Sci 11:1729–1737
75. Hetényi C, van der Spoel D (2006) Blind docking of drug-sized compounds to proteins with up to a thousand residues. FEBS Lett 580:1447–1450
76. Ghersi D, Sanchez R (2009) Improving accuracy and efficiency of blind protein-ligand docking by focusing on predicted binding sites. Proteins 74:417–424
77. Pérot S, Sperandio O, Miteva MA, Camproux A-C, Villoutreix BO (2010) Druggable pockets and binding site centric chemical space: a paradigm shift in drug discovery. Drug Discov Today 15:656–667
78. Leis S, Schneider S, Zacharias M (2010) In silico prediction of binding sites on proteins. Curr Med Chem 17:1550–1562
79. Harris R, Olson AJ, Goodsell DS (2008) Automated prediction of ligand-binding sites in proteins. Proteins 70:1506–1517
80. Cosconati S, Forli S, Perryman AL, Harris R, Goodsell DS, Olson AJ (2010) Virtual screening with AutoDock: theory and practice. Expert Opin Drug Discovery 5:597–607
81. Forli S, Huey R, Pique ME, Sanner MF, Goodsell DS, Olson AJ (2016) Computational protein-ligand docking and virtual drug screening with the AutoDock suite. Nat Protoc 11:905–919
82. Bajorath J (2002) Integration of virtual and high-throughput screening. Nat Rev Drug Discov 1:882–894
83. Ghosh S, Nie A, An J, Huang Z (2006) Structure-based virtual screening of chemical libraries for drug discovery. Curr Opin Chem Biol 10:194–202

84. Lipinski CA, Lombardo F, Dominy BW, Feeney PJ (1997) Experimental and computational approaches to estimate solubility and permeability in drug discovery and development settings. Adv Drug Deliv Rev 23:3–25
85. Totrov M, Abagyan R (2008) Flexible ligand docking to multiple receptor conformations: a practical alternative. Curr Opin Struct Biol 18:178–184
86. Ferrari AM, Wei BQ, Costantino L, Shoichet BK (2004) Soft docking and multiple receptor conformations in virtual screening. J Med Chem 47:5076
87. Källblad P, Dean PM (2003) Efficient conformational sampling of local side-chain flexibility. J Mol Biol 326:1651–1665
88. Lexa KW, Carlson HA (2012) Protein flexibility in docking and surface mapping. Q Rev Biophys 45:301–343
89. Jorgensen WL, Tirado-Rives J (1988) The OPLS [optimized potentials for liquid simulations] potential functions for proteins, energy minimizations for crystals of cyclic peptides and crambin. J Am Chem Soc 110:1657–1666
90. MacKerell AD Jr, Bashford D, Bellott M, Dunbrack RL Jr, Evanseck JD, Field MJ et al (1998) All-atom empirical potential for molecular modeling and dynamics studies of proteins. J Phys Chem B 102:3586–3616
91. Oostenbrink C, Villa A, Mark AE, Van Gunsteren WF (2004) A biomolecular force field based on the free enthalpy of hydration and solvation: the GROMOS force-field parameter sets 53A5 and 53A6. J Comput Chem 25:1656–1676
92. Wang J, Wolf RM, Caldwell JW, Kollman PA, Case DA (2004) Development and testing of a general amber force field. J Comput Chem 25:1157–1174
93. Jorgensen WL, Maxwell DS, Tirado-Rives J (1996) Development and testing of the OPLS all-atom force field on conformational energetics and properties of organic liquids. J Am Chem Soc 118:11225–11236
94. Vanommeslaeghe K, MacKerell AD Jr (2012) Automation of the CHARMM General Force Field (CGenFF) I: bond perception and atom typing. J Chem Inf Model 52:3144
95. Vanommeslaeghe K, Raman EP, MacKerell AD Jr (2012) Automation of the CHARMM general force field (CGenFF) II: assignment of bonded parameters and partial atomic charges. J Chem Inf Model 52:3155
96. McCammon JA, Gelin BR, Karplus M (1977) Dynamics of folded proteins. Nature 267:585
97. Abrams C, Bussi G (2013) Enhanced sampling in molecular dynamics using metadynamics, replica-exchange, and temperature-acceleration. Entropy 16:163–199
98. Jorgensen WL, Thomas LL (2008) Perspective on free-energy perturbation calculations for chemical equilibria. J Chem Theory Comput 4:869
99. Laio A, Parrinello M (2002) Escaping free-energy minima. Proc Natl Acad Sci U S A 99:12562–12566
100. Isralewitz B, Gao M, Schulten K (2001) Steered molecular dynamics and mechanical functions of proteins. Curr Opin Struct Biol 11:224–230
101. Hamelberg D, Mongan J, McCammon JA (2004) Accelerated molecular dynamics: a promising and efficient simulation method for biomolecules. J Chem Phys 120:11919–11929
102. Torrie GM, Valleau JP (1977) Nonphysical sampling distributions in Monte Carlo free-energy estimation: Umbrella sampling. J Comput Phys 23:187–199
103. Sugita Y, Okamoto Y (1999) Replica-exchange molecular dynamics method for protein folding. Chem Phys Lett 314:141–151
104. Lindahl E, Hess B, Van Der Spoel D (2001) GROMACS 3.0: a package for molecular simulation and trajectory analysis. J Mol Model 7:306–317
105. Van Der Spoel D, Lindahl E, Hess B, Groenhof G, Mark AE, Berendsen HJ (2005) GROMACS: fast, flexible, and free. J Comput Chem 26:1701–1718
106. Gullingsrud J, Saam J, Phillips J (2006) psfgen User's Guide, vol 51. Theoretical and Computational Biophysics Group, University of Illinois and Beckman Institute, Urbana, p 61801
107. Case DA, Cheatham TE, Darden T, Gohlke H, Luo R, Merz KM et al (2005) The Amber biomolecular simulation programs. J Comput Chem 26:1668–1688
108. Jorgensen WL, Chandrasekhar J, Madura JD, Impey RW, Klein ML (1983) Comparison of simple potential functions for simulating liquid water. J Chem Phys 79:926–935
109. Berendsen HJ, Postma JP, van Gunsteren WF, & Hermans J (1981) Interaction models for water in relation to protein hydration. In Intermolecular forces (pp. 331–342). Springer, Dordrecht

Chapter 3

How Docking Programs Work

Gabriela Bitencourt-Ferreira and Walter Filgueira de Azevedo Jr.

Abstract

Protein–ligand docking simulations are of central interest for computer-aided drug design. Docking is also of pivotal importance to understand the structural basis for protein–ligand binding affinity. In the last decades, we have seen an explosion in the number of three-dimensional structures of protein–ligand complexes available at the Protein Data Bank. These structures gave further support for the development and validation of in silico approaches to address the binding of small molecules to proteins. As a result, we have now dozens of open source programs and web servers to carry out molecular docking simulations. The development of the docking programs and the success of such simulations called the attention of a broad spectrum of researchers not necessarily familiar with computer simulations. In this scenario, it is essential for those involved in experimental studies of protein–ligand interactions and biophysical techniques to have a glimpse of the basics of the protein–ligand docking simulations. Applications of protein–ligand docking simulations to drug development and discovery were able to identify hits, inhibitors, and even drugs. In the present chapter, we cover the fundamental ideas behind protein–ligand docking programs for non-specialists, which may benefit from such knowledge when studying molecular recognition mechanism.

Key words Docking, Protein, Ligand, Drug design, Molecular recognition

1 Introduction

Protein–ligand docking simulation is a computational methodology that primarily seeks to find the position for a ligand in the binding site of a protein target [1, 2]. This type of computational analysis of protein–ligand interactions plays a vital role in computer-aided drug design as well as to the understanding of fundamental biochemical processes [3–10]. Although not strictly correct from the enzymology point of view, the simplification of the classic key–lock theory of enzyme specificity [11, 12] is a naïve model that we can use to understand the basics of the protein–ligand docking simulations or as said by Koshland [13],

> "I was also particularly intrigued with his classic key-lock (or template) theory of enzyme specificity, which like all great theories seemed so obvious once one understood it."

Walter Filgueira de Azevedo Jr. (ed.), *Docking Screens for Drug Discovery*, Methods in Molecular Biology, vol. 2053,
https://doi.org/10.1007/978-1-4939-9752-7_3, © Springer Science+Business Media, LLC, part of Springer Nature 2019

The classic key–lock theory of enzyme specificity will guide us through the exploration of this fascinating world protein–ligand docking simulations [14]. However, we will not restrict ourselves to docking of enzymes since it is possible to explore the basic idea of key fitting in the lock for any protein.

We can visualize the protein–ligand docking problem as an optimization problem, where we try to find the optimal location for a small-molecule ligand into the protein structure. Protein–ligand docking approach is the most common method for computer-aided drug design. This approach has been extensively applied to drug discovery ever since the early 1980s [15], and the increase of the computational power and the availability of protein structures have been the major factors for the development of the field.

It is customary with a modest workstation to carry out simulations of thousands of potential ligands against a protein structure. The availability of open source docking programs [16–22] made it possible to perform protein–ligand docking projects [23–39] with a low budget. Moreover, the integration of the docking programs in a workflow allows us to carry out docking simulations in a unified way that facilitates the simulations and the analysis of the docking results [40].

If we consider current protein–ligand docking programs, they all share a universal design that is independent of the choice of algorithms implemented in a specific application. Any protein–ligand docking program is composed of at least a search algorithm and a scoring function. Many programs make available more than one search algorithm, for instance, AutoDock4 that makes available four search algorithms: genetic algorithm, Lamarckian genetic algorithm, local search, and simulated annealing [16–19]. On the other hand, a docking program as Glide [41–43] makes available more than one scoring function. Some programs make possible a combination of search algorithms and scoring functions, for instance, the program Molegro Virtual Docker (MVD) [44, 45].

In the program MVD, we have four search algorithms (differential evolution, simplex evolution, iterated simplex, and ant colony optimization) and four scoring functions (MolDock Score, MolDock Score with GRID, Plants Score, and Plants Score with GRID). The grid-based scoring functions available in the program MVD are faster than MolDock and Plants Scores since they calculate potential-energy values on a cubic grid [44] before the docking simulation.

2 Analogy with the Key–Lock Theory

As we anticipated in the introduction of this chapter, we will treat protein–ligand docking simulations as a key–lock problem. Let us see the ligand of a protein target as a key and the binding site of the

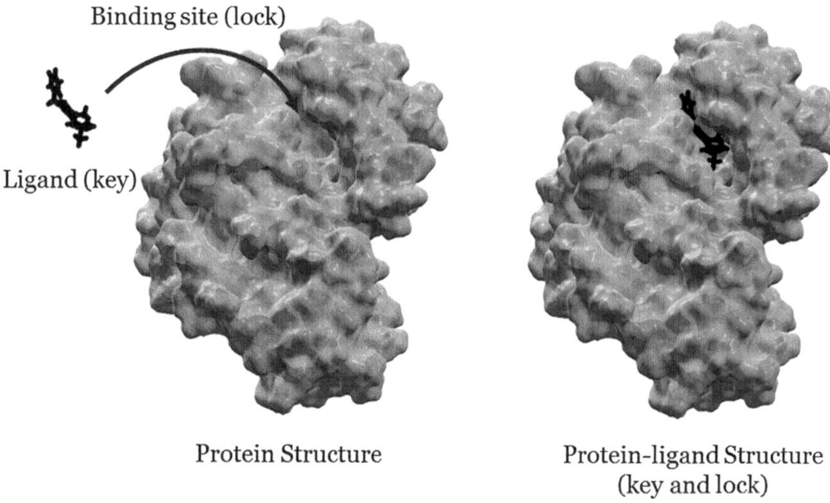

Fig. 1 Protein–ligand complex formation under the view of the key–lock paradigm. Here we show the protein surface and the ligand. We used the program MVD [44] to generate this figure

protein structure as a lock. Figure 1 shows protein–ligand interactions under the view of the key–lock theory. It is possible to visualize the whole idea of protein–ligand docking simulations through the analogy with the key–lock theory. It is as simple as to try to fit the key in the lock.

From a realistic point of view, it is necessary to consider that the experimenter who is trying to put the key into the lock is blindfolded. Let us also think that the experimenter is close to the door. Holding the key with his/her right hand at first, he/she tries with the left hand to locate the position of the lock. From the knowledge of the location of the lock, the moving of the key may be able to get close to the lock and then with small adjustments the experimenter can put it in the lock.

This analogy does not take into consideration the fine details of the internal mechanism of the lock, which is analogous to induced fit [14] of the binding site due to the interaction with the ligand. It is possible to simulate small adjustments due to the ligand binding into the protein structure through the flexibility of amino acid side chain.

It is clear that the key–lock approximation to protein–ligand docking simulation is a simple paradigm. Nevertheless, it is adequate for a crude view of what is going on during the protein–ligand docking simulation. We play around with a key and move it toward the lock mimicking the dock of the ligand in the binding site. We may act as the search algorithm of a docking program with our hand holding the key where we play with this key trying to find its position onto the lock. It is quite straightforward and in some fundamental aspects of the simulation, a valid approximation to the real problem. For instance, any search algorithm tries to accommodate the ligand into the designated binding site of the protein structure.

3 Docking Not as a Key–Lock Problem

As for protein–ligand interaction, the key–lock theory is an oversimplification of the real problem. For instance, proteins are not rigid structures. The same is true for its binding site. For a realistic view, we consider the flexibility of the amino acid side chains. To illustrate, let us analyze the rotatable angles in the side chain of the tryptophan (Fig. 2), we have two additional rotatable angles (φ_1 and φ_2): the angle ω involves main-chain atoms that we do not typically consider in protein–ligand docking simulations.

Adding the flexibility of the side chains of amino acids increases the computer demands for a given simulation substantially. One possibility to reduce the complexity of the protein–ligand system is to focus on the amino acids of the binding site. For instance, in Fig. 3, we have a docking sphere centered at the ATP-binding pocket of the protein cyclin-dependent kinase 2. For this protein–-ligand docking simulation, we restrict the flexibility to the side chains inside the docking sphere. We keep the protein system external to the docking sphere as a rigid body.

Each additional rotatable angle to be added is an extra degree of freedom for the system being simulated. We know that proteins are not dry entities; they interact with solvent and co-factors that we do not add to the key–lock approximation. Finally, the ligand itself is not necessarily a rigid structure. Rotatable angles should be included as additional degrees of freedom. Figure 4 shows a typical ligand where we highlight the rotatable angles in the structure.

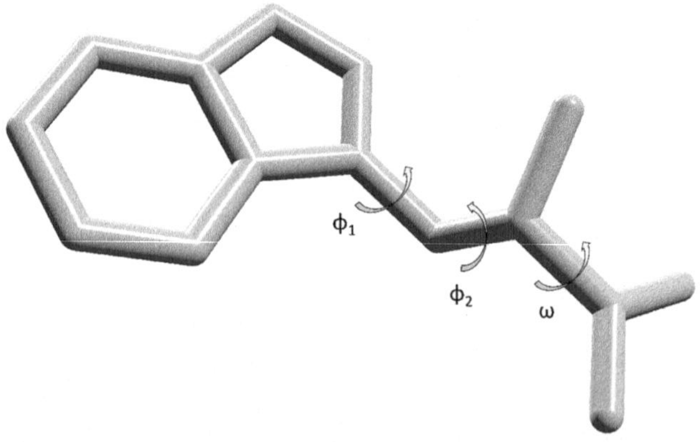

Fig. 2 Rotatable angles (φ_1 and φ_2) in the amino acid tryptophan. We used the program MVD [44] to generate this figure

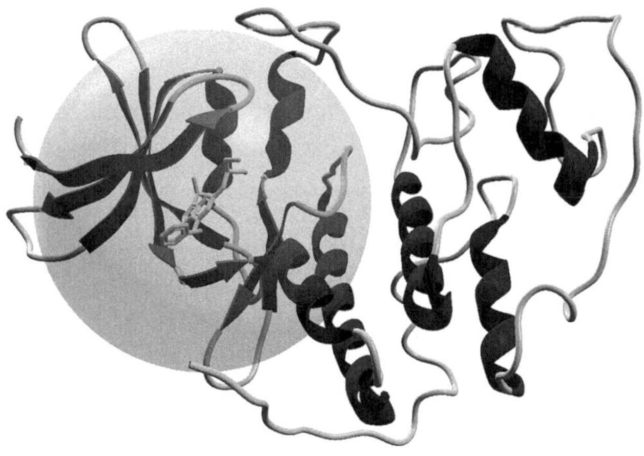

Fig. 3 Docking sphere centered at the active site of the cyclin-dependent kinase 2. We used the program MVD [44] to generate this figure

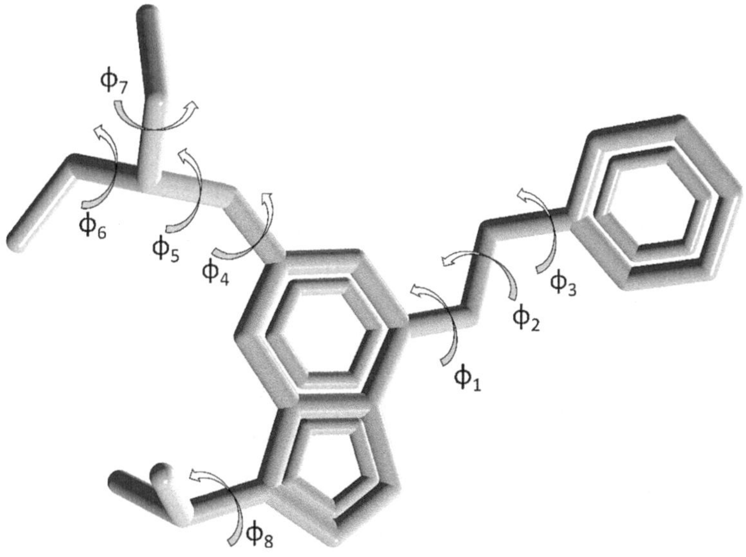

Fig. 4 Rotatable angles (φs) in a ligand. We used the program MVD [44] to generate this figure

In summary, key–lock analogies are useful for explanation of the overall process that occurs during docking simulations. However, we should keep in mind that protein–ligand structures are complex biomolecular systems that need to be carefully analyzed if we expect to generate a reliable computational model for them.

4 Search Algorithms

The whole idea behind the search algorithm in any protein–ligand docking simulation is to provide a computational technique to explore the relative orientation of the ligand into the binding pocket. Such algorithms must allow full scanning of the binding pocket and also consider the flexibility of the ligand and, sometimes, of the side chains of the amino acids found in the binding pocket. Increasing the complexity of the system being simulated with higher degrees of freedom (number of rotatable angles) has, as a result, a rise in the simulation time.

The most successful search algorithms for docking simulation are those based on evolutionary computing such as genetic algorithm available in the program AutoDock and differential evolution available in the program MVD. Such heuristic methods have the advantage of being faster than search algorithms such as exhaustive search. On the other hand, since these biologically inspired algorithms are stochastic, application of them should always be undertaken with care, as they are all dependent on the random seed used to generate the initial population of the evolutionary algorithms.

5 Scoring Functions

Scoring functions are computational approximations to predict protein–ligand binding affinity. Most of the modern development of scoring function for prediction of protein–ligand binding affinity, and their application to the selection of candidate poses generated by the search algorithms started with the pioneering work of Böhm in the early 1990s [46–51]. Docking programs such as AutoDock, AutoDock Vina, and MVD make use of empirical scoring functions that somehow work very similar to the ideas proposed by Böhm.

Let us consider that we express the protein–ligand binding affinity by the Gibbs free energy of binding for protein–ligand complexes (ΔG). The empirical scoring function tries to approximate experimental binding affinity (ΔG_e) through a regression model where we used the experimental data to determine the relative weights of each term in the regression equation. Below we have a generic empirical scoring function to illustrate the fundamental issues behind its development,

$$\Delta G_t = \alpha_0 + \alpha_1 \sum_{i=1}^{N} \sum_{j=1}^{M} V_{\mathrm{vdw},i,j} + \alpha_2 \sum_{i=1}^{N} \sum_{j=1}^{M} V_{\mathrm{Hbond},i,j} + \alpha_3 \sum_{i=1}^{N} \sum_{j=1}^{M} V_{\mathrm{elec},i,j} + \alpha_4 \sum_{i=1}^{N} \sum_{j=1}^{M} V_{\mathrm{desol},i,j} \quad (1)$$

where ΔG_t is the theoretical binding affinity, α_0 is the regression constant, α_1 is the relative weight of the van der Waals interaction term (V_{vdw}), α_2 is the relative weight of the hydrogen bond term (V_{Hbond}), α_3 is the relative weight of the electrostatic potential term (V_{elec}), and α_4 is the relative weight of the desolvation potential term (V_{desol}). It is feasible to add many other energy terms to the regression model, but the idea is the same. The protein–ligand docking program AutoDock4 [19] uses an additional variable in Eq. (1) to evaluate the number to rotatable angles ($N_{Torsion}$) in the ligand.

In protein–ligand docking, it is customary to consider the amount of torsions angles related to the entropic energy term. The summations are taken for atoms from the ligand (i) and protein (j) inside a predefined cutoff radius. In the above equation, N indicates the number of ligand atoms and M the number of protein atoms. We may apply these scoring functions to select the best pose generated by the search algorithm or evaluate binding affinity for any protein–ligand complex.

6 Overview

To have an integrated view of how protein–ligand docking programs work, we are going to consider the ideal situation where we have an ensemble of crystallographic structures for which experimental binding affinity is available. The atomic coordinates for the receptor–ligand complexes are available at the Protein Data Bank (PDB) [52–54], and the binding affinity data are available at MOAD [55], BindingDB [56], and PDBbind [57]. Figure 5 illustrates the primary steps involved in this docking project.

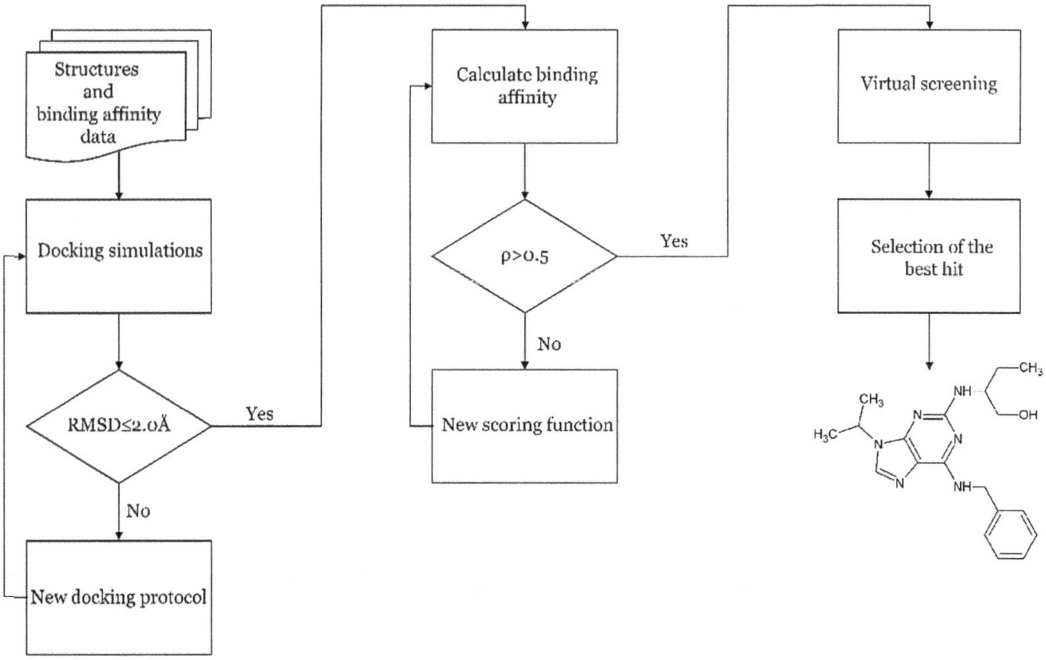

Fig. 5 This flowchart highlights all the steps of a modern approach to protein–ligand docking simulations. Here, ρ indicates Spearman's correlation coefficient

The first step in our docking project is the selection of the structures to be used in the project. Most of the docking simulations are based on structures determined by X-ray diffraction crystallography and nuclear magnetic resonance techniques. Moreover, we can employ homology models based on experimental structures in such simulations. It is even possible to use ab initio structures as a receptor. Nevertheless, since the docking itself is a computational methodology, it is safer to rely on experimental structures for docking simulations. Once a structure or an ensemble of structures has been selected, the next steps involve validation of the docking protocol. These steps should be carefully executed to give support to further analysis of protein–ligand complexes generated in docking simulations.

Initially, we have to answer critical questions to assess the performance of a docking program. (1) Is the docking program able to recover the crystallographic position of a ligand? (2) Is the docking program able to predict ligand binding affinity with reasonable performance?

For the first question, we generally evaluate the docking root-mean-square deviation (RMSD), calculated by Eq. (2)

$$\text{RMSD} = \sqrt{\frac{\sum_{i=1}^{N}\left[(x_{x,i} - x_{p,i})^2 + (y_{x,i} - y_{p,i})^2 + (z_{x,i} - z_{p,i})^2\right]}{N}} \quad (2)$$

where x_x, y_x, and z_x are the experimental coordinates for the ligand, and x_p, y_p, and z_p are the atomic coordinates for the position generated by the docking simulation. We call pose the computer-generated position for the ligand. When we calculate the summation, we consider the N nonhydrogen atoms in the ligand structure. So, it is clear that the ideal would be an RMSD = 0.0 Å. Most of the researchers involved in the development of docking programs consider that RMSD ≤ 2.0 Å is acceptable [40].

Since the majority of the docking programs generate more than one pose, it is customary to evaluate the docking accuracy of all poses created for a docking simulation. The following equation defines docking accuracy (DA) as follows:

$$\text{DA} = f_l + 0.5(f_l - f_h) \quad (3)$$

where f_l is the fraction poses for which the docking RMSD is less than l and f_h is the fraction of poses for which the docking RMSD is less than h, where $l < h$ [58, 59].

After selecting the best docking protocol, we can answer the second question, when considering the predictive performance of docking programs to calculate ligand binding affinity, the evaluation relay mostly on statistical analysis of correlation coefficients. For instance, Spearman's correlation coefficient (ρ) calculated between predicted and calculated binding affinities [60]. To assess the predictive performance of a scoring function, we estimate the binding affinity for all PDB files in the ensemble of structures. The correlation coefficient between the predicted and experimental binding affinities determines the success of a computational approach. It is expected to have a $\rho > 0.5$.

Once we defined a docking protocol, it is possible to apply it to identify a new potential ligand, named here as a hit. To find a hit, we usually try to dock small molecules available in databases such as ZINC [61, 62]. The process of scanning a database of small molecules using docking simulations is called virtual screening [7, 8]. It is possible to test thousands or even millions of molecules to try to find the potential new binder to the protein target. It is common to focus on virtual screening simulations of promising candidates using natural product datasets or trying drug repurposing to reduce computer usage. This procedure attempts to use an already approved drug to treat a different disease [63], for instance, use of aspirin to treat cancer [26].

7 Docking Exercise

To highlight the main concepts described in this chapter, we will consider a protein–ligand docking simulation of a protein target. We take as an example the study of cyclin-dependent kinase 2. This enzyme is an essential target for the development of anticancer drugs [64–74]. To run our simulations, we use the program MVD [44]. The first step in any docking simulation is the validation of the docking protocol; as we explained in the previous sections, we may evaluate the docking performance using the RMSD and the DA.

We considered the crystallographic structure of CDK2 in complex with roscovitine (PDB access code: 2A4L) [75]. We used a combination of differential evolution search algorithm with MolDock scoring function [44]. In the redocking simulation, the docking simulation to recover the crystallographic position of the ligand, we generated 50 poses. We show the lowest score pose in Fig. 6.

In Fig. 6, we see that the pose (dark gray) is close to the crystallographic position of the ligand (light gray). For this simulation, we have an RMSD of 0.97 Å, which is a value below the recommended limit of 2.0 Å. We could reach further validation through the application of this docking protocol to additional crystallographic structures of CDK2 in complex with different ligands. Such a procedure is called ensemble docking [40]. Such a set of docking RMSD's could be used to calculate the docking accuracy as indicated in Eq. (3). Ideal values of docking accuracy should be higher than 50%. Once validated this docking protocol, we may use an organic molecule dataset to investigate the binding

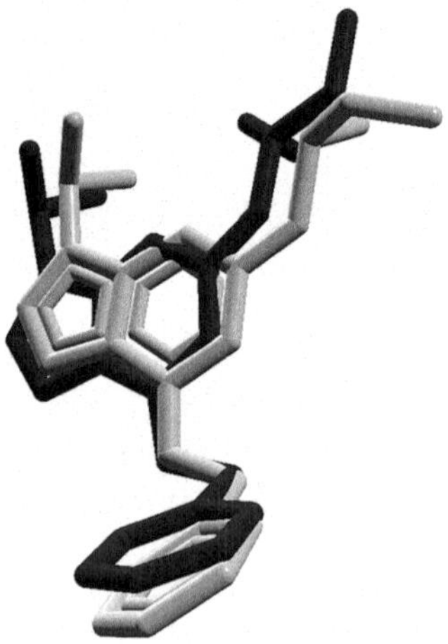

Fig. 6 Redocking results for the structure 2A4L using the program MVD [44]

of new potential inhibitors. To do so, we apply the approved protocol and use the scoring function values to evaluate the best hits among all entries available in the dataset.

8 Colophon

We employed the program MVD [44] to generate Figs. 1–4 and 6. We created Fig. 5 using Microsoft PowerPoint 2016. We performed the protein–ligand docking simulations reported on this chapter using a Desktop PC with 4GB of memory, a 1 TB hard disk, and an Intel® Core® i3-2120 @ 3.30 GHz processor running Windows 8.1.

9 Final Remarks

Protein–ligand docking simulations have been extensively used in the last three decades and have become the main computational approach in the computer-aided drug design. Considering the explosion in the number of protein structures available at the PDB, we may say that we live the golden age for molecular docking simulations. The atomic coordinates of the protein–ligand complexes along the experimental binding affinity data available from isothermal titration calorimetry (ITC) [76–78] make possible to develop and train a new generation of scoring functions and also to test the docking accuracy of the search algorithms extensively. To have a reliable docking simulation validation is mandatory. Therefore, we should take the flowchart described in Fig. 1 as a rule-of-thumb for anyone undertaking docking simulation. Particular attention should be devoted to biological systems for which structural and binding affinity information is available [79–109], which allows us to explore different scoring functions and docking protocols and validate them using the experimental data as a guide. Recent development in the machine learning techniques gave new tools to the community interested in docking studies [23–32]. Through the application of supervised machine learning techniques, we can develop scoring functions targeted to the biological systems of interest. For instance, we could train a scoring function as described by Eq. (1) to have their predictive performance optimized for a protein–ligand system of interest. Such approaches have shown superior predictive performance when compared with traditional scoring functions [40].

Most of the docking simulations consider the receptor as a rigid body, ignoring conformational changes due to ligand binding. To overcome this problem, we may combine protein–ligand docking with molecular dynamics simulations [110–114], where the initial structure for a molecular dynamics study came from docking simulation. Such a combination of computational methodologies not only addresses the flexibility of the protein–ligand complexes but also investigates the stability of the ligand during the molecular dynamics simulations, corroborating the structure obtained by molecular docking.

Acknowledgments

This work was supported by grants from CNPq (Brazil) (308883/2014-4). This study was financed in part by the Coordenação de Aperfeiçoamento de Pessoal de Nivel Superior—Brasil (CAPES)—Finance Code 001. GB-F acknowledges support from PUCRS/BPA fellowship. WFA is a senior researcher for CNPq (Brazil) (Process Numbers: 308883/2014-4 and 309029/2018-0).

References

1. Azevedo LS, Moraes FP, Xavier MM, Pantoja EO, Villavicencio B, Finck JA et al (2012) Recent progress of molecular docking simulations applied to development of drugs. Curr Bioinforma 7:352–365
2. Lengauer T, Rarey M (1996) Computational methods for biomolecular docking. Curr Opin Struct Biol 6:402–406
3. Breda A, Basso LA, Santos DS, de Azevedo Jr WF (2008) Virtual screening of drugs: score functions, docking, and drug design. Curr Comput Aided Drug Des 4:265–272
4. de Azevedo WF Jr (2008) Protein-drug interactions. Curr Drug Targets 9:1030
5. de Azevedo WF Jr, Dias R (2008) Computational methods for calculation of ligand-binding affinity. Curr Drug Targets 9:1031–1039
6. Dias R, de Azevedo WF Jr (2008) Molecular docking algorithms. Curr Drug Targets 9:1040–1047
7. de Azevedo WF Jr (2010) MolDock applied to structure-based virtual screening. Curr Drug Targets 11:327–334
8. de Azevedo WF Jr (2010) Structure-based virtual screening. Curr Drug Targets 11:261–263
9. de Avila MB, de Azevedo WF (2014) Data mining of docking results. Application to

3-dehydroquinate dehydratase. Curr Bioinf 9:361–379

10. Kitchen DB, Decornez H, Furr JR, Bajorath J (2004) Docking and scoring in virtual screening for drug discovery: methods and applications. Nat Rev Drug Discov 3:935–949
11. Fischer E (1890) Ueber die optischen Isomeren des Traubezuckers, der Gluconsäure und der Zuckersäure. Ber Dtsch Chem Ges 23:2611–2624
12. Fischer E (1894) Einfluss der Configuration auf die Wirkung der Enzyme. Ber Dtsch Chem Ges 27:2985–2993
13. Koshland DE Jr (1994) The key-lock theory and the induced fit theory. Angew Chem Int Ed Engl 33:2375–2378
14. Jorgensen WL (1991) Rusting of the lock and key model for protein-ligand binding. Science 254:954–955
15. Kuntz ID, Blaney JM, Oatley SJ, Langridge R, Ferrin TE (1982) A geometric approach to macromolecule-ligand interactions. J Mol Biol 161:269–288
16. Goodsell DS, Olson AJ (1990) Automated docking of substrates to proteins by simulated annealing. Proteins 8:195–202
17. Morris GM, Goodsell DS, Huey R, Olson AJ (1996) Distributed automated docking of flexible ligands to proteins: parallel applications of AutoDock 2.4. J Comput Aided Mol Des 10:293–304
18. Morris GM, Goodsell DS, Halliday RS, Huey R, Hart WE, Belew RK et al (1998) Automated docking using a lamarckian genetic algorithm and empirical binding free energy function. J Comput Chem 19:1639–1662
19. Morris GM, Huey R, Lindstrom W, Sanner MF, Belew RK, Goodsell DS et al (2009) AutoDock4 and AutoDockTools4: automated docking with selective receptor flexibility. J Comput Chem 30:2785–2791
20. Trott O, Olson AJ (2010) AutoDock Vina: improving the speed and accuracy of docking with a new scoring function, efficient optimization, and multithreading. J Comput Chem 31:455–461
21. Yang JM, Chen CC (2004) GEMDOCK: a generic evolutionary method for molecular docking. Proteins 55:288–304
22. Yang JM, Shen TW (2005) A pharmacophore-based evolutionary approach for screening selective estrogen receptor modulators. Proteins 59:205–220
23. Bitencourt-Ferreira G, de Azevedo Jr WF (2018) Development of a machine-learning model to predict Gibbs free energy of binding for protein-ligand complexes. Biophys Chem 240:63–69
24. de Ávila MB, de Azevedo WF Jr (2018) Development of machine learning models to predict inhibition of 3-dehydroquinate dehydratase. Chem Biol Drug Des 92:1468–1474
25. Russo S, de Azevedo WF (2019) Advances in the understanding of the cannabinoid receptor 1—focusing on the inverse agonists interactions. Curr Med Chem. https://doi.org/10.2174/0929867325666180417165247
26. Amaral MEA, Nery LR, Leite CE, de Azevedo Junior WF, Campos MM (2018) Pre-clinical effects of metformin and aspirin on the cell lines of different breast cancer subtypes. Invest New Drugs 36:782–796
27. Levin NMB, Pintro VO, Bitencourt-Ferreira G, Mattos BB, Silvério AC, de Azevedo Jr WF (2018) Development of CDK-targeted scoring functions for prediction of binding affinity. Biophys Chem 235:1–8
28. Freitas PG, Elias TC, Pinto IA, Costa LT, de Carvalho PVSD, Omote DQ et al (2018) Computational approach to the discovery of phytochemical molecules with therapeutic potential targets to the PKCZ protein. Lett Drug Des Discovery 15:488–499
29. Pintro VO, Azevedo WF (2017) Optimized virtual screening workflow. Towards target-based polynomial scoring functions for HIV-1 protease. Comb Chem High Throughput Screen 20:820–827
30. de Ávila MB, Xavier MM, Pintro VO, de Azevedo WF (2017) Supervised machine learning techniques to predict binding affinity. A study for cyclin-dependent kinase 2. Biochem Biophys Res Commun 494:305–310
31. Heck GS, Pintro VO, Pereira RR, de Ávila MB, Levin NMB, de Azevedo WF (2017) Supervised machine learning methods applied to predict ligand-binding affinity. Curr Med Chem 24:2459–2470
32. Levin NM, Pintro VO, de Ávila MB, de Mattos BB, De Azevedo WF Jr (2017) Understanding the structural basis for inhibition of cyclin-dependent kinases. New pieces in the molecular puzzle. Curr Drug Targets 18:1104–1111
33. Teles CB, Moreira-Dill LS, Silva Ade A, Facundo VA, de Azevedo WF Jr, da Silva LH et al (2015) A Lupane-triterpene isolated from *Combretum leprosum* Mart. Fruit extracts that interferes with the intracellular development of *Leishmania* (L.) *amazonensis in vitro*. BMC Complement Altern Med 15:165

34. Coracini JD, de Azevedo WF Jr (2014) Shikimate kinase, a protein target for drug design. Curr Med Chem 21:592–604
35. Moraes FP, de Azevedo WF Jr (2012) Targeting imidazoline site on monoamine oxidase B through molecular docking simulations. J Mol Model 18:3877–3886
36. Soares MB, Silva CV, Bastos TM, Guimarães ET, Figueira CP, Smirlis D et al (2012) Anti-*Trypanosoma cruzi* activity of nicotinamide. Acta Trop 12:224–229
37. Vianna CP, de Azevedo WF Jr (2012) Identification of new potential *Mycobacterium tuberculosis* shikimate kinase inhibitors through molecular docking simulations. J Mol Model 18:755–764
38. Sá MS, de Menezes MN, Krettli AU, Ribeiro IM, Tomassini TC, Ribeiro dos Santos R et al (2011) Antimalarial activity of physalins B, D, F, and G. J Nat Prod 74:2269–2272
39. Canduri F, Perez PC, Caceres RA, de Azevedo WF Jr (2008) CDK9 a potential target for drug development. Med Chem 4:210–218
40. Xavier MM, Heck GS, de Avila MB, Levin NM, Pintro VO, Carvalho NL et al (2016) SAnDReS a computational tool for statistical analysis of docking results and development of scoring functions. Comb Chem High Throughput Screen 19:801–812
41. Friesner RA, Murphy RB, Repasky MP, Frye LL, Greenwood JR, Halgren TA et al (2006) Extra precision glide: docking and scoring incorporating a model of hydrophobic enclosure for protein-ligand complexes. J Med Chem 49:6177–6196
42. Halgren TA, Murphy RB, Friesner RA, Beard HS, Frye LL, Pollard WT et al (2004) Glide: a new approach for rapid, accurate docking and scoring. 2. Enrichment factors in database screening. J Med Chem 47:1750–1759
43. Friesner RA, Banks JL, Murphy RB, Halgren TA, Klicic JJ, Mainz DT et al (2004) Glide: a new approach for rapid, accurate docking and scoring. 1. Method and assessment of docking accuracy. J Med Chem 47:1739–1749
44. Thomsen R, Christensen MH (2006) MolDock: a new technique for high-accuracy molecular docking. J Med Chem 49:3315–3321
45. Heberlé G, de Azevedo WF Jr (2011) Bio-inspired algorithms applied to molecular docking simulations. Curr Med Chem 18:1339–1352
46. Böhm HJ (1993) A novel computational tool for automated structure-based drug design. J Mol Recognit 6:131–137
47. Böhm HJ (1994) The development of a simple empirical scoring function to estimate the binding constant for a protein-ligand complex of known three-dimensional structure. J Comput Aided Mol Des 8:243–256
48. Böhm HJ (1996) Towards the automatic design of synthetically accessible protein ligands: peptides, amides and peptidomimetics. J Comput Aided Mol Des 10:265–272
49. Stahl M, Böhm HJ (1998) Development of filter functions for protein-ligand docking. J Mol Graph Model 16:121–132
50. Klebe G, Böhm HJ (1997) Energetic and entropic factors determining binding affinity in protein-ligand complexes. J Recept Signal Transduct Res 17:459–473
51. Böhm HJ, Banner DW, Weber L (1999) Combinatorial docking and combinatorial chemistry: design of potent non-peptide thrombin inhibitors. J Comput Aided Mol Des 13:51–56
52. Berman HM, Westbrook J, Feng Z, Gilliland G, Bhat TN, Weissig H et al (2000) The protein data bank. Nucleic Acids Res 28:235–242
53. Berman HM, Battistuz T, Bhat TN, Bluhm WF, Bourne PE, Burkhardt K et al (2002) The protein data bank. Acta Crystallogr D Biol Crystallogr 58:899–907
54. Westbrook J, Feng Z, Chen L, Yang H, Berman HM (2003) The protein data Bank and structural genomics. Nucleic Acids Res 31:489–491
55. Hu L, Benson ML, Smith RD, Lerner MG, Carlson HA (2005) Binding MOAD (mother of all databases). Proteins 60:333–340
56. Liu T, Lin Y, Wen X, Jorrisen RN, Gilson MK (2007) BindingDB: a web-accessible database of experimentally determined protein-ligand binding affinities. Nucleic Acids Res 35:198–201
57. Wang R, Fang X, Lu Y, Wang S (2004) The PDBbind database: collection of binding affinities for protein-ligand complexes with known three-dimensional structures. J Med Chem 47:2977–2980
58. Ballante F, Marshall GR (2016) An automated strategy for binding-pose selection and docking assessment in structure-based drug design. J Chem Inf Model 56:54–72
59. Vieth M, Hirst JD, Kolinski A, Brooks CL III (1998) Assessing energy functions for flexible docking. J Comput Chem 19:1612–1622
60. Zar JH (1972) Significance testing of the spearman rank correlation coefficient. J Am Stat Assoc 67:578–580

61. Irwin JJ, Shoichet BK (2005) ZINC--a free database of commercially available compounds for virtual screening. J Chem Inf Model 45:177–182
62. Irwin JJ, Sterling T, Mysinger MM, Bolstad ES, Coleman RG (2012) ZINC: a free tool to discover chemistry for biology. J Chem Inf Model 52:1757–1768
63. Ashburn TT, Thor KB (2004) Drug repositioning: identifying and developing new uses for existing drugs. Nat Rev Drug Discov 3:673–683
64. Morgan DO (1995) Principles of CDK regulation. Nature 374:131–134
65. Murray AW (1994) Cyclin-dependent kinases: regulators of the cell cycle and more. Chem Biol 1:191–195
66. Kim SH, Schulze-Gahmen U, Brandsen J, de Azevedo Júnior WF (1996) Structural basis for chemical inhibition of CDK2. Prog Cell Cycle Res 2:137–145
67. De Azevedo WF Jr, Mueller-Dieckmann HJ, Schulze-Gahmen U, Worland PJ, Sausville E, Kim SH (1996) Structural basis for specificity and potency of a flavonoid inhibitor of human CDK2, a cell cycle kinase. Proc Natl Acad Sci U S A 93:2735–2740
68. Canduri F, de Azevedo WF Jr (2005) Structural basis for interaction of inhibitors with cyclin-dependent kinase 2. Curr Comput Aided Drug Des 1:53–64
69. Krystof V, Cankar P, Frysová I, Slouka J, Kontopidis G, Dzubák P et al (2006) 4-arylazo-3,5-diamino-^{1}H-pyrazole CDK inhibitors: SAR study, crystal structure in complex with CDK2, selectivity, and cellular effects. J Med Chem 49:6500–6509
70. De Bondt HL, Rosenblatt J, Jancarik J, Jones HD, Morgan DO, Kim SH (1993) Crystal structure of cyclin-dependent kinase 2. Nature 363:595–602
71. Schulze-Gahmen U, De Bondt HL, Kim SH (1996) High-resolution crystal structures of human cyclin-dependent kinase 2 with and without ATP: bound waters and natural ligand as guides for inhibitor design. J Med Chem 39:4540–4546
72. de Azevedo WF Jr (2016) Opinion paper: targeting multiple cyclin-dependent kinases (CDKs): a new strategy for molecular docking studies. Curr Drug Targets 17:2
73. Leopoldino AM, Canduri F, Cabral H, Junqueira M, de Marqui AB, Apponi LH et al (2006) Expression, purification, and circular dichroism analysis of human CDK9. Protein Expr Purif 47:614–620
74. Volkart PA, Bitencourt-Ferreira G, Souto AA, de Azevedo WF (2019) Cyclin-dependent kinase 2 in cellular senescence and cancer. A structural and functional review. Curr Drug Targets 20(7):716–726. https://doi.org/10.2174/1389450120666181204165344
75. De Azevedo WF, Leclerc S, Meijer L, Havlicek L, Strnad M, Kim SH (1997) Inhibition of cyclin-dependent kinases by purine analogues: crystal structure of human cdk2 complexed with roscovitine. Eur J Biochem 243:518–526
76. de Azevedo WF Jr, Dias R (2008) Experimental approaches to evaluate the thermodynamics of protein-drug interactions. Curr Drug Targets 9:1071–1076
77. Ma W, Yang L, He L (2018) Overview of the detection methods for equilibrium dissociation constant KD of drug-receptor interaction. J Pharm Anal 8:147–152
78. Falconer RJ (2016) Applications of isothermal titration calorimetry—the research and technical developments from 2011 to 2015. J Mol Recognit 29:504–515
79. de Ávila MB, Bitencourt-Ferreira G, de Azevedo WF Jr (2019) Structural basis for inhibition of Enoyl-[Acyl Carrier Protein] Reductase (InhA) from *Mycobacterium tuberculosis*. Curr Med Chem. https://doi.org/10.2174/0929867326666181203125229
80. Pereira JH, Canduri F, de Oliveira JS, da Silveira NJ, Basso LA, Palma MS et al (2003) Structural bioinformatics study of EPSP synthase from *Mycobacterium tuberculosis*. Biochem Biophys Res Commun 312:608–614
81. Borges JC, Pereira JH, Vasconcelos IB, dos Santos GC, Olivieri JR, Ramos CH et al (2006) Phosphate closes the solution structure of the 5-enolpyruvylshikimate-3-phosphate synthase (EPSPS) from *Mycobacterium tuberculosis*. Arch Biochem Biophys 452:156–164
82. Marques MR, Pereira JH, Oliveira JS, Basso LA, de Azevedo WF Jr, Santos DS et al (2007) The inhibition of 5-enolpyruvylshikimate-3-phosphate synthase as a model for development of novel antimicrobials. Curr Drug Targets 8:445–457
83. Marques MR, Vaso A, Neto JR, Fossey MA, Oliveira JS, Basso LA et al (2008) Dynamics of glyphosate-induced conformational changes of *Mycobacterium tuberculosis* 5-enolpyruvylshikimate-3-phosphate synthase (EC 2.5.1.19) determined by hydrogen-deuterium exchange and electrospray mass spectrometry. Biochemistry 47:7509–7522

84. de Azevedo WF Jr, Canduri F, dos Santos DM, Silva RG, de Oliveira JS, de Carvalho LP et al (2003) Crystal structure of human purine nucleoside phosphorylase at 2.3A resolution. Biochem Biophys Res Commun 308:545–552
85. dos Santos DM, Canduri F, Pereira JH, Vinicius Bertacine Dias M, Silva RG et al (2003) Crystal structure of human purine nucleoside phosphorylase complexed with acyclovir. Biochem Biophys Res Commun 308:553–559
86. Filgueira de Azevedo W Jr, Canduri F, Marangoni dos Santos D, Pereira JH, Dias MV, Silva RG et al (2003) Structural basis for inhibition of human PNP by immucillin-H. Biochem Biophys Res Commun 309:917–922
87. Filgueira de Azevedo W Jr, dos Santos GC, dos Santos DM, Olivieri JR, Canduri F, Silva RG et al (2003) Docking and small angle X-ray scattering studies of purine nucleoside phosphorylase. Biochem Biophys Res Commun 309:923–928
88. de Azevedo WF Jr, Canduri F, dos Santos DM, Pereira JH, Bertacine Dias MV, Silva RG et al (2003) Crystal structure of human PNP complexed with guanine. Biochem Biophys Res Commun 312:767–772
89. da Silveira NJ, Uchôa HB, Canduri F, Pereira JH, Camera JC Jr, Basso LA et al (2004) Structural bioinformatics study of PNP from *Schistosoma mansoni*. Biochem Biophys Res Commun 322:100–104
90. Canduri F, dos Santos DM, Silva RG, Mendes MA, Basso LA, Palma MS et al (2004) Structures of human purine nucleoside phosphorylase complexed with inosine and ddI. Biochem Biophys Res Commun 313:907–914
91. Nolasco DO, Canduri F, Pereira JH, Cortinóz JR, Palma MS, Oliveira JS et al (2004) Crystallographic structure of PNP from *Mycobacterium tuberculosis* at 1.9A resolution. Biochem Biophys Res Commun 324:789–794
92. Canduri F, Fadel V, Dias MV, Basso LA, Palma MS, Santos DS et al (2005) Crystal structure of human PNP complexed with hypoxanthine and sulfate ion. Biochem Biophys Res Commun 326:335–338
93. Canduri F, Fadel V, Basso LA, Palma MS, Santos DS, de Azevedo WF Jr (2005) New catalytic mechanism for human purine nucleoside phosphorylase. Biochem Biophys Res Commun 327:646–649
94. Canduri F, Silva RG, dos Santos DM, Palma MS, Basso LA, Santos DS et al (2005) Structure of human PNP complexed with ligands. Acta Crystallogr D Biol Crystallogr 61:856–862
95. Silva RG, Pereira JH, Canduri F, de Azevedo WF Jr, Basso LA, Santos DS (2005) Kinetics and crystal structure of human purine nucleoside phosphorylase in complex with 7-methyl-6-thio-guanosine. Arch Biochem Biophys 442:49–58
96. de Azevedo WF Jr, Canduri F, Basso LA, Palma MS, Santos DS (2006) Determining the structural basis for specificity of ligands using crystallographic screening. Cell Biochem Biophys 44:405–411
97. Ducati RG, Basso LA, Santos DS, de Azevedo WF Jr (2010) Crystallographic and docking studies of purine nucleoside phosphorylase from *Mycobacterium tuberculosis*. Bioorg Med Chem 18:4769–4774
98. Pereira JH, Vasconcelos IB, Oliveira JS, Caceres RA, de Azevedo WF Jr, Basso LA et al (2007) Shikimate kinase: a potential target for development of novel lectiagents. Curr Drug Targets 8:459–468
99. Delatorre P, Rocha BA, Souza EP, Oliveira TM, Bezerra GA, Moreno FB et al (2007) Structure of a lectin from *Canavalia gladiata* seeds: new structural insights for old molecules. BMC Struct Biol 7:52
100. Canduri F, de Azevedo WF (2008) Protein crystallography in drug discovery. Curr Drug Targets 9:1048–1053
101. Canduri F, Perez PC, Caceres RA, de Azevedo WF Jr (2007) Protein kinases as targets for antiparasitic chemotherapy drugs. Curr Drug Targets 8:389–398
102. Dias MV, Borges JC, Ely F, Pereira JH, Canduri F, Ramos CH et al (2006) Structure of chorismate synthase from *Mycobacterium tuberculosis*. J Struct Biol 154:130–143
103. Dias MV, Ely F, Palma MS, de Azevedo WF Jr, Basso LA, Santos DS (2007) Chorismate synthase: an attractive target for drug development against orphan diseases. Curr Drug Targets 8:437–444
104. de Azevedo WF Jr (2011) Protein targets for development of drugs against *Mycobacterium tuberculosis*. Curr Med Chem 18:1255–1257
105. Dias MV, Faím LM, Vasconcelos IB, de Oliveira JS, Basso LA, Santos DS et al (2007) Effects of the magnesium and chloride ions and shikimate on the structure of shikimate kinase from *Mycobacterium tuberculosis*. Acta Crystallogr Sect F Struct Biol Cryst Commun 63:1–6
106. de Azevedo WF Jr, Ward RJ, Canduri F, Soares A, Giglio JR, Arni RK (1998) Crystal structure of piratoxin-I: a calcium-

independent, myotoxic phospholipase A2-homologue from *Bothrops pirajai* venom. Toxicon 36:1395–1406

107. Bezerra GA, Oliveira TM, Moreno FB, de Souza EP, da Rocha BA, Benevides RG et al (2007) Structural analysis of *Canavalia maritima* and *Canavalia gladiata* lectins complexed with different dimannosides: new insights into the understanding of the structure-biological activity relationship in legume lectins. J Struct Biol 160:168–176
108. Delatorre P, Rocha BA, Gadelha CA, Santi-Gadelha T, Cajazeiras JB, Souza EP et al (2006) Crystal structure of a lectin from *Canavalia maritima* (ConM) in complex with trehalose and maltose reveals relevant mutation in ConA-like lectins. J Struct Biol 154:280–286
109. Rádis-Baptista G, Moreno FB, de Lima Nogueira L, Martins AM, de Oliveira TD, Toyama MH et al (2006) Crotacetin, a novel snake venom C-type lectin homolog of convulxin, exhibits an unpredictable antimicrobial activity. Cell Biochem Biophys 44:412–423
110. de Azevedo WF Jr (2011) Molecular dynamics simulations of protein targets identified in *Mycobacterium tuberculosis*. Curr Med Chem 18:1353–1366
111. Sforça ML, Oyama S Jr, Canduri F, Lorenzi CC, Pertinhez TA, Konno K et al (2004) How C-terminal carboxyamidation alters the biological activity of peptides from the venom of the eumenine solitary wasp. Biochemistry 43:5608–5617
112. de Azevedo WF Jr, Canduri F, Fadel V, Teodoro LG, Hial V, Gomes RA (2001) Molecular model for the binary complex of uropepsin and pepstatin. Biochem Biophys Res Commun 287:277–281
113. Salmaso V, Moro S (2018) Bridging molecular docking to molecular dynamics in exploring ligand-protein recognition process: an overview. Front Pharmacol 9:923
114. Kontoyianni M, Lacy B (2018) Toward computational understanding of molecular recognition in the human metabolizing cytochrome P450s. Curr Med Chem 25:3353–3373

Chapter 4

SAnDReS: A Computational Tool for Docking

Gabriela Bitencourt-Ferreira and Walter Filgueira de Azevedo Jr.

Abstract

Since the early 1980s, we have witnessed considerable progress in the development and application of docking programs to assess protein–ligand interactions. Most of these applications had as a goal the identification of potential new binders to protein targets. Another remarkable progress is taking place in the determination of the structures of protein–ligand complexes, mostly using X-ray diffraction crystallography. Considering these developments, we have a favorable scenario for the creation of a computational tool that integrates into one workflow all steps involved in molecular docking simulations. We had these goals in mind when we developed the program SAnDReS. This program allows the integration of all computational features related to modern docking studies into one workflow. SAnDReS not only carries out docking simulations but also evaluates several docking protocols allowing the selection of the best approach for a given protein system. SAnDReS is a free and open-source (GNU General Public License) computational environment for running docking simulations. Here, we describe the combination of SAnDReS and AutoDock4 for protein–ligand docking simulations. AutoDock4 is a free program that has been applied to over a thousand receptor–ligand docking simulations. The dataset described in this chapter is available for downloading at https://github.com/azevedolab/sandres

Key words SAnDReS, AutoDock4, Docking, Binding affinity, Drug design, Molecular recognition

1 Introduction

Since the mid-1980s and the early 1990s, many research groups have successfully reported structure-based drug design studies [1–3]. These pioneering studies used X-ray diffraction crystallographic structures of the complexes involving a protein target and a small organic molecule bound to it. Analysis of this experimental information allowed researchers to identify the structural basis for the protein–ligand interactions. As computational power increased, it was also feasible to carry out analysis of potential new drugs with a protein target through in silico approaches. Among the computational tools used to address the drug design and development, protein–ligand docking simulation is one of the most used methods. In this technique, we simulate the joining of a small molecule against the binding site of a protein structure.

Walter Filgueira de Azevedo Jr. (ed.), *Docking Screens for Drug Discovery*, Methods in Molecular Biology, vol. 2053, https://doi.org/10.1007/978-1-4939-9752-7_4,

The development of protein–ligand docking methods started in the early 1980s [4]. Once computational tools became available, in silico techniques were successfully applied to develop many approved drugs including HIV-1 protease inhibitors [5–10]. In general, we may say that drug design has advanced substantially from the use of in silico approaches, which nowadays is the first approach in drug discovery [11, 12].

Furthermore, application of docking simulation was able to identify binders to a wide spectrum of protein targets [13–23]. In parallel with the development of docking technology, we have also witnessed an explosion in the number of protein complexes available in the Protein Data Bank [24–26]. Moreover, the availability of experimental information on inhibition constant (K_i), dissociation constant (K_d), half maximal inhibitory concentration (IC_{50}), and Gibbs free energy of binding (ΔG) provide a solid framework of structural and binding affinity data that allows us to investigate the structural basis for inhibition of enzymes. Experimental binding affinity data are available at MOAD [27], BindingDB [28], and PDBbind [29].

This favorable scenario made possible the development of the program SAnDReS [30], which provides an integrated computational environment for carrying out docking simulations. SAnDReS is an acronym for Statistical Analysis of Docking Results and Scoring Functions and takes a different approach to molecular docking studies; it focuses on the simulation of a system composed of an ensemble of crystallographic structures for which ligand binding affinity data are available. Here, we named this ensemble of crystallographic structures with binding affinity data as a biological system. SAnDReS is also a tool for statistical analysis of docking simulations and evaluation of the predictive performance of computational models developed to calculate binding affinity [30]. SAnDReS was developed in Python 3, using the SciPy, NumPy, scikit-learn [31], and Matplotlib libraries. In this chapter, we focus on the combined use of SAnDReS-AutoDock4 for docking simulations. AutoDock is a robust protein–ligand docking program [32–35]. There are 1160 studies about the application of AutoDock to docking simulations (search carried out on October 26, 2018, using the keyword "autodock" in PubMed).

Integration of AutoDock4 in the program SAnDReS makes it possible to carry out docking simulations in an elegant and fast computational tool. We have successfully employed SAnDReS to study coagulation factor Xa [30], cyclin-dependent kinases [36, 37], HIV-1 protease [38], estrogen receptor [39], cannabinoid receptor 1 [40], and 3-dehydroquinate dehydratase [41]. Also, we used SAnDReS to develop a machine-learning model to predict the Gibbs free energy of binding for protein–ligand complexes [42]. In the following sections, we describe the application of SAnDReS to an ensemble of cyclin-dependent kinases and highlight the main integrated tools available for docking simulations and analysis of the predictive performance of this in silico methodology.

2 Dataset

To explain how to apply the combined use of SAnDReS-AutoDock4 for docking simulations, we chose a dataset composed of cyclin-dependent kinase 2 (CDK2) for which IC_{50} data were available. We considered here a dataset with 89 CDK structures solved at a crystallographic resolution higher than 2.0 Å. This dataset will be referred to as HR-CDK2-IC_{50} dataset (high-resolution CDK2 structures with IC_{50} data). We previously described the application of SAnDReS to a larger dataset consisting of 170 structures [37]. Table 1 shows the PDB access codes for all structures in the dataset. This enzyme has been studied as a protein target, mainly because of its role in controlling cell cycle progression and the potential use of CDK inhibitors as anticancer drugs [43, 44]. For recent reviews, see de Azevedo 2016 [45] and Levin et al. 2016 [46]. All inhibitors in the HR-CDK2-IC_{50} dataset are bound to the ATP-binding pocket of CDK2.

3 Installing SAnDReS on Windows

SAnDReS is a free and open-source (GNU General Public License) program. You may download SAnDReS code from GitHub (https://github.com/azevedolab/sandres). You need to have Python 3 installed on your computer to run SAnDReS. Also, you need to install NumPy, Matplotlib, scikit-learn, and SciPy. You can make the installation process more accessible by installing Anaconda. To install SAnDReS, we follow these steps:

1. Install Anaconda 32 bits (https://www.anaconda.com/download/).
2. Download SAnDReS 1.1.0 from GitHub (https://github.com/azevedolab/sandres).
3. Unzip the zipped file (sandres.zip).
4. Copy sandres directory to c:\.
5. Open a command prompt window and type: cd c:\sandres
   then type: python sandres1_GUI.py

In Fig. 1, we have SAnDReS main GUI interface. From this interface, we can easily set up all necessary files to run protein–ligand simulations.

Table 1
PDB access codes for all structures in the HR-CDK2-IC_{50} dataset

PDB access codes	Protein identification
1H00, 1H01, 1H07, 1JVP, 1OIR, 1OIT, 1PXI, 1URW, 1YKR, 2A0C, 2B52, 2B54, 2B55, 2BHE, 2BTR, 2BTS, 2C68, 2C6I, 2C6K, 2C6M, 2CLX, 2R3F, 2R3G, 2R3H, 2R3J, 2R3K, 2R3L, 2R3M, 2R3N, 2R3O, 2R3P, 2VTH, 2VTQ, 2VTR, 2VTS, 2VTT, 2VU3, 2VV9, 2 W05, 3EZR, 3EZV, 3FZ1, 3IG7, 3IGG, 3NS9, 3PJ8, 3PXZ, 3PY0, 3QQK, 3QTQ, 3QTR, 3QTS, 3QTU, 3QTW, 3QTX, 3QU0, 3R8V, 3R8Z, 3R9D, 3R9N, 3R9O, 3RAH, 3RAL, 3RJC, 3RK7, 3RK9, 3RMF, 3RNI, 3RPR, 3RPV, 3RPY, 3RZB, 3S00, 3S1H, 3SQQ, 3TI1, 3TIY, 3UNJ, 4BGH, 4FKI, 4NJ3, 4RJ3, 5D1J, 2R3I, 2R3R, 4FKL, 4GCJ	Human cyclin-dependent kinase 2
1V0O	Cell division control protein 2 homolog from *Plasmodium falciparum*
3DDQ	Human cyclin-dependent kinase 2 in complex with cyclin A

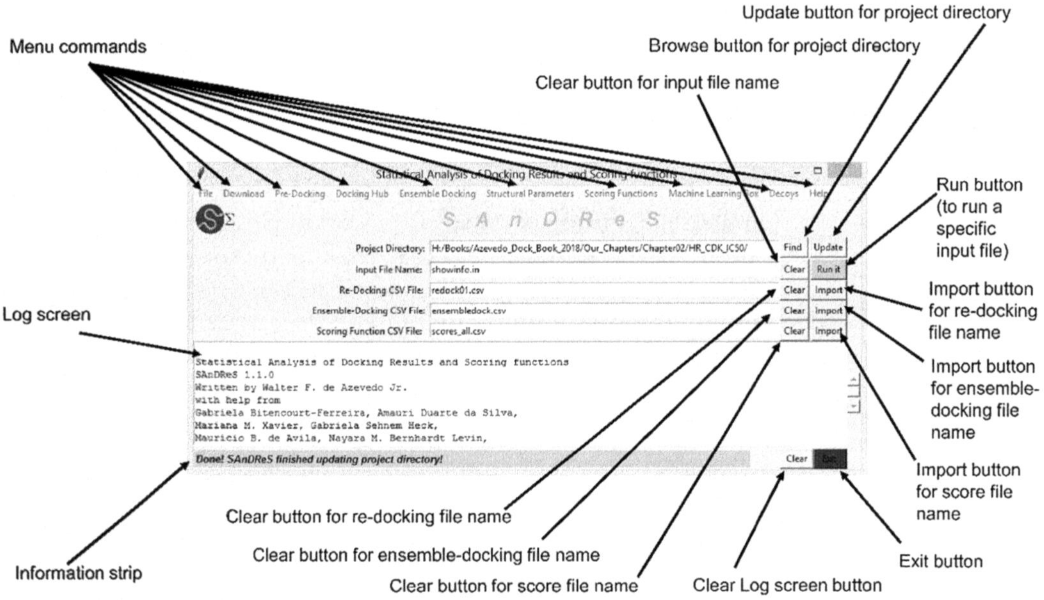

Fig. 1 SAnDReS GUI interface. Here we describe the main buttons used to carry out docking simulations using SAnDReS. For docking simulations using SAnDReS, the user must paste the PDB access codes for the crystallographic structures using the Download button (Download→Input PDB Access Codes). Then the user downloads the structures (Download→Structures). After downloading the structures, we download binding affinity data (Download→Binding Affinity). In the next step, we filter out dataset in using the Pre-Docking button. Finally, we employ Docking Hub to carry out docking simulations. We use the Ensemble Docking button to evaluate docking performance

4 Overview of the Use of SAnDReS-AutoDock4 for Docking

Our goal in developing SAnDReS was to have an integrated tool for docking simulations and for the development of machine-learning models to predict binding affinity. Here our focus is on the docking tools of SAnDReS. We may say that there are thousands of approaches [47] to protein–ligand docking simulations, but if we consider the choice of the biomolecular system, protein–ligand docking simulations, and the validation methods, they all share a common framework described below, independent of the programs used in the protein–ligand docking simulations. This common core found on all docking programs was explored in the development of SAnDReS.

We designed SAnDReS to handle PDB files of crystallographic structures. It has been decided to focus on the crystallographic information because of the majority of the structural information available for protein–ligand complexes for which there are experimental binding details come from the X-ray crystallography technique [48]. SAnDReS was designed to analyze data from any protein–ligand docking program; the only requisite is to have protein structures in Protein Data Bank (PDB) format, ligands in Structure Data Format (SDF), docking and scoring function data in comma-separated values (CSV) format. Figure 2 illustrates all steps necessary to carry out molecular docking simulation of a biological system using the combination of SAnDReS-AutoDock4 programs.

We consider as a biological system an ensemble of structures for which ligand binding affinity data are available. In our example, the HR-CDK2-IC_{50} dataset. In the flowchart, the first step is the download of the biological systems (PDB and CSV files). In the following, SAnDReS filters the dataset, in a step named here pre-docking. The filtered data are submitted to docking simulations. The current version of SAnDReS automatically generates inputs necessary to run AutoDock4 except for the conversion from the PDB to the PDBQT format. We used AutoDockTools4 [49–51] to carry out this conversion. The user has to convert PDB files to the PDBQT format before running AutoDock4. The rest of the AutoDock4 running is fully automated through SAnDReS. In the next step, SAnDReS carries out docking running AutoDock4; this phase is named docking hub. The docking results are submitted to statistical analysis to evaluate the docking performance of different protocols.

4.1 Downloading Biological System

Once we have chosen the PDB access codes that comprise the dataset, we insert the codes separated by commas and SAnDReS carries out a download of the structures and the binding affinity data from the PDB.

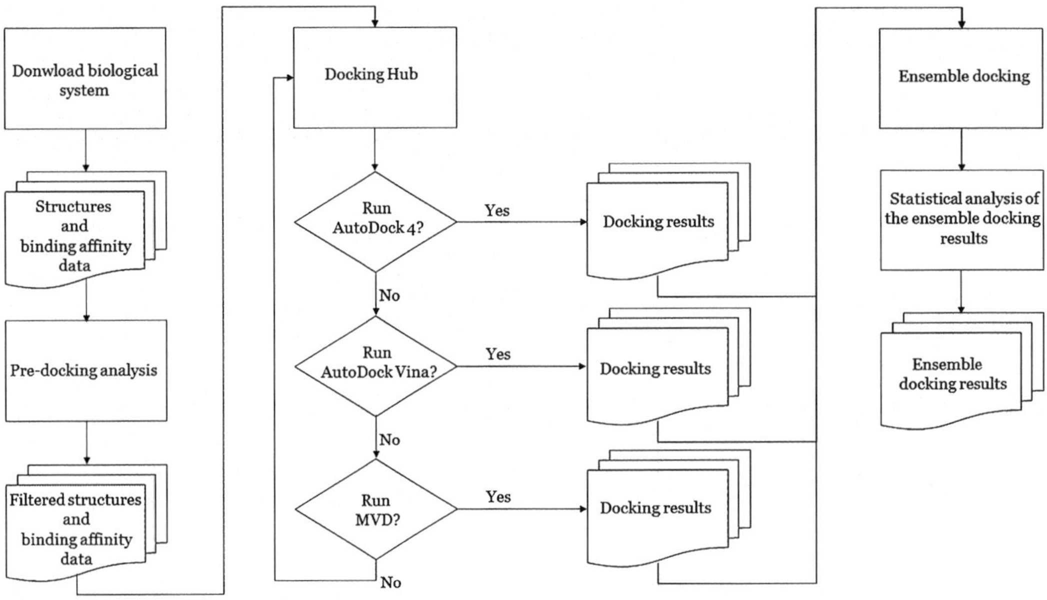

Fig. 2 Protein–ligand docking simulation with SAnDReS. This flowchart describes all steps necessary to carry out docking simulations with the combination of SAnDReS–AutoDock4

4.2 Predocking

In the pre-docking phase, we intend to prepare the PDB and CSV files for docking simulations. At first, SAnDReS checks the integrality of the structural and binding data. Although PDB has been doing a great job integrating structural and binding–affinity data, a search carried out using the advanced tool option may return the PDB access codes for which no binding affinity data are available.

SAnDReS checks whether the binding information is available for all structures in the dataset or not. It is also possible to filter out the dataset and eliminate repeated ligands. In doing so, we expect to have a dataset with no repeated ligands, which improves the chemical diversity of the dataset. It is also possible to evaluate the overall quality of the crystallographic information of our dataset. Furthermore, SAnDReS can analyze protein–ligand interactions for all structures in the dataset. Figure 3 shows the number of intermolecular contacts per residue using a cutoff distance of 4.5 Å. The top contact amino acid is the Leu-83, an interaction point identified in the molecular fork of CDK structures [52–59].

4.3 Docking Hub

SAnDReS allows running AutoDock4, AutoDock Vina, and Molegro Virtual Docker (MVD). This interface facilitates docking running which reduces the overall time of the analysis since SAnDReS generates all necessary input files to run the previously mention docking programs. Here we carried out docking simulations using AutoDock4 through the docking hub interface of SAnDReS. We may choose among the all available protocols of AutoDock4. Figure 4 shows the docking-set up interface, where the users may

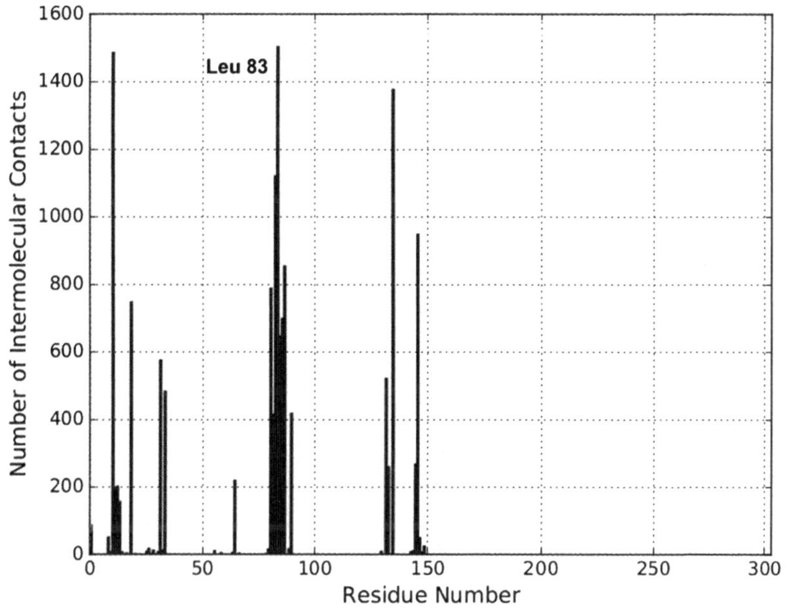

Fig. 3 Protein–ligand interactions for all structures in the HR-CDK2-IC_{50} dataset

set the different docking options. For instance, we may run docking simulations using the four search algorithms: Lamarckian genetic algorithm (LGA), genetic algorithm (GA), local search (LS), and simulated annealing (SA). SAnDReS may also calculate the AutoDock scoring function values for the crystallographic position of the ligand using the energy of the PDB structure (EPDB) option.

In summary, to run AutoDock4 using SAnDReS we click on the sequence: *AutoGrid→Set up DPF.* The setup DPF (docking parameter files) window generates the necessary input files to run AutoDock4. We have to choose the docking protocol in the Setup DPF menu and then click on the Save DPF button. To run the AutoDock4, we click on the sequence: *AutoDock→Analysis.* Once finished the docking simulations, SAnDReS may merge all output files in one file that brings docking results or energy of the crystallographic position of the ligand for all structures in the dataset.

4.4 Ensemble Docking

In this step, we may evaluate docking performance. SAnDReS investigates two significant features of the docking simulations: docking RMSD and docking accuracy. SAnDReS has previously assessed the docking root-mean-squared deviation for every structure in the dataset. We calculate the docking RMSD as follows:

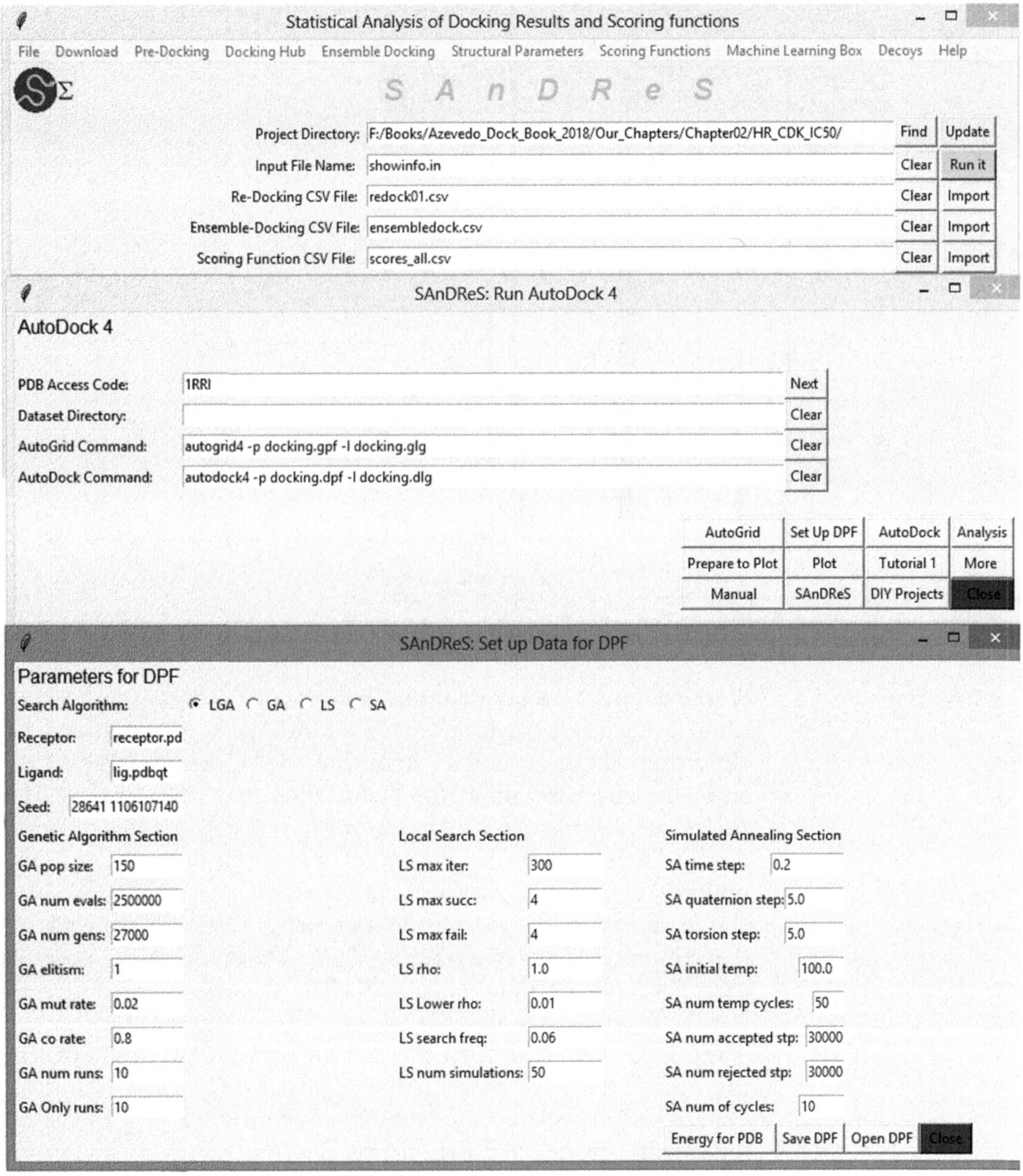

Fig. 4 Docking-set up, the interface of SAnDReS

$$\mathrm{RMSD} = \sqrt{\frac{\sum_{i=1}^{N}\left[\left(x_{x,i}-x_{p,i}\right)^2+\left(y_{x,i}-y_{p,i}\right)^2+\left(z_{x,i}-z_{p,i}\right)^2\right]}{N}} \quad (1)$$

where x_x, y_x, and z_x are the experimental coordinates for the ligand and x_p, y_p, and z_p are the atomic coordinates for the position generated by the docking simulation.

Then SAnDReS also calculates the docking accuracy (DA). The equation below defines docking accuracy (DA) as follows:

$$DA = f_l + 0.5(f_l - f_h) \tag{2}$$

where f_l is the fraction of poses for which the docking RMSD is less than l and f_h is the fraction of poses for which the docking RMSD is less than h, where $l < h$ [60, 61]. SAnDReS calculates two correlation coefficients, squared correlation coefficient (R^2) and Spearman's rank correlation coefficient (ρ). We define R^2 by the following equation:

$$R^2 = 1 - \frac{\text{RSS}}{\text{TSS}} \tag{3}$$

We calculate the terms residual sum of squares (RSS) and the total sum of squares (TSS) as follows:

$$\text{RSS} = \sum_{i=1}^{N} \left(y_i - y_{\text{calc},i}\right)^2 \tag{4}$$

and

$$\text{TSS} = \sum_{i=1}^{N} \left(y_i - \langle y \rangle\right)^2 \tag{5}$$

where $y_{\text{calc},i}$ are the values obtained by feeding independent variables into the regression equation obtained using supervised machine learning techniques available in the scikit-learn library [31]. The variables y_i are the experimental observations, for instance, log(IC_{50}), $\langle y \rangle$ is the mean value for y, and N the number of observations. We define the Spearman's rank correlation coefficient (ρ) by the following expression:

$$\rho = 1 - \frac{6\sum_{i=1}^{N} d_i^2}{N\left(N^2 - 1\right)} \tag{6}$$

In the above equation, the term d_i indicates the difference in the ranks for a given observation [31].

Statistical analysis of docking performance of AutoDock4 running LGA for all structures in the HR-CDK2-IC_{50} dataset indicates that Spearman's rank correlation coefficient ranges from −0.139 to 0.245 between the docking RMSD and the scoring function values. Analysis of DA shows a percentage of 88.764. Nearly 90% of the HR-CDK2-IC_{50} dataset shows docking RMSD below 2.0 Å, which strongly indicates that AutoDock4 is adequate to analyze CDK2–ligand interactions.

5 Availability

The program SAnDReS is implemented in Python 3 and available to download under the GNU (General Public License) license at https://github.com/azevedolab/sandres.

6 Colophon

We employed the program SAnDReS to generate Figs. 1, 3, and 4. We created Fig. 2 using Microsoft PowerPoint 2016. We performed the protein–ligand docking simulations reported on this chapter using a Desktop PC with 4GB of memory, a 1 TB hard disk, and an Intel® Core® i3-2120 @ 3.30 GHz processor running Windows 8.1.

7 Final Remarks

SAnDReS allows fast, integrated, and reliable docking simulations. Its development had as a goal to make available an integrated computational tool to carry out docking simulations, analysis of these simulations, and creation of machine learning models to predict binding affinity. In this chapter, we described SAnDReS application to protein–ligand docking simulations. One of the basic concepts behind SAnDReS is the biological system [30, 38, 62–74]. SAnDReS seeks to perform docking for an ensemble of crystallographic structures for which binding affinity data are available. Here we call a set of crystallographic structures along with binding affinity data as a biological system. With this approach, SAnDReS is adequate for biological systems with at least 30 crystallographic structures.

As a proof of concept, we investigated CDK2 biological system using an ensemble of structures composed of 89 entries (Table 1). Application of AutoDock4 through SAnDReS interface was able to generate results with a docking accuracy close to 90%. Also, the integrated interface of SAnDReS allowed us to efficiently perform molecular docking simulations, without the need for editing the input files necessary to run AutoDock4. In summary, SAnDReS is an integrated tool that facilitates protein–ligand simulations and incorporates a systems approach to the analysis of docking simulations which adds flexibility and increase the reliability of docking simulations. The development of the program SAnDReS is the direct result of our combined structural and computational studies of protein–ligand interactions [75–114]. We can use SAnDReS to study any receptor–ligand system; the only conditions are the availability of crystallographic structures and ligand binding information.

Acknowledgments

This work was supported by grants from CNPq (Brazil) (308883/2014-4). This study was financed in part by the Coordenação de Aperfeiçoamento de Pessoal de Nivel Superior—Brasil (CAPES)—Finance Code 001. GB-F acknowledges support from PUCRS/BPA fellowship. WFA is a senior researcher for CNPq (Brazil) (Process Numbers: 308883/2014-4 and 309029/2018-0).

References

1. Roberts NA, Martin JA, Kinchington D, Broadhurst AV, Craig JC, Duncan IB et al (1990) Rational design of peptide-based HIV proteinase inhibitors. Science 248:358–361
2. Erickson J, Neidhart DJ, VanDrie J, Kempf DJ, Wang XC, Norbeck DW et al (1990) Design, activity, and 2.8 a crystal structure of a C2 symmetric inhibitor complexed to HIV-1 protease. Science 249:527–533
3. Dorsey BD, Levin RB, McDaniel SL, Vacca JP, Guare JP, Darke PL et al (1994) L-735,524: the design of a potent and orally bioavailable HIV protease inhibitor. J Med Chem 37:3443–3451
4. Kuntz ID, Blaney JM, Oatley SJ, Langridge R, Ferrin TE (1982) A geometric approach to macromolecule-ligand interactions. J Mol Biol 161:269–288
5. DesJarlais RL, Dixon JS (1994) A shape- and chemistry-based docking method and its use in the design of HIV-1 protease inhibitors. J Comput Aided Mol Des 8:231–242
6. Lunney EA, Hagen SE, Domagala JM, Humblet C, Kosinski J, Tait BD et al (1994) A novel nonpeptide HIV-1 protease inhibitor: elucidation of the binding mode and its application in the design of related analogs. J Med Chem 37:2664–2677
7. Vaillancourt M, Cohen E, Sauvé G (1995) Characterization of dynamic state inhibitors of HIV-1 protease. J Enzym Inhib 9:217–233
8. Gehlhaar DK, Verkhivker GM, Rejto PA, Sherman CJ, Fogel DB, Fogel LJ et al (1995) Molecular recognition of the inhibitor AG-1343 by HIV-1 protease: conformationally flexible docking by evolutionary programming. Chem Biol 2:317–324
9. King BL, Vajda S, DeLisi C (1996) Empirical free energy as a target function in docking and design: application to HIV-1 protease inhibitors. FEBS Lett 384:87–91
10. Wang S, Milne GW, Yan X, Posey IJ, Nicklaus MC, Graham L et al (1996) Discovery of novel, non-peptide HIV-1 protease inhibitors by pharmacophore searching. J Med Chem 39:2047–2054
11. Muegge I, Bergner A, Kriegl JM (2017) Computer-aided drug design at Boehringer ingelheim. J Comput Aided Mol Des 31:275–285
12. Hillisch A, Heinrich N, Wild H (2015) Computational chemistry in the pharmaceutical industry: from childhood to adolescence. Chem Med Chem 10:1958–1962
13. Kuntz ID (1992) Structure-based strategies for drug design and discovery. Science 257:1078–1082
14. Shoichet BK, Stroud RM, Santi DV, Kuntz ID, Perry KM (1993) Structure-based discovery of inhibitors of thymidylate synthase. Science 259:1445–1450
15. Rutenber E, Fauman EB, Keenan RJ, Fong S, Furth PS, Ortiz de Montellano PR et al (1993) Structure of a non-peptide inhibitor complexed with HIV-1 protease. Developing a cycle of structure-based drug design. J Biol Chem 268:15343–15346
16. Zheng Q, Kyle DJ (1996) Computational screening of combinatorial libraries. Bioorg Med Chem 4:631–638
17. Gschwend DA, Good AC, Kuntz ID (1996) Molecular docking towards drug discovery. J Mol Recognit 9:175–186
18. Finn PW (1996) Computer-based screening of compound databases for the identification of novel leads. Drug Discov Today 1:363–370
19. Horvath D (1997) A virtual screening approach applied to the search for trypanothione reductase inhibitors. J Med Chem 40:2412–2423
20. Toyoda T, Brobey RKB, Sano G, Horii T, Tomioka N, Itai A (1997) Lead discovery of inhibitors of the dihydrofolate reductase domain of *Plasmodium Falciparum* dihydrofolate reductase-thymidylate synthase. Biochem Biophys Res Commun 235:515–519

21. Olson AJ, Goodsell DS (1998) Automated docking and the search for HIV protease inhibitors. SAR QSAR Environ Res 8:273–285
22. Walters WP, Stahl MT, Murcko MA (1998) Virtual screening—an overview. Drug Discov Today 3:160–178
23. Toney JH, Fitzgerald PMD, Groversharma N, Olson SH, May WJ, Sundelof JG et al (1998) Antibiotic sensitization using biphenyl Tetrazoles as potent inhibitors of *Bacteroides fragilis* Metallo-BetaLactamase. Chem Biol 5:185–196
24. Berman HM, Westbrook J, Feng Z, Gilliland G, Bhat TN, Weissig H et al (2000) The protein data bank. Nucleic Acids Res 28:235–242
25. Berman HM, Battistuz T, Bhat TN, Bluhm WF, Bourne PE, Burkhardt K et al (2002) The protein data bank. Acta Crystallogr D Biol Crystallogr 58:899–907
26. Westbrook J, Feng Z, Chen L, Yang H, Berman HM (2003) The protein data Bank and structural genomics. Nucleic Acids Res 31:489–491
27. Hu L, Benson ML, Smith RD, Lerner MG, Carlson HA (2005) Binding MOAD (mother of all databases). Proteins 60:333–340
28. Liu T, Lin Y, Wen X, Jorrisen RN, Gilson MK (2007) BindingDB: a web-accessible database of experimentally determined protein-ligand binding affinities. Nucleic Acids Res 35:198–201
29. Wang R, Fang X, Lu Y, Wang S (2004) The PDBbind database: collection of binding affinities for protein-ligand complexes with known three-dimensional structures. J Med Chem 47:2977–2980
30. Xavier MM, Heck GS, de Avila MB, Levin NM, Pintro VO, Carvalho NL et al (2016) SAnDReS a computational tool for statistical analysis of docking results and development of scoring functions. Comb Chem High Throughput Screen 19:801–812
31. Pedregosa F, Varoquaux G, Gramfort A, Michel V, Thirion B, Grisel O et al (2011) Scikit-learn: machine learning in python. J Mach Learn Res 12:2825–2830
32. Goodsell DS, Olson AJ (1990) Automated docking of substrates to proteins by simulated annealing. Proteins 8:195–202
33. Goodsell DS, Morris GM, Olson AJ (1996) Docking of flexible ligands: applications of AutoDock. J Mol Recognit 9:1–5
34. Morris GM, Goodsell DS, Huey R, Olson AJ (1996) Distributed automated docking of flexible ligands to proteins: parallel applications of AutoDock 2.4. J Comput Aided Mol Des 10:293–304
35. Morris GM, Goodsell DS, Halliday RS, Huey R, Hart WE, Belew RK et al (1998) Automated docking using a Lamarckian genetic algorithm and and empirical binding free energy function. J Comput Chem 19:1639–1662
36. de Ávila MB, Xavier MM, Pintro VO, de Azevedo WF (2017) Supervised machine learning techniques to predict binding affinity. A study for cyclin-dependent kinase 2. Biochem Biophys Res Commun 494:305–310
37. Levin NMB, Pintro VO, Bitencourt-Ferreira-G, Mattos BB, Silvério AC, de Azevedo Jr WF (2018) Development of CDK-targeted scoring functions for prediction of binding affinity. Biophys Chem 235:1–8
38. Pintro VO, Azevedo WF (2017) Optimized virtual screening workflow. Towards target-based polynomial scoring functions for HIV-1 protease. Comb Chem High Throughput Screen 20:820–827
39. Amaral MEA, Nery LR, Leite CE, de Azevedo Junior WF, Campos MM (2018) Pre-clinical effects of metformin and aspirin on the cell lines of different breast cancer subtypes. Investig New Drugs 36:782–796
40. Russo S, de Azevedo WF (2018) Advances in the understanding of the cannabinoid receptor 1—focusing on the inverse agonists interactions. Curr Med Chem. https://doi.org/10.2174/0929867325666180417165247
41. de Ávila MB, de Azevedo WF Jr (2018) Development of machine learning models to predict inhibition of 3-dehydroquinate dehydratase. Chem Biol Drug Des 92:1468–1474
42. Bitencourt-Ferreira G, de Azevedo Jr WF (2018) Development of a machine-learning model to predict Gibbs free energy of binding for protein-ligand complexes. Biophys Chem 240:63–69
43. Morgan DO (1995) Principles of CDK regulation. Nature 374:131–134
44. Murray AW (1994) Cyclin-dependent kinases: regulators of the cell cycle and more. Chem Biol 1:191–195
45. de Azevedo WF Jr (2016) Opinion paper: targeting multiple cyclin-dependent kinases (CDKs): a new strategy for molecular docking studies. Curr Drug Targets 17:2
46. Levin NM, Pintro VO, de Ávila MB, de Mattos BB, De Azevedo WF Jr (2017) Understanding the structural basis for inhibition of cyclin-dependent kinases. New pieces in the molecular puzzle. Curr Drug Targets 18:1104–1111

47. Jaghoori MM, Bleijlevens B, Olabarriaga SD (2016) 1001 ways to run AutoDock Vina for virtual screening. J Comput Aided Mol Des 30:237–249
48. Heck GS, Pintro VO, Pereira RR, de Ávila MB, Levin NMB, de Azevedo WF (2017) Supervised machine learning methods applied to predict ligand-binding affinity. Curr Med Chem 24:2459–2470
49. Morris GM, Huey R, Lindstrom W, Sanner MF, Belew RK, Goodsell DS, Olson AJ (2009) AutoDock4 and AutoDockTools4: automated docking with selective receptor flexibility. J Comput Chem 30:2785–2791
50. Morris GM, Huey R, Olson AJ (2008) Using AutoDock for ligand-receptor docking. Curr Protoc bioinformatics. Chapter 8:unit 8.14
51. El-Hachem N, Haibe-Kains B, Khalil A, Kobeissy FH, Nemer G (2017) AutoDock and AutoDockTools for protein-ligand docking: Beta-site amyloid precursor protein cleaving enzyme 1(BACE1) as a case study. Methods Mol Biol 1598:391–403
52. Kim SH, Schulze-Gahmen U, Brandsen J, de Azevedo Júnior WF (1996) Structural basis for chemical inhibition of CDK2. Prog Cell Cycle Res 2:137–145
53. de Azevedo WF Jr, Mueller-Dieckmann HJ, Schulze-Gahmen U, Worland PJ, Sausville E, Kim SH (1996) Structural basis for specificity and potency of a flavonoid inhibitor of human CDK2, a cell cycle kinase. Proc Natl Acad Sci U S A 93:2735–2740
54. de Azevedo WF, Leclerc S, Meijer L, Havlicek L, Strnad M, Kim SH (1997) Inhibition of cyclin-dependent kinases by purine analogues: crystal structure of human CDK2 complexed with roscovitine. Eur J Biochem 243:518–526
55. de Azevedo WF Jr, Canduri F, da Silveira NJ (2002) Structural basis for inhibition of cyclin-dependent kinase 9 by flavopiridol. Biochem Biophys Res Commun 293:566–571
56. Filgueira de Azevedo W Jr, Gaspar RT, Canduri F, Camera JC Jr, Freitas da Silveira NJ (2002) Molecular model of cyclin-dependent kinase 5 complexed with roscovitine. Biochem Biophys Res Commun 297:1154–1158
57. Canduri F, Uchoa HB, de Azevedo WF Jr (2004) Molecular models of cyclin-dependent kinase 1 complexed with inhibitors. Biochem Biophys Res Commun 324:661–666
58. Canduri F, de Azevedo WF Jr (2005) Structural basis for interaction of inhibitors with Cyclin-dependent kinase 2. Curr Comput Aided Drug Des 1:53–64
59. Krystof V, Cankar P, Frysová I, Slouka J, Kontopidis G, Dzubák P et al (2006) 4-arylazo-3,5-diamino-^{1}H-pyrazole CDK inhibitors: SAR study, crystal structure in complex with CDK2, selectivity, and cellular effects. J Med Chem 49:6500–6509
60. Vieth M, Hirst JD, Kolinski A, Brooks CL III (1998) Assessing energy functions for flexible docking. J Comput Chem 19:1612–1622
61. Ballante F, Marshall GR (2016) An automated strategy for binding-pose selection and docking assessment in structure-based drug design. J Chem Inf Model 56:54–72
62. Azevedo LS, Moraes FP, Xavier MM, Pantoja EO, Villavicencio B, Finck JA et al (2012) Recent Progress of molecular docking simulations applied to development of drugs. Curr Bioinf 7:352–365
63. Heberlé G, de Azevedo WF Jr (2011) Bio-inspired algorithms applied to molecular docking simulations. Curr Med Chem 18:1339–1352
64. Vianna CP, de Azevedo WF Jr (2012) Identification of new potential *Mycobacterium tuberculosis* shikimate kinase inhibitors through molecular docking simulations. J Mol Model 18:755–764
65. Moraes FP, de Azevedo WF Jr (2012) Targeting imidazoline site on monoamine oxidase B through molecular docking simulations. J Mol Model 18:3877–3886
66. Coracini JD, de Azevedo WF Jr (2014) Shikimate kinase, a protein target for drug design. Curr Med Chem 21:592–604
67. Teles CB, Moreira-Dill LS, Silva Ade A, Facundo VA, de Azevedo WF Jr, da Silva LH et al (2015) A Lupane-triterpene isolated from *Combretum leprosum* Mart. fruit extracts that interferes with the intracellular development of *Leishmania* (L.) *amazonensis in vitro*. BMC Complement Altern Med 15:165
68. Freitas PG, Elias TC, Pinto IA, Costa LT, de Carvalho PVSD, Omote DQ et al (2018) Computational approach to the discovery of phytochemical molecules with therapeutic potential targets to the PKCZ protein. Lett Drug Des Discovery 15:488–499
69. de Azevedo WF Jr (2010) Structure-based virtual screening. Curr Drug Targets 11:261–263
70. de Azevedo WF Jr (2010) MolDock applied to structure-based virtual screening. Curr Drug Targets 11:327–334

71. Dias R, de Azevedo WF Jr (2008) Molecular docking algorithms. Curr Drug Targets 9:1040–1047
72. de Azevedo WF Jr, Dias R (2008) Computational methods for calculation of ligand-binding affinity. Curr Drug Targets 9:1031–1039
73. de Ávila MB, Bitencourt-Ferreira G, de Azevedo WF Jr (2018) Structural basis for inhibition of Enoyl-[acyl carrier protein] Reductase (InhA) from *Mycobacterium tuberculosis*. Curr Med Chem. https://doi.org/10.2174/0929867326666181203125229
74. Volkart PA, Bitencourt-Ferreira G, Souto AA, de Azevedo WF (2019) Cyclin-dependent kinase 2 in cellular senescence and cancer. A structural and functional review. Curr Drug Targets 20(7):716–726. https://doi.org/10.2174/1389450120666181204165344
75. Canduri F, Fadel V, Basso LA, Palma MS, Santos DS, de Azevedo WF Jr (2005) New catalytic mechanism for human purine nucleoside phosphorylase. Biochem Biophys Res Commun 327:646–649
76. Filgueira de Azevedo W Jr, Canduri F, Simões de Oliveira J, Basso LA, Palma MS, Pereira JH et al (2002) Molecular model of shikimate kinase from *Mycobacterium tuberculosis*. Biochem Biophys Res Commun 295:142–148
77. Canduri F, Teodoro LG, Fadel V, Lorenzi CC, Hial V, Gomes RA et al (2001) Structure of human uropepsin at 2.45 a resolution. Acta Crystallogr D Biol Crystallogr 57:1560–1570
78. Pereira JH, Canduri F, de Oliveira JS, da Silveira NJ, Basso LA, Palma MS et al (2003) Structural bioinformatics study of EPSP synthase from *Mycobacterium tuberculosis*. Biochem Biophys Res Commun 312:608–614
79. de Azevedo WF Jr, Dias R (2008) Experimental approaches to evaluate the thermodynamics of protein-drug interactions. Curr Drug Targets 9:1071–1076
80. Delatorre P, Rocha BA, Souza EP, Oliveira TM, Bezerra GA, Moreno FB et al (2007) Structure of a lectin from *Canavalia gladiata* seeds: new structural insights for old molecules. BMC Struct Biol 7:52
81. de Azevedo WF Jr, Canduri F, dos Santos DM, Pereira JH, Bertacine Dias MV, Silva RG et al (2003) Crystal structure of human PNP complexed with guanine. Biochem Biophys Res Commun 312:767–772
82. Canduri F, de Azevedo WF (2008) Protein crystallography in drug discovery. Curr Drug Targets 9:1048–1053
83. Marques MR, Pereira JH, Oliveira JS, Basso LA, de Azevedo WF Jr, Santos DS et al (2007) The inhibition of 5-enolpyruvylshikimate-3-phosphate synthase as a model for development of novel antimicrobials. Curr Drug Targets 8:445–457
84. Filgueira de Azevedo W Jr, dos Santos GC, dos Santos DM, Olivieri JR, Canduri F, Silva RG et al (2003) Docking and small angle X-ray scattering studies of purine nucleoside phosphorylase. Biochem Biophys Res Commun 309:923–928
85. Canduri F, Perez PC, Caceres RA, de Azevedo WF Jr (2007) Protein kinases as targets for antiparasitic chemotherapy drugs. Curr Drug Targets 8:389–398
86. Dias MV, Borges JC, Ely F, Pereira JH, Canduri F, Ramos CH et al (2006) Structure of chorismate synthase from *Mycobacterium tuberculosis*. J Struct Biol 154:130–143
87. Dias MV, Ely F, Palma MS, de Azevedo WF Jr, Basso LA, Santos DS (2007) Chorismate synthase: an attractive target for drug development against orphan diseases. Curr Drug Targets 8:437–444
88. Silva RG, Pereira JH, Canduri F, de Azevedo WF Jr, Basso LA, Santos DS (2005) Kinetics and crystal structure of human purine nucleoside phosphorylase in complex with 7-methyl-6-thio-guanosine. Arch Biochem Biophys 442:49–58
89. Timmers LF, Caceres RA, Vivan AL, Gava LM, Dias R, Ducati RG et al (2008) Structural studies of human purine nucleoside phosphorylase: towards a new specific empirical scoring function. Arch Biochem Biophys 479:28–38
90. de Azevedo WF Jr (2011) Molecular dynamics simulations of protein targets identified in *Mycobacterium tuberculosis*. Curr Med Chem 18:1353–1366
91. de Azevedo WF Jr (2011) Protein targets for development of drugs against *Mycobacterium tuberculosis*. Curr Med Chem 18:1255–1257
92. Caceres RA, Saraiva Timmers LF, Dias R, Basso LA, Santos DS, de Azevedo WF Jr (2008) Molecular modeling and dynamics simulations of PNP from *Streptococcus agalactiae*. Bioorg Med Chem 16:4984–4993
93. Dias MV, Faím LM, Vasconcelos IB, de Oliveira JS, Basso LA, Santos DS et al (2007) Effects of the magnesium and chloride ions and shikimate on the structure of shikimate kinase from *Mycobacterium tuberculosis*. Acta Crystallogr Sect F Struct Biol Cryst Commun 63:1–6

94. de Azevedo WF Jr, Ward RJ, Canduri F, Soares A, Giglio JR, Arni RK (1998) Crystal structure of piratoxin-I: a calcium-independent, myotoxic phospholipase A2-homologue from *Bothrops pirajai* venom. Toxicon 36:1395–1406
95. Dias R, Timmers LF, Caceres RA, de Azevedo WF Jr (2008) Evaluation of molecular docking using polynomial empirical scoring functions. Curr Drug Targets 9:1062–1070
96. da Silveira NJ, Uchôa HB, Canduri F, Pereira JH, Camera JC Jr, Basso LA et al (2004) Structural bioinformatics study of PNP from *Schistosoma mansoni*. Biochem Biophys Res Commun 322:100–104
97. de Azevedo WF Jr, Dias R (2008) Evaluation of ligand-binding affinity using polynomial empirical scoring functions. Bioorg Med Chem 16:9378–9382
98. Bezerra GA, Oliveira TM, Moreno FB, de Souza EP, da Rocha BA, Benevides RG et al (2007) Structural analysis of *Canavalia maritima* and *Canavalia gladiata* lectins complexed with different dimannosides: new insights into the understanding of the structure-biological activity relationship in legume lectins. J Struct Biol 160:168–176
99. Canduri F, Fadel V, Dias MV, Basso LA, Palma MS, Santos DS et al (2005) Crystal structure of human PNP complexed with hypoxanthine and sulfate ion. Biochem Biophys Res Commun 326:335–338
100. Timmers LF, Pauli I, Caceres RA, de Azevedo WF Jr (2008) Drug-binding databases. Curr Drug Targets 9:1092–1099
101. Delatorre P, Rocha BA, Gadelha CA, Santi-Gadelha T, Cajazeiras JB, Souza EP et al (2006) Crystal structure of a lectin from *Canavalia maritima* (ConM) in complex with trehalose and maltose reveals relevant mutation in ConA-like lectins. J Struct Biol 154:280–286
102. Rádis-Baptista G, Moreno FB, de Lima NL, Martins AM, de Oliveira TD, Toyama MH et al (2006) Crotacetin, a novel snake venom C-type lectin homolog of convulxin, exhibits an unpredictable antimicrobial activity. Cell Biochem Biophys 44:412–423
103. Breda A, Basso LA, Santos DS, de Azevedo Jr WF (2008) Virtual screening of drugs: score functions, docking, and drug design. Curr Comput Aided Drug Des 4:265–272
104. Nolasco DO, Canduri F, Pereira JH, Cortinóz JR, Palma MS, Oliveira JS et al (2004) Crystallographic structure of PNP from *Mycobacterium tuberculosis* at 1.9A resolution. Biochem Biophys Res Commun 324:789–794
105. Arcuri HA, Canduri F, Pereira JH, da Silveira NJ, Camera Júnior JC, de Oliveira JS et al (2004) Molecular models for shikimate pathway enzymes of *Xylella fastidiosa*. Biochem Biophys Res Commun 320:979–991
106. Soares MB, Silva CV, Bastos TM, Guimarães ET, Figueira CP, Smirlis D et al (2012) Anti-*Trypanosoma cruzi* activity of nicotinamide. Acta Trop 12:224–229
107. Rocha BA, Delatorre P, Oliveira TM, Benevides RG, Pires AF, Sousa AA et al (2011) Structural basis for both pro- and anti-inflammatory response induced by mannose-specific legume lectin from *Cymbosema roseum*. Biochimie 93:806–816
108. Ducati RG, Basso LA, Santos DS, de Azevedo WF Jr (2010) Crystallographic and docking studies of purine nucleoside phosphorylase from *Mycobacterium tuberculosis*. Bioorg Med Chem 18:4769–4774
109. Manhani KK, Arcuri HA, da Silveira NJ, Uchôa HB, de Azevedo WF Jr, Canduri F (2005) Molecular models of protein kinase 6 from *Plasmodium falciparum*. J Mol Model 12:42–48
110. Arcuri HA, Borges JC, Fonseca IO, Pereira JH, Neto JR, Basso LA et al (2008) Structural studies of shikimate 5-dehydrogenase from *Mycobacterium tuberculosis*. Proteins 72:720–730
111. Marques MR, Vaso A, Neto JR, Fossey MA, Oliveira JS, Basso LA et al (2008) Dynamics of glyphosate-induced conformational changes of *Mycobacterium tuberculosis* 5-enolpyruvylshikimate-3-phosphate synthase (EC 2.5.1.19) determined by hydrogen-deuterium exchange and electrospray mass spectrometry. Biochemistry 47:7509–7522
112. Cavada BS, Moreno FB, da Rocha BA, de Azevedo WF Jr, Castellón RE, Goersch GV et al (2006) cDNA cloning and 1.75 a crystal structure determination of PPL2, an endochitinase and N-acetylglucosamine-binding hemagglutinin from *Parkia platycephala* seeds. FEBS J 273:3962–3974
113. Arcuri HA, Zafalon GF, Marucci EA, Bonalumi CE, da Silveira NJ, Machado JM (2010) SKPDB: a structural database of shikimate pathway enzymes. BMC Bioinformatics 11:12
114. Moreno FB, de Oliveira TM, Martil DE, Viçoti MM, Bezerra GA, Abrego JR et al (2008) Identification of a new quaternary association for legume lectins. J Struct Biol 161:133–143

Chapter 5

Electrostatic Energy in Protein–Ligand Complexes

Gabriela Bitencourt-Ferreira, Martina Veit-Acosta, and Walter Filgueira de Azevedo Jr.

Abstract

Computational analysis of protein–ligand interactions is of pivotal importance for drug design. Assessment of ligand binding energy allows us to have a glimpse of the potential of a small organic molecule as a ligand to the binding site of a protein target. Considering scoring functions available in docking programs such as AutoDock4, AutoDock Vina, and Molegro Virtual Docker, we could say that they all rely on equations that sum each type of protein–ligand interactions to model the binding affinity. Most of the scoring functions consider electrostatic interactions involving the protein and the ligand. In this chapter, we present the main physics concepts necessary to understand electrostatics interactions relevant to molecular recognition of a ligand by the binding pocket of a protein target. Moreover, we analyze the electrostatic potential energy for an ensemble of structures to highlight the main features related to the importance of this interaction for binding affinity.

Key words Electrostatic interactions, Binding affinity, Drug design, Shikimate pathway, Molecular recognition

1 Introduction

The availability of experimental data about dissociation constant (K_d), Gibbs free energy of binding (ΔG), inhibition constant (K_i), half maximal inhibitory concentration (IC_{50}), provide a solid base for the development of computational models to predict binding affinity. Experimental binding affinity data are available at databases such as MOAD [1], BindingDB [2], and PDBbind [3]. Moreover, the richness of structural data available in the Protein Data Bank (PDB) [4–6] and the previously mentioned binding data can be used to create empirical scoring functions to predict binding affinity for protein-ligand complexes based on their atomic coordinates.

Scoring functions are computational approximations to predict protein–ligand binding affinity. Most of the modern development of scoring function for prediction of protein–ligand binding affinity started with the pioneering work of Böhm in the early 1990s

Walter Filgueira de Azevedo Jr. (ed.), *Docking Screens for Drug Discovery*, Methods in Molecular Biology, vol. 2053, https://doi.org/10.1007/978-1-4939-9752-7_5,

[7–12]. Docking programs such as AutoDock [13–16], AutoDock Vina [17, 18], and Molegro Virtual Docker (MVD) [19–21] make use of empirical scoring functions that somehow work very similar to the ideas proposed by Böhm.

One of the most used scoring functions to assess receptor–ligand binding affinity is the AutoDock4 semi-empirical free energy-force field scoring function [13–16]. Several studies showed that this scoring function could carry out a reliable evaluation of the binding energies of ligands to receptors [22, 23]. Briefly, AutoDock4 applies this force field through a two-step calculation. Firstly, AutoDock4 assesses the intramolecular energetics of the conversion from the unbound to the bound structures of the receptor–ligand complexes and then calculates the intermolecular energetics of the system. Let us consider that we express the binding affinity for receptor–ligand complexes as $pK_{\mathrm{i}} = \log(K_{\mathrm{i}})$, where K_{i} is the inhibition constant. Below we have the AutoDock4 semi-empirical free energy-force field scoring function,

$$pK_{\mathrm{i}} = \left(V_{\mathrm{bound}}^{\mathrm{L-L}} - V_{\mathrm{unbound}}^{\mathrm{L-L}}\right) + \left(V_{\mathrm{bound}}^{\mathrm{R-R}} - V_{\mathrm{unbound}}^{\mathrm{R-R}}\right) + \left(V_{\mathrm{bound}}^{\mathrm{R-L}} - V_{\mathrm{unbound}}^{\mathrm{R-L}}\right) + \Delta S_{\mathrm{system}} \quad (1)$$

The above equation includes an evaluation of the loss of torsional entropy upon binding ($\Delta S_{\mathrm{system}}$) and six pairwise atomic terms (V) where the L and the R, respectively, refer to the "ligand" and the "receptor" in a receptor–ligand complex. The expression for the conformational entropy lost upon binding of equation ($\Delta S_{\mathrm{system}}$) is as follows:

$$\Delta S_{\mathrm{system}} = \alpha_0 N_{\mathrm{tors}} \quad (2)$$

where N_{tors} represents the number of rotatable bonds in the ligand and α_0 the relative weight of this term.

The empirical scoring function tries to approximate the calculated binding affinity (V) to the experimental binding affinity ($pK_{\mathrm{i,\,exp}}$) through a regression model where we used the experimental data to determine the relative weights of each term in the regression equation. We calculate the pairwise energetic terms of Eq. (1) as follows:

$$V = \alpha_1 \sum_{i,j} \left(\frac{A_{ij}}{r_{ij}^{12}} - \frac{B_{ij}}{r_{ij}^{6}}\right) + \alpha_2 \sum_{i,j} E(t)\left(\frac{C_{ij}}{r_{ij}^{12}} - \frac{D_{ij}}{r_{ij}^{10}}\right) + \alpha_3 \sum_{i,j} \times \frac{q_i q_j}{\varepsilon(r_{ij}) r_{ij}} + \alpha_4 \sum_{i,j} \left(S_i V_j + S_j V_i\right) e^{-r_{ij}^2/2\sigma^2} \quad (3)$$

In the above equation, the αs represent the regression weights of the energy terms. The first term of the above equation calculates the dispersal/repulsion interactions, which is the equation of the Lennard-Jones potential [24]. The second term is a modification of the expression of the Lennard-Jones potential based on a 10/12 potential. It estimates the intermolecular hydrogen bonding interaction energy. The next term is the electrostatic potential, and the final one accounts for the desolvation potential. This last potential considers the volume of atoms (V_i or V_j) multiplied by a solvation parameter (S_i or S_j) and an exponential function with a distance weight of $\sigma = 3.5$ Å. In the above equation, the summations operate over all pairs of ligand atoms (i) and receptor atoms (j) besides all pairs of atoms in the ligand that are apart by three or more bonds.

It is feasible to add many other energy terms to Eq. (3), for instance, contact area and dipole energy, but the idea is the same. The summations are taken for atoms from the ligand and protein inside a predefined cutoff radius. We may apply these scoring functions to select the best pose generated by a search algorithm of a docking program or evaluate binding affinity based on the crystallographic structure for any protein–ligand complex. One key feature of the development of any scoring function is the assessment of electrostatic interactions for the protein–ligand system. In this chapter, we will give a broader view of the electrostatic interactions.

2 Coulomb's Law

To have a physical interpretation of the electrostatic interactions present in protein–ligand complexes, let us consider a system composed of two point charges q_1 and q_2 as shown in Fig. 1. The charge q_1 is at position $\vec{r}_1$ and the charge q_2 is at position $\vec{r}_2$. The term point charge used here is a mathematical abstraction; the protons and electrons have finite volumes. We see point charges as one whose dimensions are small compared with the distance between them. From the vector analysis of Fig. 1, we have the vector $\vec{r}_{12}$ as follows:

$$\vec{r}_{12} = \vec{r}_2 - \vec{r}_1$$

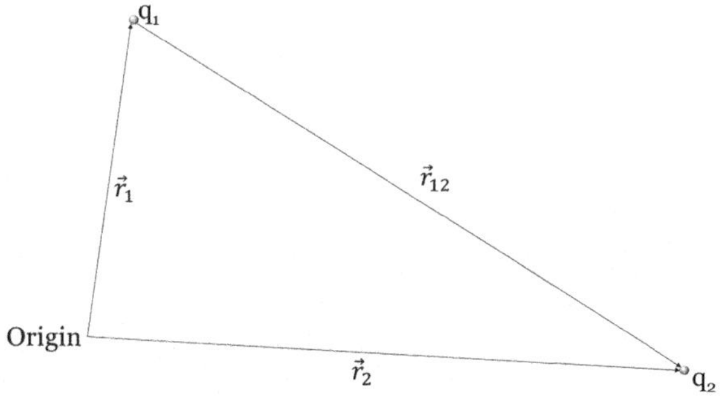

Fig. 1 A system composed of two point charges

The vector $\vec{r}_{12}$ joins q_1 and q_2 and points from q_1 to q_2. In the international system of units, electric charges are measured in Coulombs (C).

The force $\left(\vec{F}_{12}\right)$ exerted by q_1 on q_2 is given by Coulomb's law as follows:

$$\vec{F}_{12} = \frac{1}{4\pi\varepsilon_0}\frac{q_1 q_2}{r_{12}^3}\vec{r}_{12} \tag{4}$$

where ε_0 is permittivity of vacuum, and its value is approximately 8.854.10–12 C^2N^{-1} m^{-2}. The above equation is called Coulomb's law and is valid in the free space. Considering that we take punctual charges immersed in different media, we have that Coulomb's law still holds but with a different proportionality constant, as follows:

$$\vec{F}_{12} = \frac{1}{4\pi\varepsilon_r\varepsilon_0}\frac{q_1 q_2}{r_{12}^3}\vec{r}_{12} \tag{5}$$

where the quantity ε_r is called the relative permittivity of a material. The ε_r of water is 80.2 at a temperature of 20 °C. Therefore, we observe a reduction in the force between charges when immersed in water.

Let us consider a system composed of three point charges as shown in Fig. 2. Addition of a third charge (q_3) does not modify the force between charges q_1 and q_2. The resultant force that acts upon charge q_2 has now two components, namely, the force due to charge q_1 and the additional force due to q_3. The vector summation of the two forces acting on charge q_2 (F_2) has the following expression:

$$\vec{F}_2 = \frac{1}{4\pi\varepsilon_r\varepsilon_0}\left(\frac{q_1 q_2}{r_{12}^3}\vec{r}_{12} + \frac{q_2 q_3}{r_{32}^3}\vec{r}_{32}\right)$$

Rearranging the terms, we have the equation for a system composed of two point charges acting on a third charge as follows:

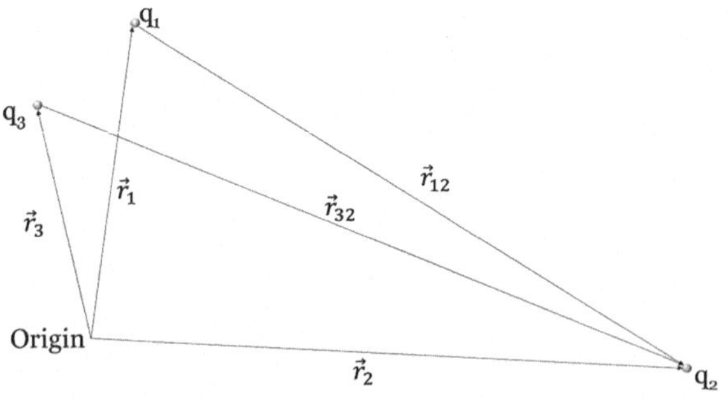

Fig. 2 A system composed of three point charges

$$\vec{F}_2 = \frac{1}{4\pi\varepsilon_r\varepsilon_0} q_2 \left(\frac{q_1}{r_{12}^3} \vec{r}_{12} + \frac{q_3}{r_{32}^3} \vec{r}_{32} \right)$$

In general, we may say that forces involving point electric charges are pairwise additive; therefore, if we consider a system composed of N charges, with $N - 1$ charges acting on charge i, we have the following expression for the force working on point charge i,

$$\vec{F}_i = \frac{1}{4\pi\varepsilon_r\varepsilon_0} q_i \left(\sum_{j \neq i}^{N} \frac{q_j}{r_{ij}^3} \vec{r}_{ij} \right) \tag{6}$$

3 Electrostatic Potential Energy

The electrostatic force is a conservative force since it only depends on the initial and final positions. Let us consider a system composed of two point charges q and Q in which the positive test charge q moves toward the stationary point charge Q. In the previous section, we saw that the magnitude of the force on a positive test charge as calculated by Coulomb's law is given by Eq. (4). Electrostatic potential energy (U) of a point charge q at position r from a charge Q, is defined as the negative work (W) done by electrostatic force to bring from a position r_{ref} to r position as follows:

$$U = -\int_{r_{\text{ref}}}^{r} \vec{F} \cdot \mathrm{d}\vec{r}$$

where $\mathrm{d}\vec{r}$ is the displacement vector from the reference point r_{ref} where $U = 0$ J and the position r of point charge q. The dot product (.) means that we take the component of the force acting along the displacement $\mathrm{d}r$. Substituting Eq (5) in the above expression, we have,

$$U = -\int_{r_{\text{ref}}}^{r} \vec{F} \cdot \mathrm{d}\vec{r}$$

$$= -\int_{r_{\text{ref}}}^{r} \frac{1}{4\pi\varepsilon_r\varepsilon_0} \frac{qQ}{r^3} \vec{r} \cdot \mathrm{d}\vec{r} = -\frac{qQ}{4\pi\varepsilon_r\varepsilon_0} \int_{r_{\text{ref}}}^{r} \frac{r}{r^3} \mathrm{d}r = -\frac{qQ}{4\pi\varepsilon_r\varepsilon_0} \int_{r_{\text{ref}}}^{r} \frac{1}{r^2} \mathrm{d}r$$

Considering the reference point for which $U = 0$ J at ∞ we have,

$$U = -\frac{qQ}{4\pi\varepsilon_r\varepsilon_0} \left(-\frac{1}{r} \right) \Bigg|_{\infty}^{r} = \frac{qQ}{4\pi\varepsilon_r\varepsilon_0} \left(\frac{1}{r} \right)$$

So the electrostatic potential energy (U) for a system composed of two charges q and Q is given by the following equation:

$$U = \frac{qQ}{4\pi\varepsilon_r\varepsilon_0 r} \tag{7}$$

For a system composed of N point charges, the electrostatic potential energy ($U_{\text{electrostatic}}$) is given by the following expression:

$$U_{\text{electrostatic}} = \sum_{i,j} \frac{q_i q_j}{4\pi\varepsilon_r\varepsilon_0 r_{ij}} \tag{8}$$

The above equation is the electrostatic term of the AutoDock4 empirical scoring function, where we consider that $\varepsilon(r_{ij})$ is $4\pi\varepsilon_r\varepsilon_0$. Evaluation of $\varepsilon(r_{ij})$ for biomolecules is a challenge from the computational point of view. Specifically for the AutoDock4, $\varepsilon(r_{ij})$ is approximated by a sigmoidal distance-dependent permittivity function, based on the work of Mehler and Solmajer [25].

$$\varepsilon(r) = A + \frac{B}{1 + ke^{-\lambda Br}} \tag{9}$$

In the above equation, the constants have the following values: $B = \varepsilon_r - A$; $\varepsilon_r =$ the relative permittivity constant of bulk water at 25 °C = 78.4; $A = -8.5525$, $\lambda = 0.003627$ and $k = 7.7839$.

In biological systems such as proteins and nucleic acids, we find fully charged atoms. Nevertheless, most of atoms show only partial charges. For this reason, the variable for charges in the previously explained equations could mean partial charges. There are several algorithms to calculate partial charges for biological systems. Amongst the most used approaches, we could highlight the Partial Equalization of Orbital Electronegativity (PEOE) method [26]. AutoDockTools4 [22] uses this algorithm to estimate partial charges. In the next section, we discuss the application of Eqs. (8 and 9) to determine the electrostatic potential energy of protein-–ligand complexes.

4 Calculating Electrostatic Potential for Protein–Ligand Complexes

To illustrate the calculations of electrostatic interactions of protein–ligand complexes, we took a biological system composed of enzymes of the shikimate pathway. This metabolic route is a target for the development of herbicides and antibacterial drugs [27]. Shikimate pathway has been submitted to intense structural and computational studies [28–65] due to its relevance for drug design and development.

We searched the PDB for the enzymes 3-deoxy-D-arabino-heptulosonate 7-phosphate (DAHP) synthase (EC 2.5.1.54), shikimate kinase (EC 2.7.1.71), and 3-dehydroquinate dehydratase (EC 4.2.1.10) of this metabolic route for which inhibition constant (K_i) data are available. We found a total of 24 crystallographic

Table 1
Shikimate pathway enzymes used in this study

Enzyme classification	PDB access codes
2.5.1.54	4UMA, 4UMB, 4UMC
2.7.1.71	4BQS
4.2.1.10	1H0R, 1GU1, 1V1J, 2BT4, 2C4V, 2C4W, 2XB8, 2XB9, 3N76, 3N7A, 3N86, 3N87, 3N8K, 3N8N, 4B6O, 4B6P, 4B6R, 4B6S, 4CIW, 4CIY

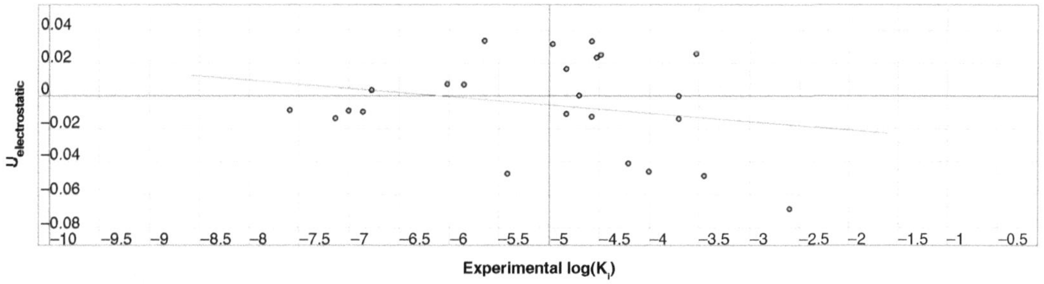

Fig. 3 Scattering plot for experimental log(K_i) and theoretical $U_{electrostatic}$. We generated this plot with the program Molegro Data Modeller (MDM) [19]

structures for which K_i data are available (search carried out on December 18, 2018). Table 1 shows the PDB access codes for all structures identified in the PDB.

We implemented Eqs. (8 and 9) in Python (program SFSXplorer) and considered the partially charged charges calculated using AutoDockTools4 [22]. The scattering plot for experimental binding affinity (log(K_i)) and the calculated electrostatic potential energy $U_{electrostatic}$ is shown in Fig. 3. Spearman's rank correlation between experimental log(K_i) and $U_{electrostatic}$ is −0.22. This level of correlation is not significant. Nevertheless, electrostatic interactions have been shown of pivotal importance for ligand binding affinity in recent studies focused on specific enzymes [66–75]. The low level of significance may be due to the application of a pure electrostatic potential without consideration of additional interactions such as the Lennard-Jones potential and intermolecular hydrogen bonds.

5 Colophon

We created Figs. 1 and 2 using Microsoft PowerPoint 2016. We generated Fig. 3 with the Molegro Data Modeller (MDM)

[19]. We performed scoring function calculation described in this chapter using a Desktop PC with 4GB of memory, a 1 TB hard disk, and an Intel® Core® i3-2120 @ 3.30 GHz processor running Windows 8.1.

6 Availability

SFSXplorer is implemented in Python and available to download under the GNU license at https://github.com/azevedolab/SFSXplorer. The shikimate dataset is available for downloading at https://azevedolab.net/receptor-ligand-systems-database.php.

7 Final Remarks

In summary, we can easily calculate electrostatic interactions using classical electromagnetism (Eq. (8)) and implement this equation in a high-level computer language such as Python. The availability of experimental information for structures and binding affinity opens the possibility to generate enzyme-targeted scoring functions for prediction of binding affinity where we employ the experimental data to calibrate a complete scoring function for a specific biological system.

Acknowledgments

This work was supported by grants from CNPq (Brazil) (308883/2014-4). This study was financed in part by the Coordenação de Aperfeiçoamento de Pessoal de Nivel Superior—Brasil (CAPES)—Finance Code 001. GB-F acknowledges support from PUCRS/BPA fellowship. MV-A acknowledges support from PUCRS/IC Jr. WFA is a senior researcher for CNPq (Brazil) (Process Numbers: 308883/2014-4 and 309029/2018-0).

References

1. Hu L, Benson ML, Smith RD, Lerner MG, Carlson HA (2005) Binding MOAD (Mother Of All Databases). Proteins 60:333–340
2. Liu T, Lin Y, Wen X, Jorrisen RN, Gilson MK (2007) BindingDB: a web-accessible database of experimentally determined protein-ligand binding affinities. Nucleic Acids Res 35:198–201
3. Wang R, Fang X, Lu Y, Wang S (2004) The PDBbind database: collection of binding affinities for protein-ligand complexes with known three-dimensional structures. J Med Chem 47:2977–2980
4. Berman HM, Westbrook J, Feng Z, Gilliland G, Bhat TN, Weissig H et al (2000) The Protein Data Bank. Nucleic Acids Res 28:235–242
5. Berman HM, Battistuz T, Bhat TN, Bluhm WF, Bourne PE, Burkhardt K et al (2002) The Protein Data Bank. Acta Crystallogr D Biol Crystallogr 58:899–907

6. Westbrook J, Feng Z, Chen L, Yang H, Berman HM (2003) The Protein Data Bank and structural genomics. Nucleic Acids Res 31:489–491
7. Böhm HJ (1993) A novel computational tool for automated structure-based drug design. J Mol Recognit 6:131–137
8. Böhm HJ (1994) The development of a simple empirical scoring function to estimate the binding constant for a protein-ligand complex of known three-dimensional structure. J Comput Aided Mol Des 8:243–256
9. Böhm HJ (1996) Towards the automatic design of synthetically accessible protein ligands: peptides, amides and peptidomimetics. J Comput Aided Mol Des 10:265–272
10. Stahl M, Böhm HJ (1998) Development of filter functions for protein-ligand docking. J Mol Graph Model 16:121–132
11. Klebe G, Böhm HJ (1997) Energetic and entropic factors determining binding affinity in protein-ligand complexes. J Recept Signal Transduct Res 17:459–473
12. Böhm HJ, Banner DW, Weber L (1999) Combinatorial docking and combinatorial chemistry: design of potent non-peptide thrombin inhibitors. J Comput Aided Mol Des 13:51–56
13. Goodsell DS, Olson AJ (1990) Automated docking of substrates to proteins by simulated annealing. Proteins 8:195–202
14. Goodsell DS, Morris GM, Olson AJ (1996) Docking of flexible ligands: applications of AutoDock. J Mol Recognit 9:1–5
15. Morris GM, Goodsell DS, Huey R, Olson AJ (1996) Distributed automated docking of flexible ligands to proteins: Parallel applications of AutoDock 2.4. J Comput Aided Mol Des 10:293–304
16. Morris GM, Goodsell DS, Halliday RS, Huey R, Hart WE, Belew RK et al (1998) Automated docking using a lamarckian genetic algorithm and and empirical binding free energy function. J Comput Chem 19:1639–1662
17. Trott O, Olson AJ (2010) AutoDock Vina: improving the speed and accuracy of docking with a new scoring function, efficient optimization, and multithreading. J Comput Chem 31:455–461
18. Jaghoori MM, Bleijlevens B, Olabarriaga SD (2016) 1001 Ways to run AutoDock Vina for virtual screening. J Comput Aided Mol Des 30:237–249
19. Thomsen R, Christensen MH (2006) MolDock: a new technique for high-accuracy molecular docking. J Med Chem 49:3315–3321
20. Heberlé G, de Azevedo WF Jr (2011) Bio-inspired algorithms applied to molecular docking simulations. Curr Med Chem 18:1339–1352
21. de Azevedo WF Jr (2010) MolDock applied to structure-based virtual screening. Curr Drug Targets 11:327–334
22. Morris GM, Huey R, Lindstrom W, Sanner MF, Belew RK, Goodsell DS et al (2009) AutoDock4 and AutoDockTools4: Automated docking with selective receptor flexibility. J Comput Chem 30:2785–2791
23. Huey R, Morris GM, Olson AJ, Goodsell DS (2007) A semiempirical free energy force field with charge-based desolvation. J Comput Chem 28:1145–1152
24. Lennard-Jones JE (1931) Cohesion. Proc Phys Soc 43:461–482
25. Mehler EL, Solmajer T (1991) Electrostatic effects in proteins: comparison of dielectric and charge models. Protein Eng 4:903–910
26. Gasteiger J, Marsili M (1980) Iterative partial equalization of orbital electronegativity—a rapid access to atomic charges. Tetrahedron 36:3219–3228
27. Parish T, Stoker NG (2002) The common aromatic amino acid biosynthesis pathway is essential in *Mycobacterium tuberculosis*. Microbiology 148:3069–3077
28. Pereira JH, Canduri F, de Oliveira JS, da Silveira NJ, Basso LA, Palma MS et al (2003) Structural bioinformatics study of EPSP synthase from *Mycobacterium tuberculosis*. Biochem Biophys Res Commun 312:608–614
29. Arcuri HA, Canduri F, Pereira JH, da Silveira NJ, Camera JC Jr, de Oliveira JS et al (2004) Molecular models for shikimate pathway enzymes of *Xylella fastidiosa*. Biochem Biophys Res Commun 320:979–991
30. Dias MV, Ely F, Canduri F, Pereira JH, Frazzon J, Basso LA et al (2004) Crystallization and preliminary X-ray crystallographic analysis of chorismate synthase from *Mycobacterium tuberculosis*. Acta Crystallogr D Biol Crystallogr 60:2003–2005
31. Uchôa HB, Jorge GE, Freitas Da Silveira NJ, Camera JC Jr, Canduri F, De Azevedo WF Jr (2004) Parmodel: a web server for automated comparative modeling of proteins. Biochem Biophys Res Commun 325:1481–1486
32. Pereira JH, de Oliveira JS, Canduri F, Dias MV, Palma MS, Basso LA et al (2004) Structure of shikimate kinase from *Mycobacterium tuberculosis* reveals the binding of shikimic acid. Acta Crystallogr D Biol Crystallogr 60:2310–2319
33. Silveira NJ, Uchôa HB, Pereira JH, Canduri F, Basso LA, Palma MS et al (2005) Molecular

models of protein targets from *Mycobacterium tuberculosis.* J Mol Model 11:160–166
34. Dias MV, Borges JC, Ely F, Pereira JH, Canduri F, Ramos CH et al (2006) Structure of chorismate synthase from *Mycobacterium tuberculosis.* J Struct Biol 154:130–143
35. da Silveira NJ, Bonalumi CE, Uchôa HB, Pereira JH, Canduri F, de Azevedo WF (2006) DBMODELING: a database applied to the study of protein targets from genome projects. Cell Biochem Biophys 44:366–374
36. Borges JC, Pereira JH, Vasconcelos IB, dos Santos GC, Olivieri JR, Ramos CH et al (2006) Phosphate closes the solution structure of the 5-enolpyruvylshikimate-3-phosphate synthase (EPSPS) from *Mycobacterium tuberculosis.* Arch Biochem Biophys 452:156–164
37. da Silveira NJF, Bonalumi CE, Arcuri HA, de Azevedo WF Jr (2007) Molecular modeling databases: a new way in the search of proteins targets for drug development. Curr Bioinf 2:1–10
38. Dias MV, Faím LM, Vasconcelos IB, de Oliveira JS, Basso LA, Santos DS et al (2007) Effects of the magnesium and chloride ions and shikimate on the structure of shikimate kinase from *Mycobacterium tuberculosis.* Acta Crystallogr Sect F Struct Biol Cryst Commun 63:1–6
39. Dias MV, Ely F, Palma MS, de Azevedo WF Jr, Basso LA, Santos DS (2007) Chorismate synthase: an attractive target for drug development against orphan diseases. Curr Drug Targets 8:437–444
40. Marques MR, Pereira JH, Oliveira JS, Basso LA, de Azevedo WF Jr, Santos DS et al (2007) The inhibition of 5-enolpyruvylshikimate-3-phosphate synthase as a model for development of novel antimicrobials. Curr Drug Targets 8:445–457
41. Pereira JH, Vasconcelos IB, Oliveira JS, Caceres RA, de Azevedo WF Jr, Basso LA et al (2007) Shikimate kinase: a potential target for development of novel antitubercular agents. Curr Drug Targets 8:459–468
42. Marques MR, Vaso A, Neto JR, Fossey MA, Oliveira JS, Basso LA et al (2008) Dynamics of glyphosate-induced conformational changes of *Mycobacterium tuberculosis* 5-enolpyruvylshikimate-3-phosphate synthase (EC 2.5.1.19) determined by hydrogen-deuterium exchange and electrospray mass spectrometry. Biochemistry 47:7509–7522
43. Arcuri HA, Borges JC, Fonseca IO, Pereira JH, Neto JR, Basso LA et al (2008) Structural studies of shikimate 5-dehydrogenase from *Mycobacterium tuberculosis.* Proteins 72:720–730
44. Pauli I, Caceres RA, de Azevedo WF Jr (2008) Molecular modeling and dynamics studies of Shikimate Kinase from *Bacillus anthracis.* Bioorg Med Chem 16:8098–8108
45. de Azevedo WF Jr (2008) Protein-drug interactions. Curr Drug Targets 9:1030
46. de Azevedo WF Jr, Dias R (2008) Computational methods for calculation of ligand-binding affinity. Curr Drug Targets 92:1031–1039
47. Dias R, de Azevedo WF Jr (2008) Molecular docking algorithms. Curr Drug Targets 9:1040–1047
48. Canduri F, de Azevedo WF (2008) Protein crystallography in drug discovery. Curr Drug Targets 9:1048–1053
49. Pauli I, Timmers LF, Caceres RA, Soares MB, de Azevedo WF Jr (2008) *In silico* and *in vitro*: identifying new drugs. Curr Drug Targets 9:1054–1061
50. Dias R, Timmers LF, Caceres RA, de Azevedo WF Jr (2008) Evaluation of molecular docking using polynomial empirical scoring functions. Curr Drug Targets 9:1062–1070
51. de Azevedo WF Jr, Dias R (2008) Experimental approaches to evaluate the thermodynamics of protein-drug interactions. Curr Drug Targets 9:1071–1076
52. Caceres RA, Pauli I, Timmers LF, de Azevedo WF Jr (2008) Molecular recognition models: a challenge to overcome. Curr Drug Targets 9:1077–1083
53. Barcellos GB, Caceres RA, de Azevedo WF Jr (2009) Structural studies of shikimate dehydrogenase from *Bacillus anthracis* complexed with cofactor NADP. J Mol Model 15:147–155
54. de Azevedo WF Jr, Dias R, Timmers LF, Pauli I, Caceres RA, Soares MB (2009) Bioinformatics tools for screening of antiparasitic drugs. Curr Drug Targets 10:232–239
55. Arcuri HA, Zafalon GF, Marucci EA, Bonalumi CE, da Silveira NJ, Machado JM et al (2010) SKPDB: a structural database of shikimate pathway enzymes. BMC Bioinformatics 11:12
56. Hernandes MZ, Cavalcanti SM, Moreira DR, de Azevedo WF Jr, Leite AC (2010) Halogen atoms in the modern medicinal chemistry: hints for the drug design. Curr Drug Targets 11:303–314
57. De Azevedo WF Jr (2010) Structure-based virtual screening. Curr Drug Targets 11:261–263
58. de Azevedo WF Jr (2011) Molecular dynamics simulations of protein targets identified in *Mycobacterium tuberculosis.* Curr Med Chem 18:1353–1366

59. de Azevedo WF Jr (2011) Protein targets for development of drugs against *Mycobacterium tuberculosis*. Curr Med Chem 18:1255–1257
60. Vianna CP, de Azevedo WF Jr (2012) Identification of new potential *Mycobacterium tuberculosis* shikimate kinase inhibitors through molecular docking simulations. J Mol Model 18:755–764
61. Azevedo LS, Moraes FP, Xavier MM, Pantoja EO, Villavicencio B, Finck JA et al (2012) Recent progress of molecular docking simulations applied to development of drugs. Curr Bioinf 7:352–365
62. Coracini JD, de Azevedo WF Jr (2014) Shikimate kinase, a protein target for drug design. Curr Med Chem 21:592–604
63. de Avila MB, de Azevedo WF (2014) Data mining of docking results. application to 3-dehydroquinate dehydratase. Curr Bioinf 9:361–379
64. Heck GS, Pintro VO, Pereira RR, de Ávila MB, Levin NMB, de Azevedo WF (2017) Supervised machine learning methods applied to predict ligand-binding affinity. Curr Med Chem 24:2459–2470
65. de Ávila MB, Bitencourt-Ferreira G, de Azevedo WF Jr (2018) Structural basis for inhibition of Enoyl-[Acyl Carrier Protein] reductase (InhA) from *Mycobacterium tuberculosis*. Curr Med Chem. https://doi.org/10.2174/0929867326666181203125229
66. Xavier MM, Heck GS, de Avila MB, Levin NM, Pintro VO, Carvalho NL et al (2016) SAnDReS a computational tool for statistical analysis of docking results and development of scoring functions. Comb Chem High Throughput Screen 19:801–812
67. Levin NM, Pintro VO, de Ávila MB, de Mattos BB, De Azevedo WF Jr (2017) Understanding the structural basis for inhibition of Cyclin-dependent kinases. new pieces in the molecular puzzle. Curr Drug Targets 18:1104–1111
68. de Ávila MB, Xavier MM, Pintro VO, de Azevedo WF (2017) Supervised machine learning techniques to predict binding affinity. A study for cyclin-dependent kinase 2. Biochem Biophys Res Commun 494:305–310
69. Pintro VO, Azevedo WF (2017) Optimized virtual screening workflow. towards target-based polynomial scoring functions for HIV-1 protease. Comb Chem High Throughput Screen 20:820–827
70. Freitas PG, Elias TC, Pinto IA, Costa LT, de Carvalho PVSD, Omote DQ et al (2018) Computational approach to the discovery of phytochemical molecules with therapeutic potential targets to the PKCZ protein. Lett Drug Des Discovery 15:488–499
71. Levin NMB, Pintro VO, Bitencourt-Ferreira G, Mattos BB, Silvério AC, de Azevedo WF Jr (2018) Development of CDK-targeted scoring functions for prediction of binding affinity. Biophys Chem 235:1–8
72. Amaral MEA, Nery LR, Leite CE, de Azevedo WF Jr, Campos MM (2018) Pre-clinical effects of metformin and aspirin on the cell lines of different breast cancer subtypes. Invest New Drugs 36:782–796
73. de Ávila MB, de Azevedo WF Jr (2018) Development of machine learning models to predict inhibition of 3-dehydroquinate dehydratase. Chem Biol Drug Des 92:1468–1474
74. Volkart PA, Bitencourt-Ferreira G, Souto AA, de Azevedo WF (2019) Cyclin-dependent kinase 2 in cellular senescence and cancer. A structural and functional review. Curr Drug Targets 20(7):716–726. https://doi.org/10.2174/1389450120666181204165344
75. Bitencourt-Ferreira G, de Azevedo WF Jr (2018) Development of a machine-learning model to predict Gibbs free energy of binding for protein-ligand complexes. Biophys Chem 240:63–69

Chapter 6

Van der Waals Potential in Protein Complexes

Gabriela Bitencourt-Ferreira, Martina Veit-Acosta, and Walter Filgueira de Azevedo Jr.

Abstract

Van der Waals forces are determinants of the formation of protein-ligand complexes. Physical models based on the Lennard-Jones potential can estimate van der Waals interactions with considerable accuracy and with a computational complexity that allows its application to molecular docking simulations and virtual screening of large databases of small organic molecules. Several empirical scoring functions used to evaluate protein-ligand interactions approximate van der Waals interactions with the Lennard-Jones potential. In this chapter, we present the main concepts necessary to understand van der Waals interactions relevant to molecular recognition of a ligand by the binding pocket of a protein target. We describe the Lennard-Jones potential and its application to calculate potential energy for an ensemble of structures to highlight the main features related to the importance of this interaction for binding affinity.

Key words van der Waals interactions, Lennard-Jones potential, Binding affinity, Drug design, Shikimate pathway

1 Introduction

Modern computational models to predict binding affinity based on the atomic coordinates of protein-ligand complexes need to evaluate non-bonded atom-atom interactions in a physically coherent approach. For recent reviews, please *see* refs. 1–5. Considering applications to computer-aided drug design such as protein-ligand docking, the primary determinant is the computational complexity of the algorithm used to evaluate binding affinity [6–15]. Therefore, increasing the complexity of the physical model to predict binding affinity creates a theoretical model that demands more computational power. Modern methods to predict protein-ligand binding affinity have to consider the limitations of adding physical realism to a computational model.

Pioneering works of many research groups have established the experimental and theoretical framework for structure-based drug design studies [16–18]. These research initiatives employing X-ray

Walter Filgueira de Azevedo Jr. (ed.), *Docking Screens for Drug Discovery*, Methods in Molecular Biology, vol. 2053, https://doi.org/10.1007/978-1-4939-9752-7_6,

diffraction crystallography were able to solve structures of the complexes involving a protein target and a small organic molecule bound to it. A subsequent analysis of these data made it possible to identify the structural basis for the intermolecular interactions. As computational power increased, it was also possible to perform analysis of new drugs with a protein target through in silico techniques. Among these techniques employed in drug design and development, protein-ligand docking simulation is one of the most employed computational methodologies [19, 20].

The progress of molecular docking methods began in the early 1980s [21]. Once molecular docking programs became available, in silico methodologies were successfully employed to discover new drugs including HIV-1 protease (EC 3.4.23.16) inhibitors [22–27]. One key feature of all docking simulations is the evaluation of the binding affinity based on the atomic coordinates of the protein-ligand complexes. Computational tools to evaluate these complexes should be fast to allow computational assessment of thousands of positions for a given ligand. The use of quantum mechanics methods could generate coherent physical models to calculate binding affinity. On the other hand, quantum mechanics methods to handle biomolecular systems with thousands of atoms demand higher computational power [28–37] than classical approaches.

The tug-of-war between physical coherence and the computational complexity of the algorithm has a moving line that depends on the computational power available for the generation of the predictive model. As computational power increases, the complexity of the algorithm can be higher to include physical relevant interactions in the modeling of protein-ligand interactions. Nevertheless, these conflicts of interest between physics and computational complexity have some landmarks in the history of the development of computational models to predict atom-atom interactions [38]. In this chapter, our focus is on the van der Waals interactions and its approximation by the Lennard-Jones potential. To illustrate the application of the Lennard-Jones potential, we calculated the van der Waals interactions for an ensemble of crystallographic structures for which experimental binding affinity data are available.

2 van der Waals Interactions

One naïve interpretation of the van der Waals interaction is possible through a thought experiment involving two spherical gas balloons. We take these balloons initially separated by a distance $r \gg$ sum of their radii. We might consider that we hold both balloons, one in each hand. Since they are far away from each other, we can quickly move the balloons. We may say that the potential energy of this system, our two balloons, is zero when the "inter-balloon" distance is $r \gg$ sum of their radii. Consider now that we move the balloons just close enough to contact each other. From now on, if we insist on approximating them, we have to exert a force to bring them closer. Now we have positive potential energy. We could think of our balloons as atoms; when they are far apart, the potential energy of the system is zero, and as we approximate them, we reach positive potential energy. This thought experiment captures the basic idea of the interaction between two atoms.

Let us take a more realistic view of the non-bonded atom-atom interactions; we consider a system composed of two spherical atoms (atoms 1 and 2) separated by a distance r and with radii r_1 and r_2, respectively. In this situation, the positioning of the electron of atom 1 at the furthest distance from atom 2 creates in an instant the lacking of the negative charge of the atom 1 in a region close to the atom 2. We could consider this absence of negative charge as a relative positive charge.

In physical terms, we have an instant electrical dipole in atom 1. This positive charge in the atom 1 attracts the electrons from atom 2, which creates a favorable interaction if the atoms are not too close or too far apart. The closer we move both atoms, the higher is the potential energy of the system since we have to act against the repulsion of electrons in both atoms. When both atoms are at distance $r \gg$ sum of the van der Waals radii ($r_1 + r_2$) (Fig. 1a), we have a potential energy close to zero. The minimum of the potential energy is at the situation where the distance between the atoms is equal to the sum of their radii; we call this distance of equilibrium distance (r_{eqm}) (Fig. 1b). As we move the atoms closer, we have positive potential energy (Fig. 1c). Figure 1d illustrates the variation of the potential energy (V) as a function of the internuclear distance (r).

3 Lennard-Jones Potential

The original description of the Lennard-Jones potential dates back to 1931 [38]. This elegant approximation to non-bonded atom-atom interaction is present in several force fields dedicated to evaluation protein-ligand interactions, such as the functions calculated by AMBER ff99 [39, 40], AutoDock 4 [41], TreeDock [42], and ReplicOpter [43], to mention a few.

To have a deeper insight into the modeling of atom-atom interaction, let us consider a system composed of two non-bonded atoms separated by the internuclear distance r. The potential energy of this system consisting of two atoms can be expressed as a function of r, as follows:

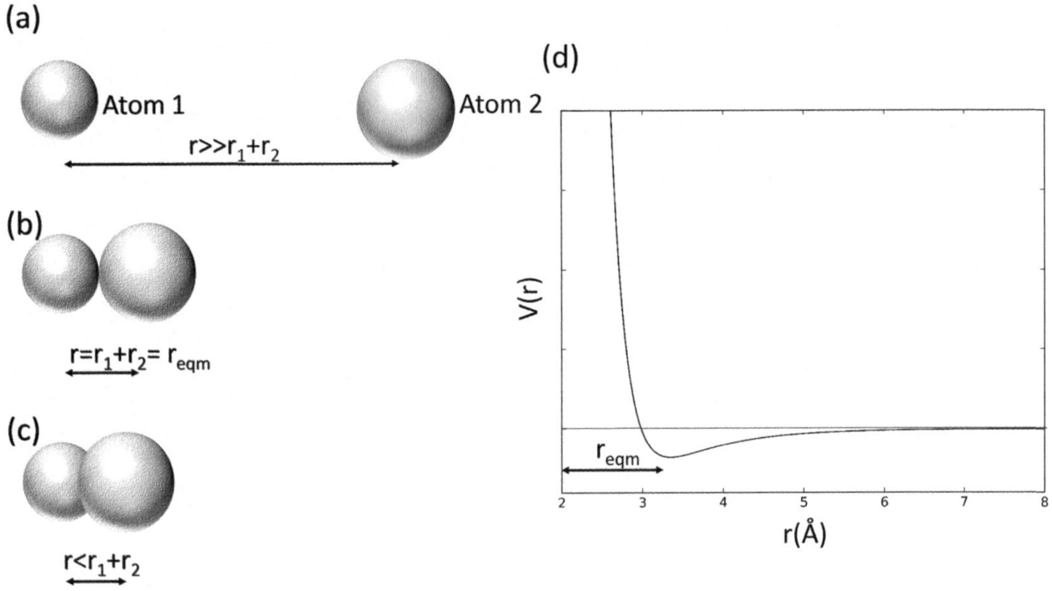

Fig. 1 Non-bonded atom-atom interactions. (**a**) In this situation, we have the internuclear distance $r \gg r_1 + r_2$. (**b**) Now we have our system separated by the equilibrium distance (r_{eqm}). (**c**) As we move the atoms closer, their electron cloud overlap, the positively charged nuclei become less shielded by the negative charges and the two atoms repel each other. (**d**) The plot of the variation of the potential energy (V) relative to the internuclear distance (r)

$$V(r) = \frac{Ae^{-br}}{r} - \frac{C_6}{r^6} \quad (1)$$

where A, b, and C_6 are parameters specific to the particular pair of atoms and have to be experimentally determined [44–46]. Eq. (1) is named Buckingham potential [44]. The first term of Eq. (1) is responsible for the repulsive exchange energy, and the $-r^{-6}$ term is related to the attractive interaction. In several empirical scoring functions, the exponential term is often approximated as follows:

$$\frac{A}{r}e^{-br} \approx \frac{C_{12}}{r^{12}} \quad (2)$$

Therefore, the potential energy can be approximated using the following expression:

$$V(r) \approx \frac{C_n}{r^n} - \frac{C_m}{r^m} = C_n r^{-n} - C_m r^{-m} \quad (3)$$

where m and n are integers, and C_n and C_m are constants whose values are based on the equilibrium separation between two atoms and the depth of the energy well.

In general, Eq. (3) is computationally implemented as follows:

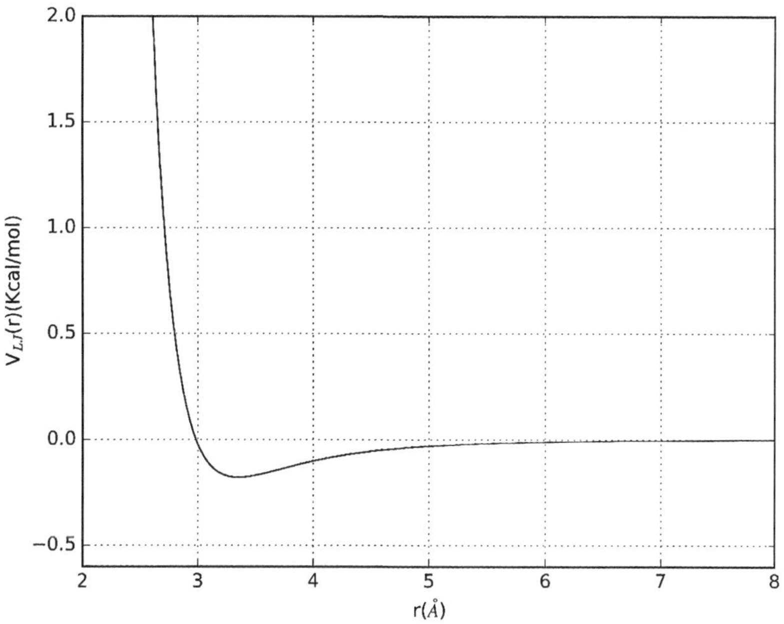

Fig. 2 Lennard-Jones 12-6 potential for nitrogen-oxygen

$$V_{\mathrm{LJ}}(r) \approx \frac{\frac{m}{n-m}\varepsilon r^{n}_{\mathrm{eqm}}}{r^{n}} - \frac{\frac{n}{n-m}\varepsilon r^{m}_{\mathrm{eqm}}}{r^{m}} \tag{4}$$

where V_{LJ} is the Lennard-Jones potential energy, ε is the well depth of the potential energy function, and r_{eqm} is the equilibrium separation between two atoms. The numbers m and n are integers taken as $n = 12$ and $m = 6$ for the original Lennard-Jones potential. Figure 2 illustrates the standard Lennard-Jones potential for N.O interaction. Although the computational form of Eq. (4) has been successfully applied to several biomolecular systems [39–43], application of Eq. (1) (exponential-6 form) has shown superior predictive performance on the evaluation of the native binding modes in biomolecular systems such as cyclin-dependent kinase (CDK) and proteases [47]. CDK and protease are both important protein targets for the development of drugs [48–61].

Such variability of predictive performance with the type of biomolecular system is in agreement with the concept of scoring function space [3]. Briefly, we see protein-ligand interaction as a result of the relation between the protein space [62] and the chemical space [63], and we propose to approach these sets as a unique complex system, where the application of computational methodologies could contribute to establishing the physical principles to understand the structural basis for the specificity of ligands for proteins. We propose to use the abstraction of a mathematical space composed of infinite computational models to predict ligand binding affinity, named here as scoring function space. By the use of supervised machine learning techniques, we can explore this scoring function space to build a computational model targeted to a specific biological system.

4 Calculating Lennard-Jones Potential for Protein-Ligand Complexes

To illustrate the calculations of van der Waals interactions of protein-ligand complexes, we took a biological system composed of enzymes of the shikimate pathway. This metabolic route is a target for the development of herbicides and antibacterial drugs [64]. The shikimate pathway has been submitted to intense structural and computational studies [65–102] due to its relevance for drug design and development.

We searched the Protein Data Bank (PDB) [103–105] for enzymes DAHP (3-Deoxy-D-arabinoheptulosonate 7-phosphate) synthase (EC 2.5.1.54), shikimate kinase (EC 2.7.1.71), and 3-dehydroquinate dehydratase (EC 4.2.1.10) of this metabolic route for which inhibition constant (K_i) data are available. We found a total of 23 crystallographic structures for which K_i data are available (search carried out on December 18, 2018). Table 1 shows the PDB access codes for all structures identified in the PDB.

We implemented Eq. (4) in Python (program SFSXplorer) and considered the self-consistent Lennard-Jones 12–6 parameters of the AutoDock 4 semi-empirical force fields [41]. The scattering plot for experimental binding affinity ($\log(K_i)$) and the calculated potential energy V_{LJ} is shown in Fig. 3. Spearman's rank correlation

Table 1
List of proteins used in this study

PDB access codes
4UMA, 4UMB, 4UMC, 4BQS, 1H0R, 1GU1, 1V1J, 2BT4, 2C4W, 2XB8, 2XB9, 3N76, 3N7A, 3N86, 3N87, 3N8K, 3N8N, 4B6O, 4B6P, 4B6R, 4B6S, 4CIW, 4CIY

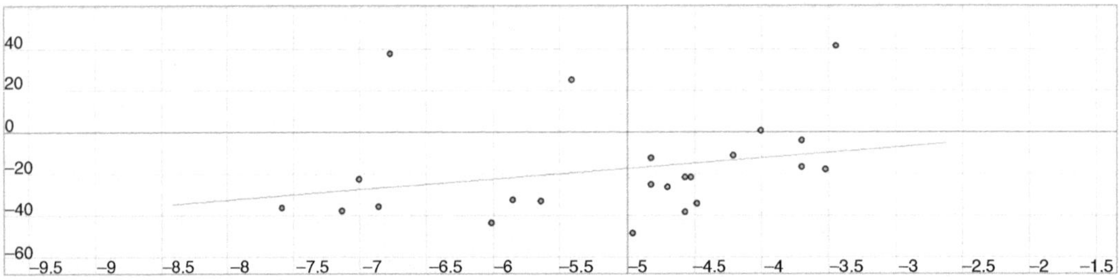

Fig. 3 Scatter plot for V_{LJ} against experimental $\log(K_i)$. We generated this plot with the program Molegro Data Modeller (MDM) [134, 135]

between experimental $\log(K_i)$ and V_{LJ} is 0.51 (p-value = 0.01). This level of correlation is significant. Furthermore, van der Waals interactions have been shown to be of pivotal importance for ligand binding affinity in several studies focused on a wide range of different proteins [106–133].

5 Availability

SFSXplorer is implemented in Python and available to download under the GNU license at https://github.com/azevedolab/SFSXplorer. The shikimate dataset is available for downloading at https://azevedolab.net/receptor-ligand-systems-database.php.

6 Colophon

We created Fig. 1 using Microsoft PowerPoint 2016. We used SFSXplorer to generate Fig. 2. We made Fig. 3 with the Molegro Data Modeller (MDM) [134, 135]. We performed scoring function calculation described in this chapter using a Desktop PC with 4 GB memory, a 1 TB hard disk, and an Intel® Core® i3-2120 @ 3.30 GHz processor running Windows 8.1.

7 Final Remarks

Van der Waals interactions can be straightforwardly computed using the Lennard-Jones potential and implemented in a high-level computer language such as Python. The availability of experimental information for structures and binding affinity opens the possibility to generate enzyme-targeted scoring functions for prediction of binding affinity where the experimental data are employed to calibrate a complete scoring function for a specific biological system.

Acknowledgments

This work was supported by grants from CNPq (Brazil) (308883/2014-4). This study was financed in part by the Coordenação de Aperfeiçoamento de Pessoal de Nivel Superior—Brasil (CAPES)—Finance Code 001. GB-F acknowledges support from PUCRS/BPA fellowship. MV-A acknowledges support from PUCRS/IC Jr. WFA is a senior researcher for CNPq (Brazil) (Process Numbers: 308883/2014-4 and 309029/2018-0).

References

1. Wang C, Greene D, Xiao L, Qi R, Luo R (2018) Recent developments and applications of the MMPBSA method. Front Mol Biosci 4:87
2. Cappel D, Sherman W, Beuming T (2017) Calculating water thermodynamics in the binding site of proteins—applications of WaterMap to drug discovery. Curr Top Med Chem 17:2586–2598
3. Bernetti M, Cavalli A, Mollica L (2017) Protein-ligand (un)binding kinetics as a new paradigm for drug discovery at the crossroad between experiments and modelling. Medchemcomm 8:534–550
4. Jaegle M, Wong EL, Tauber C, Nawrotzky E, Arkona C, Rademann J (2017) Protein-templated fragment ligations-from molecular recognition to drug discovery. Angew Chem Int Ed Engl 56:7358–7378
5. Yin J, Henriksen NM, Slochower DR, Shirts MR, Chiu MW, Mobley DL et al (2017) Overview of the SAMPL5 host-guest challenge: are we doing better? J Comput Aided Mol Des 31:1–19
6. de Azevedo WF Jr (2010) MolDock applied to structure-based virtual screening. Curr Drug Targets 11:327–334
7. Chakravarty K, Dalal DC (2018) Mathematical modelling of liposomal drug release to tumour. Math Biosci 306:82–96
8. Qi R, Luo R (2019) Robustness and efficiency of poisson-boltzmann modeling on graphics processing units. J Chem Inf Model 59:409–420
9. He X, Man VH, Ji B, Xie XQ, Wang J (2019) Calculate protein-ligand binding affinities with the extended linear interaction energy method: application on the Cathepsin S set in the D3R Grand Challenge 3. J Comput Aided Mol Des 33:105–117
10. Li A, Gilson MK (2018) Protein-ligand binding enthalpies from near-millisecond simulations: analysis of a preorganization paradox. J Chem Phys 149:072311
11. Miao Y, Huang YM, Walker RC, McCammon JA, Chang CA (2018) Ligand binding pathways and conformational transitions of the HIV protease. Biochemistry 57:1533–1541
12. Hoffer L, Muller C, Roche P, Morelli X (2018) Chemistry-driven Hit-to-lead optimization guided by structure-based approaches. Mol Inform 37:e1800059
13. Yadav BS, Tripathi V (2018) Recent advances in the system biology-based target identification and drug discovery. Curr Top Med Chem 18:1737–1744
14. Sotriffer C (2018) Docking of covalent ligands: challenges and approaches. Mol Inform 37:e1800062
15. Leelananda SP, Lindert S (2016) Computational methods in drug discovery. Beilstein J Org Chem 12:2694–2718
16. Roberts NA, Martin JA, Kinchington D, Broadhurst AV, Craig JC, Duncan IB et al (1990) Rational design of peptide-based HIV proteinase inhibitors. Science 248:358–361
17. Erickson J, Neidhart DJ, VanDrie J, Kempf DJ, Wang XC, Norbeck DW et al (1990) Design, activity, and 2.8 A crystal structure of a C2 symmetric inhibitor complexed to HIV-1 protease. Science 249:527–533
18. Dorsey BD, Levin RB, McDaniel SL, Vacca JP, Guare JP, Darke PL et al (1994) L-735,524: the design of a potent and orally bioavailable HIV protease inhibitor. J Med Chem 37:3443–3451
19. Vilar S, Sobarzo-Sanchez E, Santana L, Uriarte E (2017) Molecular docking and drug discovery in β-adrenergic receptors. Curr Med Chem 24:4340–4359
20. Xia X (2017) Bioinformatics and drug discovery. Curr Top Med Chem 17:1709–1726
21. Kuntz ID, Blaney JM, Oatley SJ, Langridge R, Ferrin TE (1982) A geometric approach to macromolecule-ligand interactions. J Mol Biol 161:269–288
22. DesJarlais RL, Dixon JS (1994) A shape- and chemistry-based docking method and its use in the design of HIV-1 protease inhibitors. J Comput Aided Mol Des 8:231–242
23. Lunney EA, Hagen SE, Domagala JM, Humblet C, Kosinski J, Tait BD et al (1994) A novel nonpeptide HIV-1 protease inhibitor: elucidation of the binding mode and its application in the design of related analogs. J Med Chem 37:2664–2677
24. Vaillancourt M, Cohen E, Sauvé G (1995) Characterization of dynamic state inhibitors of HIV-1 protease. J Enzyme Inhib 9:217–233
25. Gehlhaar DK, Verkhivker GM, Rejto PA, Sherman CJ, Fogel DB, Fogel LJ et al (1995) Molecular recognition of the inhibitor AG-1343 by HIV-1 protease: conformationally flexible docking by evolutionary programming. Chem Biol 2:317–324
26. King BL, Vajda S, DeLisi C (1996) Empirical free energy as a target function in docking and

design: application to HIV-1 protease inhibitors. FEBS Lett 384:87–91
27. Wang S, Milne GW, Yan X, Posey IJ, Nicklaus MC, Graham L et al (1996) Discovery of novel, non-peptide HIV-1 protease inhibitors by pharmacophore searching. J Med Chem 39:2047–2054
28. Adeniyi AA, Soliman MES (2017) Implementing QM in docking calculations: is it a waste of computational time? Drug Discov Today 22:1216–1223
29. Crespo A, Rodriguez-Granillo A, Lim VT (2017) Quantum-mechanics methodologies in drug discovery: applications of docking and scoring in lead optimization. Curr Top Med Chem 17:2663–2680
30. Yilmazer ND, Korth M (2016) Recent progress in treating protein-ligand interactions with quantum-mechanical methods. Int J Mol Sci 17:742
31. Cavasotto CN, Adler NS, Aucar MG (2018) Quantum chemical approaches in structure-based virtual screening and lead optimization. Front Chem 29(6):188
32. Hitzenberger M, Schuster D, Hofer TS (2017) The binding mode of the sonic hedgehog inhibitor robotnikinin, a combined docking and QM/MM MD study. Front Chem 5:76
33. Salmas RE, Is YS, Durdagi S, Stein M, Yurtsever M (2018) A QM protein-ligand investigation of antipsychotic drugs with the dopamine D2 receptor (D2R). J Biomol Struct Dyn 36:2668–2677
34. Phipps MJ, Fox T, Tautermann CS, Skylaris CK (2017) Intuitive density functional theory-based energy decomposition analysis for protein-ligand interactions. J Chem Theory Comput 13:1837–1850
35. Hylsová M, Carbain B, Fanfrlík J, Musilová L, Haldar S, Köprülüoğlu C et al (2017) Explicit treatment of active-site waters enhances quantum mechanical/implicit solvent scoring: Inhibition of CDK2 by new pyrazolo[1,5-a] pyrimidines. Eur J Med Chem 126:1118–1128
36. Pecina A, Meier R, Fanfrlík J, Lepšík M, Řezáč J, Hobza P et al (2016) The SQM/COSMO filter: reliable native pose identification based on the quantum-mechanical description of protein-ligand interactions and implicit COSMO solvation. Chem Commun (Camb) 52:3312–3315
37. Yang Z, Liu Y, Chen Z, Xu Z, Shi J, Chen K et al (2015) A quantum mechanics-based halogen bonding scoring function for protein-ligand interactions. J Mol Model 21:138
38. Lennard-Jones JE (1931) Cohesion. Proc Phys Soc 43:461–482
39. Cornell WD, Cieplak P, Bayly CI, Gould IR, Merz KM, Ferguson DM et al (1995) A second generation force field for the simulation of proteins, nucleic acids, and organic molecules. J Am Chem Soc 117:5179–5197
40. Hornak V, Abel R, Okur A, Strockbine B, Roitberg A, Simmerling C (2006) Comparison of multiple Amber force fields and development of improved protein backbone parameters. Proteins 65:712–725
41. Huey R, Morris GM, Olson AJ, Goodsell DS (2007) A semiempirical free energy force field with charge-based desolvation. J Comput Chem 28:1145–1152
42. Fahmy A, Wagner G (2002) TreeDock: a tool for protein docking based on minimizing van der Waals energies. J Am Chem Soc 124:1241–1250
43. Demerdash ON, Buyan A, Mitchell JC (2010) ReplicOpter: a replicate optimizer for flexible docking. Proteins 78:3156–3165
44. Buckingham A (1938) The classical equation of state of gaseous helium, neon and argon. Proc R Soc London Ser A 168:264–283
45. Teik-Cheng L (2007) Alternative scaling factor between Lennard-Jones and Exponential-6 potential energy functions. Mol Simul 33:1029–1032
46. Xantheas SS, Werhahn JC (2014) Universal scaling of potential energy functions describing intermolecular interactions. I. Foundations and scalable forms of new generalized Mie, Lennard-Jones, Morse, and Buckingham exponential-6 potentials. J Chem Phys 141:064117
47. Bazgier V, Berka K, Otyepka M, Banáš P (2016) Exponential repulsion improves structural predictability of molecular docking. J Comput Chem 37:2485–2494
48. Volkart PA, Bitencourt-Ferreira G, art AA, de Azevedo WF (2019) Cyclin-dependent kinase 2 in cellular senescence and cancer. A structural and functional review. Curr Drug Targets 20(7):716–726. https://doi.org/10.2174/1389450120666181204165344
49. de Azevedo WF Jr (2016) Opinion paper: targeting multiple cyclin-dependent kinases (CDKs): A new strategy for molecular docking studies. Curr Drug Targets 17:2
50. Perez PC, Caceres RA, Canduri F, de Azevedo WF Jr (2009) Molecular modeling and dynamics simulation of human cyclin-dependent kinase 3 complexed with inhibitors. Comput Biol Med 39:130–140

51. Canduri F, Perez PC, Caceres RA, de Azevedo WF Jr (2008) CDK9 a potential target for drug development. Med Chem 4:210–218
52. Krystof V, Cankar P, Frysová I, Slouka J, Kontopidis G, Dzubák P et al (2006) 4-arylazo-3,5-diamino-1H-pyrazole CDK inhibitors: SAR study, crystal structure in complex with CDK2, selectivity, and cellular effects. J Med Chem 49:6500–6509
53. Leopoldino AM, Canduri F, Cabral H, Junqueira M, de Marqui AB, Apponi LH et al (2006) Expression, purification, and circular dichroism analysis of human CDK9. Protein Expr Purif 47:614–620
54. Canduri F, de Azevedo WF Jr (2005) Structural basis for interaction of inhibitors with cyclin-dependent kinase 2. Curr Comput Aided Drug Des 1:53–64
55. Canduri F, Uchoa HB, de Azevedo WF Jr (2004) Molecular models of cyclin-dependent kinase 1 complexed with inhibitors. Biochem Biophys Res Commun 324:661–666
56. de Azevedo WF Jr, Gaspar RT, Canduri F, Camera JC Jr, da Silveira NJ (2002) Molecular model of cyclin-dependent kinase 5 complexed with roscovitine. Biochem Biophys Res Commun 297:1154–1158
57. de Azevedo WF Jr, Canduri F, da Silveira NJ (2002) Structural basis for inhibition of cyclin-dependent kinase 9 by flavopiridol. Biochem Biophys Res Commun 293:566–571
58. de Azevedo WF, Leclerc S, Meijer L, Havlicek L, Strnad M, Kim SH (1997) Inhibition of cyclin-dependent kinases by purine analogues: crystal structure of human CDK2 complexed with roscovitine. Eur J Biochem 243:518–526
59. de Azevedo WF Jr, Mueller-Dieckmann HJ, Schulze-Gahmen U, Worland PJ, Sausville E, Kim SH (1996) Structural basis for specificity and potency of a flavonoid inhibitor of human CDK2, a cell cycle kinase. Proc Natl Acad Sci U S A 93:2735–2740
60. Pang X, Liu Z, Zhai G (2014) Advances in non-peptidomimetic HIV protease inhibitors. Curr Med Chem 21:1997–2011
61. Calugi C, Guarna A, Trabocchi A (2013) Heterocyclic HIV-protease inhibitors. Curr Med Chem 20:3693–3710
62. Smith JM (1970) Natural selection and the concept of a protein space. Nature 225:563–564
63. Bohacek RS, McMartin C, Guida WC (1996) The art and practice of structure-based drug design: a molecular modeling perspective. Med Res Rev 16:3–50
64. Parish T, Stoker NG (2002) The common aromatic amino acid biosynthesis pathway is essential in *Mycobacterium tuberculosis.* Microbiology 148:3069–3077
65. Pereira JH, Canduri F, de Oliveira JS, da Silveira NJ, Basso LA, Palma MS et al (2003) Structural bioinformatics study of EPSP synthase from *Mycobacterium tuberculosis.* Biochem Biophys Res Commun 312:608–614
66. Arcuri HA, Canduri F, Pereira JH, da Silveira NJ, Camera JC Jr, de Oliveira JS et al (2004) Molecular models for shikimate pathway enzymes of *Xylella fastidiosa.* Biochem Biophys Res Commun 320:979–991
67. Dias MV, Ely F, Canduri F, Pereira JH, Frazzon J, Basso LA et al (2004) Crystallization and preliminary X-ray crystallographic analysis of chorismate synthase from *Mycobacterium tuberculosis.* Acta Crystallogr D Biol Crystallogr 60:2003–2005
68. Uchôa HB, Jorge GE, Freitas Da Silveira NJ, Camera JC Jr, Canduri F, De Azevedo WF Jr (2004) Parmodel: a web server for automated comparative modeling of proteins. Biochem Biophys Res Commun 325:1481–1486
69. Pereira JH, de Oliveira JS, Canduri F, Dias MV, Palma MS, Basso LA et al (2004) Structure of shikimate kinase from *Mycobacterium tuberculosis* reveals the binding of shikimic acid. Acta Crystallogr D Biol Crystallogr 60:2310–2319
70. Silveira NJ, Uchôa HB, Pereira JH, Canduri F, Basso LA, Palma MS et al (2005) Molecular models of protein targets from *Mycobacterium tuberculosis.* J Mol Model 11:160–166
71. Dias MV, Borges JC, Ely F, Pereira JH, Canduri F, Ramos CH et al (2006) Structure of chorismate synthase from *Mycobacterium tuberculosis.* J Struct Biol 154:130–143
72. da Silveira NJ, Bonalumi CE, Uchõa HB, Pereira JH, Canduri F, de Azevedo WF (2006) DBMODELING: a database applied to the study of protein targets from genome projects. Cell Biochem Biophys 44:366–374
73. Borges JC, Pereira JH, Vasconcelos IB, dos Santos GC, Olivieri JR, Ramos CH et al (2006) Phosphate closes the solution structure of the 5-enolpyruvylshikimate-3-phosphate synthase (EPSPS) from *Mycobacterium tuberculosis.* Arch Biochem Biophys 452:156–164
74. da Silveira NJF, Bonalumi CE, Arcuri HA, de Azevedo WF Jr (2007) Molecular modeling

databases: a new way in the search of proteins targets for drug development. Curr Bioinf 2:1–10
75. Dias MV, Faím LM, Vasconcelos IB, de Oliveira JS, Basso LA, Santos DS et al (2007) Effects of the magnesium and chloride ions and shikimate on the structure of shikimate kinase from *Mycobacterium tuberculosis*. Acta Crystallogr Sect F Struct Biol Cryst Commun 63:1–6
76. Dias MV, Ely F, Palma MS, de Azevedo WF Jr, Basso LA, Santos DS (2007) Chorismate synthase: an attractive target for drug development against orphan diseases. Curr Drug Targets 8:437–444
77. Marques MR, Pereira JH, Oliveira JS, Basso LA, de Azevedo WF Jr, Santos DS et al (2007) The inhibition of 5-enolpyruvylshikimate-3-phosphate synthase as a model for development of novel antimicrobials. Curr Drug Targets 8:445–457
78. Pereira JH, Vasconcelos IB, Oliveira JS, Caceres RA, de Azevedo WF Jr, Basso LA et al (2007) Shikimate kinase: a potential target for development of novel antitubercular agents. Curr Drug Targets 8:459–468
79. Marques MR, Vaso A, Neto JR, Fossey MA, Oliveira JS, Basso LA et al (2008) Dynamics of glyphosate-induced conformational changes of *Mycobacterium tuberculosis* 5-enolpyruvylshikimate-3-phosphate synthase (EC 2.5.1.19) determined by hydrogen-deuterium exchange and electrospray mass spectrometry. Biochemistry 47:7509–7522
80. Arcuri HA, Borges JC, Fonseca IO, Pereira JH, Neto JR, Basso LA et al (2008) Structural studies of shikimate 5-dehydrogenase from *Mycobacterium tuberculosis*. Proteins 72:720–730
81. Pauli I, Caceres RA, de Azevedo WF Jr (2008) Molecular modeling and dynamics studies of Shikimate Kinase from *Bacillus anthracis*. Bioorg Med Chem 16:8098–8108
82. de Azevedo WF Jr (2008) Protein-drug interactions. Curr Drug Targets 9:1030
83. de Azevedo WF Jr, Dias R (2008) Computational methods for calculation of ligand-binding affinity. Curr Drug Targets 92:1031–1039
84. Dias R, de Azevedo WF Jr (2008) Molecular docking algorithms. Curr Drug Targets 9:1040–1047
85. Canduri F, de Azevedo WF (2008) Protein crystallography in drug discovery. Curr Drug Targets 9:1048–1053
86. Pauli I, Timmers LF, Caceres RA, Soares MB, de Azevedo WF Jr (2008) *In silico* and *in vitro*: identifying new drugs. Curr Drug Targets 9:1054–1061
87. Dias R, Timmers LF, Caceres RA, de Azevedo WF Jr (2008) Evaluation of molecular docking using polynomial empirical scoring functions. Curr Drug Targets 9:1062–1070
88. de Azevedo WF Jr, Dias R (2008) Experimental approaches to evaluate the thermodynamics of protein-drug interactions. Curr Drug Targets 9:1071–1076
89. Caceres RA, Pauli I, Timmers LF, de Azevedo WF Jr (2008) Molecular recognition models: a challenge to overcome. Curr Drug Targets 9:1077–1083
90. Barcellos GB, Caceres RA, de Azevedo WF Jr (2009) Structural studies of shikimate dehydrogenase from *Bacillus anthracis* complexed with cofactor NADP. J Mol Model 15:147–155
91. de Azevedo WF Jr, Dias R, Timmers LF, Pauli I, Caceres RA, Soares MB (2009) Bioinformatics tools for screening of antiparasitic drugs. Curr Drug Targets 10:232–239
92. Arcuri HA, Zafalon GF, Marucci EA, Bonalumi CE, da Silveira NJ, Machado JM et al (2010) SKPDB: a structural database of shikimate pathway enzymes. BMC Bioinformatics 11:12
93. Hernandes MZ, Cavalcanti SM, Moreira DR, de Azevedo WF Jr, Leite AC (2010) Halogen atoms in the modern medicinal chemistry: hints for the drug design. Curr Drug Targets 11:303–314
94. De Azevedo WF Jr (2010) Structure-based virtual screening. Curr Drug Targets 11:261–263
95. de Azevedo WF Jr (2011) Molecular dynamics simulations of protein targets identified in *Mycobacterium tuberculosis*. Curr Med Chem 18:1353–1366
96. de Azevedo WF Jr (2011) Protein targets for development of drugs against *Mycobacterium tuberculosis*. Curr Med Chem 18:1255–1257
97. Vianna CP, de Azevedo WF Jr (2012) Identification of new potential *Mycobacterium tuberculosis* shikimate kinase inhibitors through molecular docking simulations. J Mol Model 18:755–764
98. Azevedo LS, Moraes FP, Xavier MM, Pantoja EO, Villavicencio B, Finck JA et al (2012) Recent progress of molecular docking simulations applied to development of drugs. Curr Bioinf 7:352–365
99. Coracini JD, de Azevedo WF Jr (2014) Shikimate kinase, a protein target for drug design. Curr Med Chem 21:592–604

100. de Avila MB, de Azevedo WF (2014) Data mining of docking results. Application to 3-dehydroquinate dehydratase. Curr Bioinf 9:361–379
101. Heck GS, Pintro VO, Pereira RR, de Ávila MB, Levin NMB, de Azevedo WF (2017) Supervised machine learning methods applied to predict ligand-binding affinity. Curr Med Chem 24:2459–2470
102. de Ávila MB, Bitencourt-Ferreira G, de Azevedo WF Jr (2019) Structural basis for inhibition of enoyl-[Acyl Carrier Protein] reductase (InhA) from *Mycobacterium tuberculosis*. Curr Med Chem. https://doi.org/10.2174/0929867326666181203125229
103. Berman HM, Westbrook J, Feng Z, Gilliland G, Bhat TN, Weissig H et al (2000) The protein data bank. Nucleic Acids Res 28:235–242
104. Berman HM, Battistuz T, Bhat TN, Bluhm WF, Bourne PE, Burkhardt K et al (2002) The protein data bank. Acta Crystallogr D Biol Crystallogr 58:899–907
105. Westbrook J, Feng Z, Chen L, Yang H, Berman HM (2003) The protein data bank and structural genomics. Nucleic Acids Res 31:489–491
106. Xavier MM, Heck GS, de Avila MB, Levin NM, Pintro VO, Carvalho NL et al (2016) SAnDReS a computational tool for statistical analysis of docking results and development of scoring functions. Comb Chem High Throughput Screen 19:801–812
107. Levin NM, Pintro VO, de Ávila MB, de Mattos BB, De Azevedo WF Jr (2017) Understanding the structural basis for inhibition of cyclin-dependent kinases. New pieces in the molecular puzzle. Curr Drug Targets 18:1104–1111
108. de Ávila MB, Xavier MM, Pintro VO, de Azevedo WF (2017) Supervised machine learning techniques to predict binding affinity. A study for cyclin-dependent kinase 2. Biochem Biophys Res Commun 494:305–310
109. Pintro VO, Azevedo WF (2017) Optimized virtual screening workflow. Towards target-based polynomial scoring functions for HIV-1 protease. Comb Chem High Throughput Screen 20:820–827
110. Freitas PG, Elias TC, Pinto IA, Costa LT, de Carvalho PVSD, Omote DQ et al (2018) Computational approach to the discovery of phytochemical molecules with therapeutic potential targets to the PKCZ protein. Lett Drug Des Discovery 15:488–499
111. Levin NMB, Pintro VO, Bitencourt-Ferreira-G, Mattos BB, Silvério AC, de Azevedo WF Jr (2018) Development of CDK-targeted scoring functions for prediction of binding affinity. Biophys Chem 235:1–8
112. Amaral MEA, Nery LR, Leite CE, de Azevedo WF Jr, Campos MM (2018) Pre-clinical effects of metformin and aspirin on the cell lines of different breast cancer subtypes. Invest New Drugs 36:782–796
113. de Ávila MB, de Azevedo WF Jr (2018) Development of machine learning models to predict inhibition of 3-dehydroquinate dehydratase. Chem Biol Drug Des 92:1468–1474
114. Bitencourt-Ferreira G, de Azevedo WF Jr (2018) Development of a machine-learning model to predict Gibbs free energy of binding for protein-ligand complexes. Biophys Chem 240:63–69
115. de Azevedo WF Jr, Dias R (2008) Evaluation of ligand-binding affinity using polynomial empirical scoring functions. Bioorg Med Chem 16:9378–9382
116. Delatorre P, Rocha BA, Souza EP, Oliveira TM, Bezerra GA, Moreno FB et al (2007) Structure of a lectin from *Canavalia gladiata* seeds: new structural insights for old molecules. BMC Struct Biol 7:52
117. de Azevedo WF Jr, Canduri F, dos Santos DM, Pereira JH, Bertacine Dias MV, Silva RG et al (2003) Crystal structure of human PNP complexed with guanine. Biochem Biophys Res Commun 312:767–772
118. Filgueira de Azevedo W Jr, dos Santos GC, dos Santos DM, Olivieri JR, Canduri F, Silva RG et al (2003) Docking and small angle X-ray scattering studies of purine nucleoside phosphorylase. Biochem Biophys Res Commun 309:923–928
119. Canduri F, Perez PC, Caceres RA, de Azevedo WF Jr (2007) Protein kinases as targets for antiparasitic chemotherapy drugs. Curr Drug Targets 8:389–398
120. Silva RG, Pereira JH, Canduri F, de Azevedo WF Jr, Basso LA, Santos DS (2005) Kinetics and crystal structure of human purine nucleoside phosphorylase in complex with 7-methyl-6-thio-guanosine. Arch Biochem Biophys 442:49–58
121. Timmers LF, Caceres RA, Vivan AL, Gava LM, Dias R, Ducati RG et al (2008) Structural studies of human purine nucleoside phosphorylase: towards a new specific empirical scoring function. Arch Biochem Biophys 479:28–38
122. Caceres RA, Saraiva Timmers LF, Dias R, Basso LA, Santos DS, de Azevedo WF Jr (2008) Molecular modeling and dynamics

simulations of PNP from *Streptococcus agalactiae*. Bioorg Med Chem 16:4984–4993

123. de Azevedo WF Jr, Ward RJ, Canduri F, Soares A, Giglio JR, Arni RK (1998) Crystal structure of piratoxin-I: a calcium-independent, myotoxic phospholipase A2-homologue from *Bothrops pirajai* venom. Toxicon 36:1395–1406
124. da Silveira NJ, Uchôa HB, Canduri F, Pereira JH, Camera JC Jr, Basso LA et al (2004) Structural bioinformatics study of PNP from *Schistosoma mansoni*. Biochem Biophys Res Commun 322:100–104
125. Bezerra GA, Oliveira TM, Moreno FB, de Souza EP, da Rocha BA, Benevides RG et al (2007) Structural analysis of *Canavalia maritima* and *Canavalia gladiata* lectins complexed with different dimannosides: new insights into the understanding of the structure-biological activity relationship in legume lectins. J Struct Biol 160:168–176
126. Canduri F, Fadel V, Dias MV, Basso LA, Palma MS, Santos DS et al (2005) Crystal structure of human PNP complexed with hypoxanthine and sulfate ion. Biochem Biophys Res Commun 326:335–338
127. Delatorre P, Rocha BA, Gadelha CA, Santi-Gadelha T, Cajazeiras JB, Souza EP et al (2006) Crystal structure of a lectin from *Canavalia maritima* (ConM) in complex with trehalose and maltose reveals relevant mutation in ConA-like lectins. J Struct Biol 154:280–286
128. Rádis-Baptista G, Moreno FB, de Lima Nogueira L, Martins AM, de Oliveira Toyama D, Toyama MH et al (2006) Crotacetin, a novel snake venom C-type lectin homolog of convulxin, exhibits an unpredictable antimicrobial activity. Cell Biochem Biophys 44:412–423
129. Breda A, Basso LA, Santos DS, de Azevedo WF Jr (2008) Virtual screening of drugs: score functions, docking, and drug design. Curr Comput Aided Drug Des 4:265–272
130. Nolasco DO, Canduri F, Pereira JH, Cortinóz JR, Palma MS, Oliveira JS et al (2004) Crystallographic structure of PNP from *Mycobacterium tuberculosis* at 1.9A resolution. Biochem Biophys Res Commun 324:789–794
131. Soares MB, Silva CV, Bastos TM, Guimarães ET, Figueira CP, Smirlis D et al (2012) Anti-*Trypanosoma cruzi* activity of nicotinamide. Acta Trop 12:224–229
132. Rocha BA, Delatorre P, Oliveira TM, Benevides RG, Pires AF, Sousa AA et al (2011) Structural basis for both pro- and anti-inflammatory response induced by mannose-specific legume lectin from *Cymbosema roseum*. Biochimie 93:806–816
133. Ducati RG, Basso LA, Santos DS, de Azevedo WF Jr (2010) Crystallographic and docking studies of purine nucleoside phosphorylase from *Mycobacterium tuberculosis*. Bioorg Med Chem 18:4769–4774
134. Thomsen R, Christensen MH (2006) MolDock: a new technique for high-accuracy molecular docking. J Med Chem 49:3315–3321
135. Heberlé G, de Azevedo WF Jr (2011) Bio-inspired algorithms applied to molecular docking simulations. Curr Med Chem 18:1339–1352

Chapter 7

Hydrogen Bonds in Protein-Ligand Complexes

Gabriela Bitencourt-Ferreira, Martina Veit-Acosta, and Walter Filgueira de Azevedo Jr.

Abstract

Fast and reliable evaluation of the hydrogen bond potential energy has a significant impact in the drug design and development since it allows the assessment of large databases of organic molecules in virtual screening projects focused on a protein of interest. Semi-empirical force fields implemented in molecular docking programs make it possible the evaluation of protein-ligand binding affinity where the hydrogen bond potential is a common term used in the calculation. In this chapter, we describe the concepts behind the programs used to predict hydrogen bond potential energy employing semi-empirical force fields as the ones available in the programs AMBER, AutoDock4, TreeDock, and ReplicOpter. We described here the 12-10 potential and applied it to evaluate the binding affinity for an ensemble of crystallographic structures for which experimental data about binding affinity are available.

Key words Hydrogen bond interactions, Binding affinity, Drug design, Molecular recognition, Shikimate pathway

1 Introduction

Hydrogen bonds play a pivotal role in the stabilization of the structures of proteins due to their participation in the secondary structure elements such as alpha helices and beta sheets. Since the pioneering work of Linus Pauling in the early 1950s, the central role of hydrogen bonds for protein structures was crystal clear [1–4]. It is worth noting that the determination of the alpha helix and beta sheets in protein structures was predicted before the elucidation of the first protein structure through X-ray diffraction crystallography, in 1958 [5].

Considering the role of hydrogen bond interactions for protein-ligand interactions, it is clear that among the non-bonded interactions, the hydrogen bonds are vital determinants for ligand binding affinity. As proof of concept, let us consider protein-ligand interactions for cyclin-dependent kinase (CDK). There are over 400 structures of CDK deposited in the Protein Data Bank

Walter Filgueira de Azevedo Jr. (ed.), *Docking Screens for Drug Discovery*, Methods in Molecular Biology, vol. 2053, https://doi.org/10.1007/978-1-4939-9752-7_7,

(PDB) [6–8] (search carried out on January 02, 2019). Most of these structures present competitive inhibitors bound to the ATP-binding pocket of CDK. Furthermore, many of the CDK structures present inhibitors with experimental information about the binding affinity gathered from other databases such as Binding MOAD (Mother Of All Databases) [9], BindingDB [10], and PDBbind [11]. This interest in the study of CDK has been motivated by the potential use of CDK inhibitors to treat cancer [12–30].

The binding affinity information could be accessed through the PDB. Analysis of the CDK2 structures for which IC_{50} is available indicated that ligand binding affinity is most related to intermolecular hydrogen bonds involving main chain atoms of Glu-81 and Leu-83. There is a common intermolecular hydrogen bond pattern observed in the complexes involving CDK and inhibitor. These intermolecular interactions affect the molecular fork of CDK. In summary, ligand binding specificity could be mediated by intermolecular hydrogen bonds involving key residues present in the protein target and donor and acceptor atoms in the ligand structure. Specifically, for CDK4 and CDK6, inhibitors with IC_{50} in the nanomolar range show strong intermolecular bonds involving the molecular fork. Taken together, this richness of structural and functional data made it possible to develop CDK4/6 inhibitors that reached clinical trials, for instance, palbociclib, ribociclib, and abemaciclib [31–39].

Although the precise evaluation of intermolecular hydrogen bond needs application of quantum mechanics approaches [40, 41], it is possible to generate computational models to predict hydrogen bond potential energy by means of semi-empirical force fields as the ones available in the programs AMBER ff99 [42, 43], AutoDock4 [44], TreeDock [45], and ReplicOpter [46], to mention a few. Besides the semi-empirical force fields, other programs make use of a piecewise potential function, like the ones available in the programs Molegro Virtual Docker and Plants [47–49].

In this chapter, we consider the evaluation of hydrogen bond potential as described in the AutoDock4 semi-empirical force field. To illustrate its application, we discussed the calculation of the intermolecular potential for an ensemble of protein structures for which data of inhibition constant are available.

2 Hydrogen Bond Interactions

Our focus here is the protein-ligand hydrogen bonds for protein-ligand complexes. To have a full understanding of these non-bonded interactions, let us see the typical architecture of a hydrogen bond as illustrated in Fig. 1. In a hydrogen bond interaction, we have a donor (D) and an acceptor (A) atom. Analysis of common stronger intermolecular hydrogen bonds involving

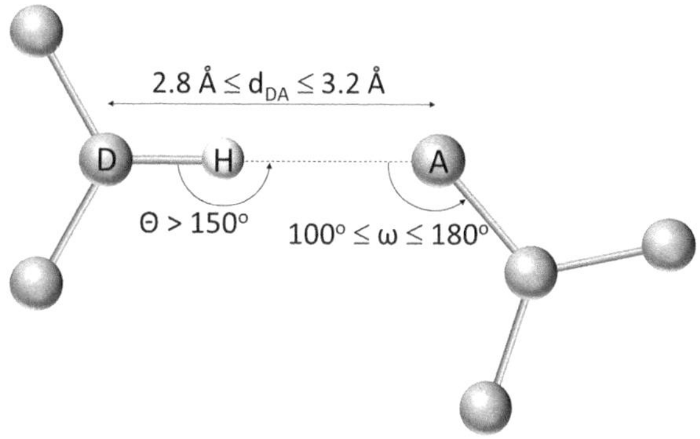

Fig. 1 Schematic of a hydrogen bond. This figure shows the interaction between the donor atom (D) and the acceptor atom (A) mediated by an atom of H

proteins and organic ligands indicates the participation of N and O of the protein structure and N, O, S, and halogen atoms from the ligand. On average, intermolecular hydrogen has a length (d_{DA}) of 3.0 Å, measure along the bond axis as illustrated in Fig. 1. The angles θ and ω assume typical values as indicated in Fig. 1. Considering protein- ligand interaction, typical energy values and distances related to hydrogen bonds are

N–H···O (1.912 kcal/mol for a $d_{DA} = 3.04$ Å)

N–H···N (3.107 kcal/mol for a $d_{DA} = 3.10$ Å)

O–H···O (5.019 kcal/mol for a $d_{DA} = 2.70$ Å)

O–H···N (6.931 kcal/mol for a $d_{DA} = 2.88$ Å)

It is also possible to have weaker intermolecular hydrogen bonds involving aromatic rings. These rings act as hydrogen bond acceptors. We have shown in Fig. 2 all 20 naturally occurring amino acids, where we highlight those for which the side chain participates in hydrogen bonds. Analysis of high-resolution crystallographic structures for protein-ligand complexes revealed that the typical hydrogen bond distance between the donor and acceptor atoms ranges from 2.5 to 3.4 Å.

The graphical representation of intermolecular hydrogen bonds for protein-ligand complexes is of pivotal importance for the evaluation of the residues responsible for ligand binding affinity. Such graphical analysis could rely on the direct representation of intermolecular hydrogen bonds available in protein such as Molegro Virtual Docker [47] and Visual Molecular Dynamics [50]. Nevertheless, such description could be troublesome, such as the one of the crystal structure of shikimate kinase from *Mycobacterium tuberculosis* in complex with ADP [51] (PDB access code: 1WE2) (Fig. 3). In Fig. 3, we have a superposition of the intermolecular

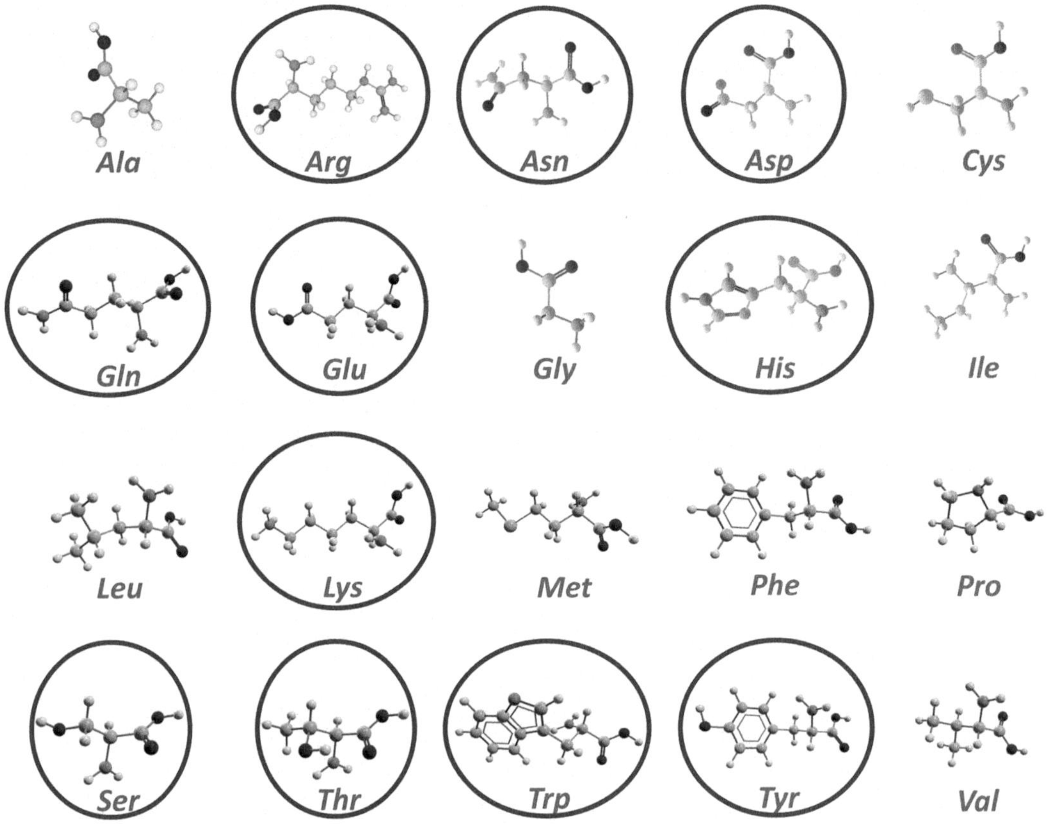

Fig. 2 This figure shows the molecular structures of all naturally occurring amino acids. We used the program Molegro Virtual Docker [47] to generate this figure. Amino acids that participate in intermolecular hydrogen bonds with ligands are circled in the figure

hydrogen bonds; in such a view, it is difficult to have a clear picture of all interactions. One way to overcome this problem of the representation is through the generation of 2D-plots of the intermolecular interactions.

One of the most successful programs to generate 2D-plots to represent protein-ligand interactions is the LigPlot [52, 53]. The program LigPlot allows determining structural criteria to assess intermolecular hydrogen bonds for protein-ligand complexes for which experimental and theoretical structures are available. This computational method brings consistency in the analysis of protein-ligand interactions since it uses the same strong structural evidence to assign a given interaction for a pair of atoms. Figure 4 shows the protein-ligand interactions for the crystal structure of shikimate kinase in complex with ADP (PDB access code: 1WE2). From Fig. 4, all intermolecular hydrogen bonds are easily identified.

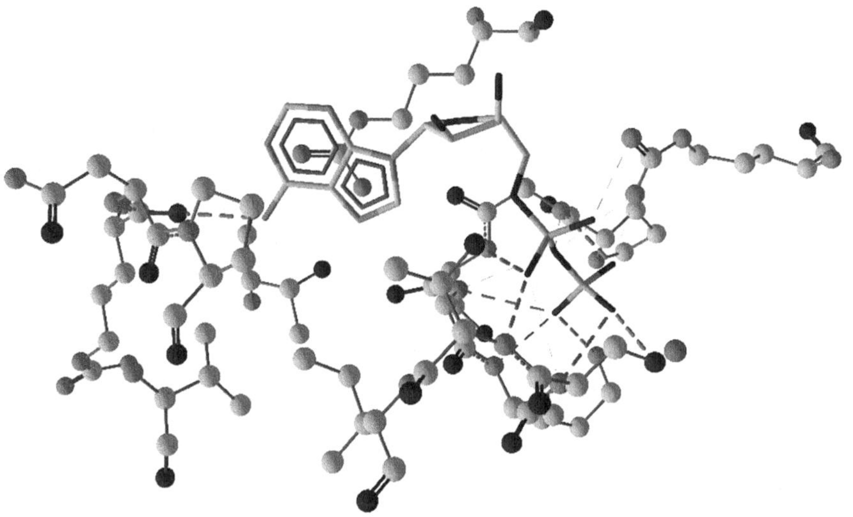

Fig. 3 Intermolecular hydrogen bonds involving shikimate kinase and ADP (PDB access code: 1WE2) [51]. We used the program Molegro Virtual Docker [47] to generate the above figure. Molegro Virtual Docker indicates hydrogen bonds as dashed lines, protein atoms as ball-and-stick, and ADP as lines

3 Hydrogen Bond Potential

In a typical semi-empirical force field equation, the term to assess intermolecular hydrogen bond potential is a modified Lennard-Jones potential. The original description of the Lennard-Jones potential dates back to 1931 [54]. We find this methodology to estimate interatomic interaction in many force fields dedicated to the evaluation of protein-ligand interactions, such as the functions calculated by AMBER ff99 [42, 43], AutoDock4 [44], TreeDock [45], and ReplicOpter [46].

In summary, the potential energy for a system composed of two atoms can be approximated using the following expression:

$$V(r) \approx \frac{C_n}{r^n} - \frac{C_m}{r^m} = C_n r^{-n} - C_m r^{-m} \tag{1}$$

where m and n are integers, and C_n and C_m are constants whose values are based on the equilibrium separation between two atoms and the depth of the energy well. The original model of the Lennard-Jones potential uses the 12-6 terms in the above equation ($n = 12$, $m = 6$) [54].

In general, Eq. (1) is computationally implemented as follows:

$$V_{\text{LJ}}(r) \approx \frac{\frac{m}{n-m}\varepsilon {r^n}_{\text{eqm}}}{r^n} - \frac{\frac{n}{n-m}\varepsilon {r^m}_{\text{eqm}}}{r^m} \tag{2}$$

where V_{LJ} is the Lennard-Jones potential energy, ε is the well depth of the potential energy function, and r_{eqm} is the equilibrium separation between two atoms. The numbers m and n are integers taken

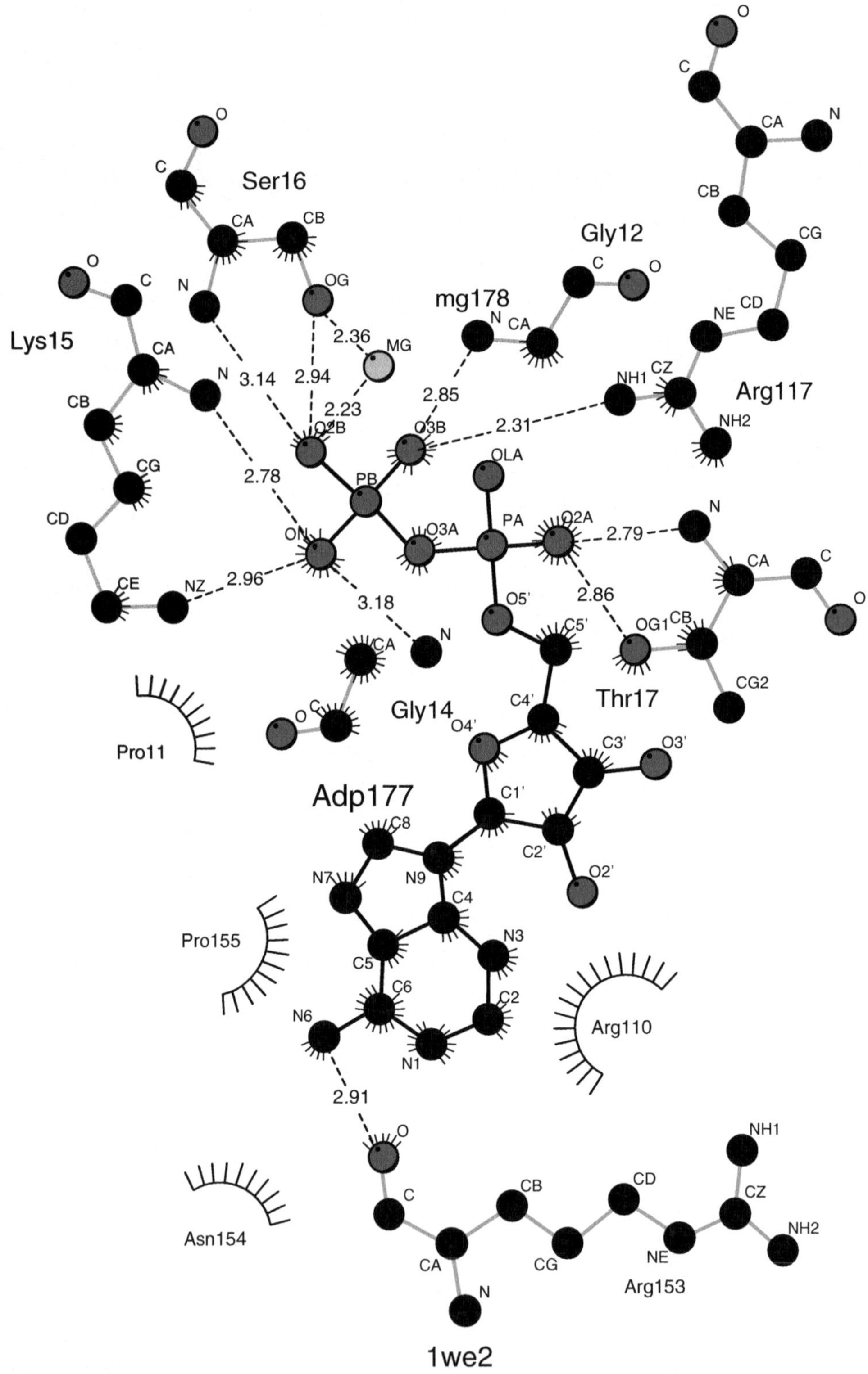

Fig. 4 Representation of protein-ligand interactions for the structure 1WE2 [51]. This figure was generated using LigPlot [52, 53]. Here we represent intermolecular hydrogen bonds as dashed lines. The program LigPlot shows the complete structures of the residues involved in the intermolecular hydrogen bonds. The program LigPlot depicts other intermolecular interactions indicating the residues as spoked arcs. The distance between acceptor and donor atoms participating in intermolecular hydrogen bonds is indicated in Å

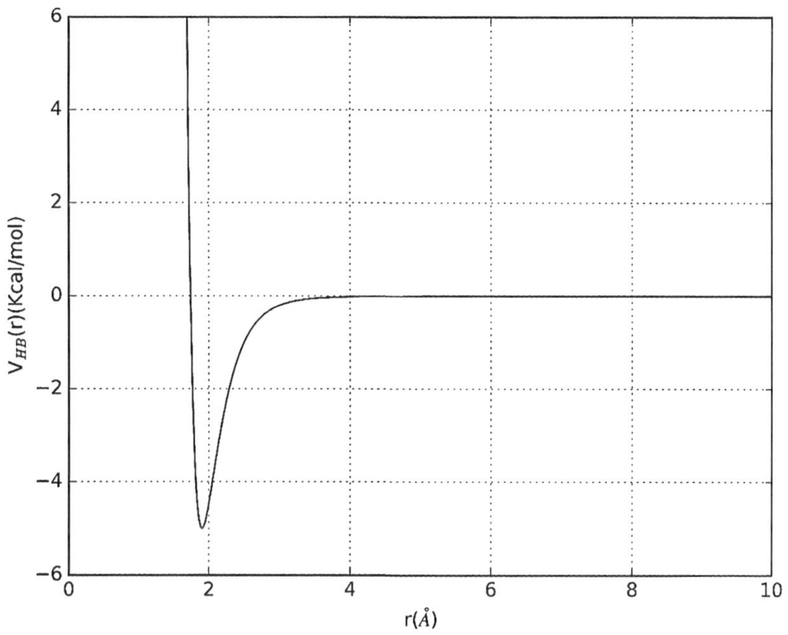

Fig. 5 Hydrogen bond potential generated using Eq. (2) for N···O pair of atoms

as $n = 12$ and $m = 6$ for the original Lennard-Jones potential. In the AutoDock4 semi-empirical force field, we employ Eq. (2) to approximate intermolecular hydrogen bond potential, where $n = 12$ and $m = 10$. Figure 5 shows the hydrogen bond potential for N···O atoms.

4 Calculating Hydrogen Bond Potential for Protein-Ligand Complexes

To illustrate the calculations of the intermolecular hydrogen bond potential of protein-ligand complexes, we considered a biological system composed of enzymes of the shikimate pathway. This metabolic route is a target for the development of herbicides and antibacterial drugs [55]. There are a substantial number of crystallographic and computational studies focused on shikimate pathway enzymes [56–89] due to their role in the development of antibacterial drugs and herbicides.

We searched the PDB for the enzymes 3-deoxy-D-arabino-heptulosonate 7-phosphate (DAHP) synthase (EC 2.5.1.54), shikimate kinase (EC 2.7.1.71), and 3-dehydroquinate dehydratase (EC 4.2.1.10) of this pathway for which inhibition constant (K_i) data are available. We found a total of 23 crystallographic structures for which K_i data are available (search carried out on December 18, 2018). Table 1 shows the PDB access codes for all structures identified in the PDB.

Table 1
Structural and binding affinity data for all structures in the dataset

PDB	Ligand	Chain	Ligand number	K_i (nM)
4UMA	GZ3	A	1351	3900
4UMC	PEQ	A	1352	360,000
4BQS	K2Q	A	1172	62,000
1V1J	FA3	A	201	15,000
2XB8	XNW	A	1144	26
2XB9	XNW	A	201	170
3N76	CA2	A	147	140
3N7A	FA1	A	147	200,000
3N86	RJP	A	147	2300
3N87	N87	A	147	11,000
3N8K	D1X	A	147	300,000
3N8N	N88	A	147	27,000
4B6O	3DQ	A	1144	100
4B6P	2HN	A	1145	74
4B6S	2HN	A	200	970
4CIW	XH2	A	1148	15,000
4CIY	NDY	A	1144	27,000
4UMB	0V5	A	1353	99,000
1GU1	FA1	A	201	30,000
1H0R	FA1	A	200	200,000
2BT4	CA2	A	160	33,000
2C4W	GAJ	A	1160	20,000
4B6R	3DQ	A	1158	1420

We implemented Eq. (2) in Python (program SFSXplorer) and considered the self-consistent Lennard-Jones 12-10 parameters of the AutoDock4 semi-empirical force fields [44]. Figure 6 shows the scatter plot for experimental binding affinity ($\log(K_i)$) and the calculated potential energy V_{HB}. Spearman's rank correlation between experimental $\log(K_i)$ and V_{HB} is 0.084. This level of correlation is not significant. Nevertheless, calculation of hydrogen bond potential using a 9–6 potential generates a Spearman's rank correlation between experimental $\log(K_i)$ and V_{HB} of 0.496 (p = value of 0.016), which is a significant correlation. Figure 6 shows the scatter plot for 9-6 potential to approximate intermolecular hydrogen bond potential.

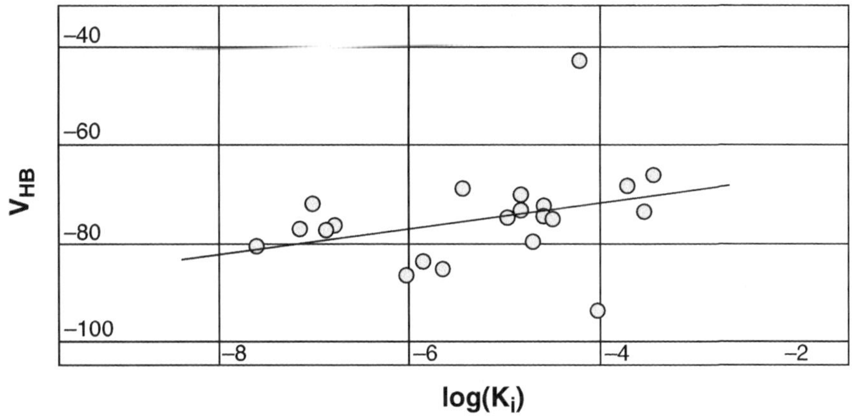

Fig. 6 Scatter plot for VHB against experimental log(K_i). We generated this plot with the program Molegro Data Modeller (MDM) [47]

5 Availability

SFSXplorer is implemented in Python and available to download under the GNU license at https://github.com/azevedolab/SFSXplorer. The shikimate dataset is available for downloading at https://azevedolab.net/receptor-ligand-systems-database.php.

6 Colophon

We created Fig. 1 using Microsoft PowerPoint 2016. We used the program Molegro Virtual Docker [47] to generate Figs. 2, 3, and 6. We made Fig. 4 using the program LigPlot [52, 53]. We used SFSXplorer to produce Fig. 5. We performed scoring function calculation described in this chapter using a Desktop PC with 4 GB memory, a 1 TB hard disk, and an Intel® Core® i3–2120 @ 3.30 GHz processor running Windows 8.1.

7 Final Remarks

Computational evaluation of binding affinity for protein-ligand complexes is an open problem in structural bioinformatics and computer-aided drug design. Among the terms usually found in the semi-empirical force fields, the hydrogen bond potential is one of the most common. Analysis of receptor-ligand interactions in different protein systems indicated that intermolecular hydrogen bonds are critical for binding affinity [90–114]. In this chapter, we see the description of the 10-6 potential for the evaluation of hydrogen bond potential for a system composed of 23 crystallographic structures. Precisely for this system, the 12-10 potential showed no significant correlation with the experimental binding affinity. On the other hand, the 9-6 potential has superior predictive performance. Taken together, we may suggest that the availability of programs where the variation for the type of *n–m* potential could be tested opens up the possibility for exploring the scoring function space and finding the type of interaction that is relevant for the biological system of interest.

Acknowledgments

This work was supported by grants from CNPq (Brazil) (308883/2014-4). This study was financed in part by the Coordenação de Aperfeiçoamento de Pessoal de Nivel Superior—Brasil (CAPES)—Finance Code 001. GB-F acknowledges support from PUCRS/BPA fellowship. MV-A acknowledges support from PUCRS/IC Jr. WFA is a senior researcher for CNPq (Brazil) (Process Numbers: 308883/2014-4 and 309029/2018-0).

References

1. Pauling L, Corey RB, Branson HR (1951) The structure of proteins: two hydrogen-bonded helical configurations of the polypeptide chain. Proc Natl Acad Sci U S A 37:205–211
2. Pauling L, Corey RB (1951) Atomic coordinates and structure factors for two helical configurations of polypeptide chains. Proc Natl Acad Sci U S A 37:235–240
3. Pauling L, Corey RB (1951) The structure of synthetic polypeptides. Proc Natl Acad Sci U S A 37:241–250
4. Pauling L, Corey RB (1951) The pleated sheet, a new layer configuration of polypeptide chains. Proc Natl Acad Sci U S A 37:251–256
5. Kendrew JC, Bodo G, Dintzis HM, Parrish RG, Wyckoff H, Phillips DC (1958) A three-dimensional model of the myoglobin molecule obtained by X-ray analysis. Nature 181:662–666
6. Berman HM, Westbrook J, Feng Z, Gilliland G, Bhat TN, Weissig H et al (2000) The Protein Data Bank. Nucleic Acids Res 28:235–242
7. Berman HM, Battistuz T, Bhat TN, Bluhm WF, Bourne PE, Burkhardt K et al (2002) The Protein Data Bank. Acta Crystallogr D Biol Crystallogr 58:899–907
8. Westbrook J, Feng Z, Chen L, Yang H, Berman HM (2003) The protein data Bank and structural genomics. Nucleic Acids Res 31:489–491
9. Hu L, Benson ML, Smith RD, Lerner MG, Carlson HA (2005) Binding MOAD (Mother Of All Databases). Proteins 60:333–340
10. Liu T, Lin Y, Wen X, Jorrisen RN, Gilson MK (2007) BindingDB: a web-accessible database of experimentally determined protein-ligand binding affinities. Nucleic Acids Res 35:198–201
11. Wang R, Fang X, Lu Y, Wang S (2004) The PDBbind database: collection of binding affinities for protein-ligand complexes with known three-dimensional structures. J Med Chem 47:2977–2980
12. Murray AW (1994) Cyclin-dependent kinases: regulators of the cell cycle and more. Chem Biol 1:191–195
13. Morgan DO (1995) Principles of CDK regulation. Nature 374:131–134
14. Volkart PA, Bitencourt-Ferreira G, Souto AA, de Azevedo WF (2019) Cyclin-dependent kinase 2 in cellular senescence and cancer. A structural and functional review. Curr Drug Targets 20(7):716–726. https://doi.org/10.2174/1389450120666181204165344
15. Levin NM, Pintro VO, de Ávila MB, de Mattos BB, De Azevedo WF Jr (2017) Understanding the structural basis for inhibition of Cyclin-dependent kinases. New pieces in the molecular puzzle. Curr Drug Targets 18:1104–1111

16. de Ávila MB, Xavier MM, Pintro VO, de Azevedo WF (2017) Supervised machine learning techniques to predict binding affinity. A study for cyclin-dependent kinase 2. Biochem Biophys Res Commun 494:305–310
17. Levin NMB, Pintro VO, Bitencourt-Ferreira-G, Mattos BB, Silvério AC, de Azevedo WF Jr (2018) Development of CDK-targeted scoring functions for prediction of binding affinity. Biophys Chem 235:1–8
18. de Azevedo WF Jr (2016) Opinion paper: targeting multiple Cyclin-dependent kinases (CDKs): a new strategy for molecular docking studies. Curr Drug Targets 17:2
19. Perez PC, Caceres RA, Canduri F, de Azevedo WF Jr (2009) Molecular modeling and dynamics simulation of human cyclin-dependent kinase 3 complexed with inhibitors. Comput Biol Med 39:130–140
20. Canduri F, Perez PC, Caceres RA, de Azevedo WF Jr (2008) CDK9 a potential target for drug development. Med Chem 4:210–218
21. Dos Santos NFP, Canduri F (2018) The emerging picture of CDK11: genetic, functional and medicinal aspects. Curr Med Chem 25:880–888
22. Paparidis NF, Durvale MC, Canduri F (2017) The emerging picture of CDK9/P-TEFb: more than 20 years of advances since PITALRE. Mol BioSyst 13:246–276
23. Leopoldino AM, Canduri F, Cabral H, Junqueira M, de Marqui AB, Apponi LH et al (2006) Expression, purification, and circular dichroism analysis of human CDK9. Protein Expr Purif 47:614–620
24. Krystof V, Cankar P, Frysová I, Slouka J, Kontopidis G, Dzubák P et al (2006) 4-arylazo-3,5-diamino-1H-pyrazole CDK inhibitors: SAR study, crystal structure in complex with CDK2, selectivity, and cellular effects. J Med Chem 49:6500–6509
25. Canduri F, de Azevedo WF Jr (2005) Structural basis for interaction of inhibitors with Cyclin-dependent kinase 2. Curr Comput Aided Drug Des 1:53–64
26. Canduri F, Uchoa HB, de Azevedo WF Jr (2004) Molecular models of cyclin-dependent kinase 1 complexed with inhibitors. Biochem Biophys Res Commun 324:661–666
27. De Azevedo WF Jr, Gaspar RT, Canduri F, Camera JC Jr, Da Silveira NJF (2002) Molecular model of cyclin-dependent kinase 5 complexed with roscovitine. Biochem Biophys Res Commun 297:1154–1158
28. de Azevedo WF Jr, Canduri F, da Silveira NJ (2002) Structural basis for inhibition of cyclin-dependent kinase 9 by flavopiridol. Biochem Biophys Res Commun 293:566–571
29. De Azevedo WF, Leclerc S, Meijer L, Havlicek L, Strnad M, Kim SH (1997) Inhibition of cyclin-dependent kinases by purine analogues: crystal structure of human cdk2 complexed with roscovitine. Eur J Biochem 243:518–526
30. De Azevedo WF Jr, Mueller-Dieckmann HJ, Schulze-Gahmen U, Worland PJ, Sausville E, Kim SH (1996) Structural basis for specificity and potency of a flavonoid inhibitor of human CDK2, a cell cycle kinase. Proc Natl Acad Sci U S A 93:2735–2740
31. Iwata H (2018) Clinical development of CDK4/6 inhibitor for breast cancer. Breast Cancer 25:402–406
32. Banys-Paluchowski M, Krawczyk N, Paluchowski P (2019) Cyclin-dependent kinase 4/6 inhibitors: what have we learnt across studies, therapy situations and substances. Curr Opin Obstet Gynecol 31:56–66
33. Roskoski R Jr (2019) Cyclin-dependent protein serine/threonine kinase inhibitors as anticancer drugs. Pharmacol Res 139:471–488
34. Kim S, Tiedt R, Loo A, Horn T, Delach S, Kovats S et al (2018) The potent and selective cyclin-dependent kinases 4 and 6 inhibitor ribociclib (LEE011) is a versatile combination partner in preclinical cancer models. Oncotarget 9:35226–35240
35. Choo JR, Lee SC (2018) CDK4-6 inhibitors in breast cancer: current status and future development. Expert Opin Drug Metab Toxicol 14:1123–1138
36. Ribnikar D, Volovat SR, Cardoso F (2018) Targeting CDK4/6 pathways and beyond in breast cancer. Breast 43:8–17
37. Martin JM, Goldstein LJ (2018) Profile of abemaciclib and its potential in the treatment of breast cancer. Onco Targets Ther 11:5253–5259
38. Robert M, Frenel JS, Bourbouloux E, Rigaud DB, Patsouris A, Augereau P et al (2018) An update on the clinical use of CDK4/6 inhibitors in breast cancer. Drugs 78:1353–1362
39. Messina C, Cattrini C, Buzzatti G, Cerbone L, Zanardi E, Messina M et al (2018) CDK4/6 inhibitors in advanced hormone receptor-positive/HER2-negative breast cancer: a systematic review and meta-analysis of randomized trials. Breast Cancer Res Treat 172:9–21
40. Cintrón MS, Johnson GP, French AD (2017) Quantum mechanics models of the methanol

dimer: OH···O hydrogen bonds of β-d-glucose moieties from crystallographic data. Carbohydr Res 443:87–94
41. Heifetz A, Chudyk EI, Gleave L, Aldeghi M, Cherezov V, Fedorov DG et al (2016) The fragment molecular orbital method reveals new insight into the chemical nature of GPCR-ligand interactions. J Chem Inf Model 56:159–172
42. Cornell WD, Cieplak P, Bayly CI, Gould IR, Merz KM, Ferguson DM et al (1995) A second generation force field for the simulation of proteins, nucleic acids, and organic molecules. J Am Chem Soc 117:5179–5197
43. Hornak V, Abel R, Okur A, Strockbine B, Roitberg A, Simmerling C (2006) Comparison of multiple Amber force fields and development of improved protein backbone parameters. Proteins 65:712–725
44. Huey R, Morris GM, Olson AJ, Goodsell DS (2007) A semiempirical free energy force field with charge-based desolvation. J Comput Chem 28:1145–1152
45. Fahmy A, Wagner G (2002) TreeDock: a tool for protein docking based on minimizing van der Waals energies. J Am Chem Soc 124:1241–1250
46. Demerdash ON, Buyan A, Mitchell JC (2010) ReplicOpter: a replicate optimizer for flexible docking. Proteins 78:3156–3165
47. Thomsen R, Christensen MH (2006) MolDock: a new technique for high-accuracy molecular docking. J Med Chem 49:3315–3321
48. de Azevedo WF Jr (2010) MolDock applied to structure-based virtual screening. Curr Drug Targets 11:327–334
49. Heberlé G, de Azevedo WF Jr (2011) Bio-inspired algorithms applied to molecular docking simulations. Curr Med Chem 18:1339–1352
50. Humphrey W, Dalke A, Schulten K (1996) VMD—visual molecular dynamics. J Mol Graph 14:33–38
51. Pereira JH, de Oliveira JS, Canduri F, Dias MV, Palma MS, Basso LA et al (2004) Structure of shikimate kinase from *Mycobacterium tuberculosis* reveals the binding of shikimic acid. Acta Crystallogr D Biol Crystallogr 60:2310–2319
52. Wallace AC, Laskowski RA, Thornton JM (1995) LIGPLOT: a program to generate schematic diagrams of protein-ligand interactions. Protein Eng 8:127–134
53. Laskowski RA, Swindells MB (2011) LigPlot+: multiple ligand-protein interaction diagrams for drug discovery. J Chem Inf Model 51:2778–2786
54. Lennard-Jones JE (1931) Cohesion. Proc Phys Soc 43:461–482
55. Parish T, Stoker NG (2002) The common aromatic amino acid biosynthesis pathway is essential in *Mycobacterium tuberculosis*. Microbiology 148:3069–3077
56. Pereira JH, Canduri F, de Oliveira JS, da Silveira NJ, Basso LA, Palma MS et al (2003) Structural bioinformatics study of EPSP synthase from *Mycobacterium tuberculosis*. Biochem Biophys Res Commun 312:608–614
57. Arcuri HA, Canduri F, Pereira JH, da Silveira NJ, Camera JC Jr, de Oliveira JS et al (2004) Molecular models for shikimate pathway enzymes of *Xylella fastidiosa*. Biochem Biophys Res Commun 320:979–991
58. Dias MV, Ely F, Canduri F, Pereira JH, Frazzon J, Basso LA et al (2004) Crystallization and preliminary X-ray crystallographic analysis of chorismate synthase from *Mycobacterium tuberculosis*. Acta Crystallogr D Biol Crystallogr 60:2003–2005
59. Uchôa HB, Jorge GE, Freitas Da Silveira NJ, Camera JC Jr, Canduri F, De Azevedo WF Jr (2004) Parmodel: a web server for automated comparative modeling of proteins. Biochem Biophys Res Commun 325:1481–1486
60. Silveira NJ, Uchôa HB, Pereira JH, Canduri F, Basso LA, Palma MS et al (2005) Molecular models of protein targets from *Mycobacterium tuberculosis*. J Mol Model 11:160–166
61. Dias MV, Borges JC, Ely F, Pereira JH, Canduri F, Ramos CH et al (2006) Structure of chorismate synthase from *Mycobacterium tuberculosis*. J Struct Biol 154:130–143
62. da Silveira NJ, Bonalumi CE, Uchõa HB, Pereira JH, Canduri F, de Azevedo WF (2006) DBMODELING: a database applied to the study of protein targets from genome projects. Cell Biochem Biophys 44:366–374
63. Borges JC, Pereira JH, Vasconcelos IB, dos Santos GC, Olivieri JR, Ramos CH et al (2006) Phosphate closes the solution structure of the 5-enolpyruvylshikimate-3-phosphate synthase (EPSPS) from *Mycobacterium tuberculosis*. Arch Biochem Biophys 452:156–164
64. da Silveira NJF, Bonalumi CE, Arcuri HA, de Azevedo WF Jr (2007) Molecular modeling databases: a new way in the search of proteins targets for drug development. Curr Bioinf 2:1–10

65. Dias MV, Faím LM, Vasconcelos IB, de Oliveira JS, Basso LA, Santos DS et al (2007) Effects of the magnesium and chloride ions and shikimate on the structure of shikimate kinase from *Mycobacterium tuberculosis*. Acta Crystallogr Sect F Struct Biol Cryst Commun 63:1–6
66. Dias MV, Ely F, Palma MS, de Azevedo WF Jr, Basso LA, Santos DS (2007) Chorismate synthase: an attractive target for drug development against orphan diseases. Curr Drug Targets 8:437–444
67. Marques MR, Pereira JH, Oliveira JS, Basso LA, de Azevedo WF Jr, Santos DS et al (2007) The inhibition of 5-enolpyruvylshikimate-3-phosphate synthase as a model for development of novel antimicrobials. Curr Drug Targets 8:445–457
68. Pereira JH, Vasconcelos IB, Oliveira JS, Caceres RA, de Azevedo WF Jr, Basso LA et al (2007) Shikimate kinase: a potential target for development of novel antitubercular agents. Curr Drug Targets 8:459–468
69. Marques MR, Vaso A, Neto JR, Fossey MA, Oliveira JS, Basso LA et al (2008) Dynamics of glyphosate-induced conformational changes of *Mycobacterium tuberculosis* 5-enolpyruvylshikimate-3-phosphate synthase (EC 2.5.1.19) determined by hydrogen-deuterium exchange and electrospray mass spectrometry. Biochemistry 47:7509–7522
70. Arcuri HA, Borges JC, Fonseca IO, Pereira JH, Neto JR, Basso LA et al (2008) Structural studies of shikimate 5-dehydrogenase from *Mycobacterium tuberculosis*. Proteins 72:720–730
71. Pauli I, Caceres RA, de Azevedo WF Jr (2008) Molecular modeling and dynamics studies of Shikimate kinase from *Bacillus anthracis*. Bioorg Med Chem 16:8098–8108
72. de Azevedo WF Jr (2008) Protein-drug interactions. Curr Drug Targets 9:1030
73. de Azevedo WF Jr, Dias R (2008) Computational methods for calculation of ligand-binding affinity. Curr Drug Targets 92:1031–1039
74. Dias R, de Azevedo WF Jr (2008) Molecular docking algorithms. Curr Drug Targets 9:1040–1047
75. Canduri F, de Azevedo WF (2008) Protein crystallography in drug discovery. Curr Drug Targets 9:1048–1053
76. Pauli I, Timmers LF, Caceres RA, Soares MB, de Azevedo WF Jr (2008) *In silico* and *in vitro*: identifying new drugs. Curr Drug Targets 9:1054–1061
77. Dias R, Timmers LF, Caceres RA, de Azevedo WF Jr (2008) Evaluation of molecular docking using polynomial empirical scoring functions. Curr Drug Targets 9:1062–1070
78. de Azevedo WF Jr, Dias R (2008) Experimental approaches to evaluate the thermodynamics of protein-drug interactions. Curr Drug Targets 9:1071–1076
79. Caceres RA, Pauli I, Timmers LF, de Azevedo WF Jr (2008) Molecular recognition models: a challenge to overcome. Curr Drug Targets 9:1077–1083
80. Barcellos GB, Caceres RA, de Azevedo WF Jr (2009) Structural studies of shikimate dehydrogenase from *Bacillus anthracis* complexed with cofactor NADP. J Mol Model 15:147–155
81. de Azevedo WF Jr, Dias R, Timmers LF, Pauli I, Caceres RA, Soares MB (2009) Bioinformatics tools for screening of antiparasitic drugs. Curr Drug Targets 10:232–239
82. Arcuri HA, Zafalon GF, Marucci EA, Bonalumi CE, da Silveira NJ, Machado JM et al (2010) SKPDB: a structural database of shikimate pathway enzymes. BMC Bioinformatics 11:12
83. Hernandes MZ, Cavalcanti SM, Moreira DR, de Azevedo WF Jr, Leite AC (2010) Halogen atoms in the modern medicinal chemistry: hints for the drug design. Curr Drug Targets 11:303–314
84. De Azevedo WF Jr (2010) Structure-based virtual screening. Curr Drug Targets 11:261–263
85. de Azevedo WF Jr (2011) Molecular dynamics simulations of protein targets identified in *Mycobacterium tuberculosis*. Curr Med Chem 18:1353–1366
86. de Azevedo WF Jr (2011) Protein targets for development of drugs against *Mycobacterium tuberculosis*. Curr Med Chem 18:1255–1257
87. Vianna CP, de Azevedo WF Jr (2012) Identification of new potential *Mycobacterium tuberculosis* shikimate kinase inhibitors through molecular docking simulations. J Mol Model 18:755–764
88. Azevedo LS, Moraes FP, Xavier MM, Pantoja EO, Villavicencio B, Finck JA et al (2012) Recent Progress of molecular docking simulations applied to development of drugs. Curr Bioinf 7:352–365
89. Coracini JD, de Azevedo WF Jr (2014) Shikimate kinase, a protein target for drug design. Curr Med Chem 21:592–604
90. Xavier MM, Heck GS, de Avila MB, Levin NM, Pintro VO, Carvalho NL et al (2016) SAnDReS a computational tool for statistical

analysis of docking results and development of scoring functions. Comb Chem High Throughput Screen 19:801–812
91. Pintro VO, Azevedo WF (2017) Optimized virtual screening workflow. Towards target-based polynomial scoring functions for HIV-1 protease. Comb Chem High Throughput Screen 20:820–827
92. Freitas PG, Elias TC, Pinto IA, Costa LT, de Carvalho PVSD, Omote DQ et al (2018) Computational approach to the discovery of phytochemical molecules with therapeutic potential targets to the PKCZ protein. Lett Drug Des Discovery 15:488–499
93. Amaral MEA, Nery LR, Leite CE, de Azevedo WF Jr, Campos MM (2018) Pre-clinical effects of metformin and aspirin on the cell lines of different breast cancer subtypes. Invest New Drugs 36:782–796
94. de Ávila MB, de Azevedo WF Jr (2018) Development of machine learning models to predict inhibition of 3-dehydroquinate dehydratase. Chem Biol Drug Des 92:1468–1474
95. Bitencourt-Ferreira G, de Azevedo WF Jr (2018) Development of a machine-learning model to predict Gibbs free energy of binding for protein-ligand complexes. Biophys Chem 240:63–69
96. de Azevedo WF Jr, Dias R (2008) Evaluation of ligand-binding affinity using polynomial empirical scoring functions. Bioorg Med Chem 16:9378–9382
97. Delatorre P, Rocha BA, Souza EP, Oliveira TM, Bezerra GA, Moreno FB et al (2007) Structure of a lectin from *Canavalia gladiata* seeds: new structural insights for old molecules. BMC Struct Biol 7:52
98. de Azevedo WF Jr, Canduri F, dos Santos DM, Pereira JH, Bertacine Dias MV, Silva RG et al (2003) Crystal structure of human PNP complexed with guanine. Biochem Biophys Res Commun 312:767–772
99. Filgueira de Azevedo W Jr, dos Santos GC, dos Santos DM, Olivieri JR, Canduri F, Silva RG et al (2003) Docking and small angle X-ray scattering studies of purine nucleoside phosphorylase. Biochem Biophys Res Commun 309:923–928
100. Canduri F, Perez PC, Caceres RA, de Azevedo WF Jr (2007) Protein kinases as targets for antiparasitic chemotherapy drugs. Curr Drug Targets 8:389–398
101. Silva RG, Pereira JH, Canduri F, de Azevedo WF Jr, Basso LA, Santos DS (2005) Kinetics and crystal structure of human purine nucleoside phosphorylase in complex with 7-methyl-6-thio-guanosine. Arch Biochem Biophys 442:49–58
102. Timmers LF, Caceres RA, Vivan AL, Gava LM, Dias R, Ducati RG et al (2008) Structural studies of human purine nucleoside phosphorylase: towards a new specific empirical scoring function. Arch Biochem Biophys 479:28–38
103. Caceres RA, Saraiva Timmers LF, Dias R, Basso LA, Santos DS, de Azevedo WF Jr (2008) Molecular modeling and dynamics simulations of PNP from *Streptococcus agalactiae*. Bioorg Med Chem 16:4984–4993
104. de Azevedo WF Jr, Ward RJ, Canduri F, Soares A, Giglio JR, Arni RK (1998) Crystal structure of piratoxin-I: a calcium-independent, myotoxic phospholipase A2-homologue from *Bothrops pirajai* venom. Toxicon 36:1395–1406
105. da Silveira NJ, Uchôa HB, Canduri F, Pereira JH, Camera JC Jr, Basso LA et al (2004) Structural bioinformatics study of PNP from *Schistosoma mansoni*. Biochem Biophys Res Commun 322:100–104
106. Bezerra GA, Oliveira TM, Moreno FB, de Souza EP, da Rocha BA, Benevides RG et al (2007) Structural analysis of *Canavalia maritima* and *Canavalia gladiata* lectins complexed with different dimannosides: new insights into the understanding of the structure-biological activity relationship in legume lectins. J Struct Biol 160:168–176
107. Canduri F, Fadel V, Dias MV, Basso LA, Palma MS, Santos DS et al (2005) Crystal structure of human PNP complexed with hypoxanthine and sulfate ion. Biochem Biophys Res Commun 326:335–338
108. Delatorre P, Rocha BA, Gadelha CA, Santi-Gadelha T, Cajazeiras JB, Souza EP et al (2006) Crystal structure of a lectin from *Canavalia maritima* (ConM) in complex with trehalose and maltose reveals relevant mutation in ConA-like lectins. J Struct Biol 154:280–286
109. Rádis-Baptista G, Moreno FB, de Lima Nogueira L, Martins AM, de Oliveira Toyama D, Toyama MH et al (2006) Crotacetin, a novel snake venom C-type lectin homolog of convulxin, exhibits an unpredictable antimicrobial activity. Cell Biochem Biophys 44:412–423
110. Breda A, Basso LA, Santos DS, de Azevedo WF Jr (2008) Virtual screening of drugs: score functions, docking, and drug design. Curr Comput Aided Drug Des 4(4):265–272
111. Nolasco DO, Canduri F, Pereira JH, Cortinóz JR, Palma MS, Oliveira JS et al (2004)

Crystallographic structure of PNP from *Mycobacterium tuberculosis* at 1.9A resolution. Biochem Biophys Res Commun 324:789–794

112. Soares MB, Silva CV, Bastos TM, Guimarães ET, Figueira CP, Smirlis D et al (2012) Anti-*Trypanosoma cruzi* activity of nicotinamide. Acta Trop 12:224–229
113. Rocha BA, Delatorre P, Oliveira TM, Benevides RG, Pires AF, Sousa AA et al (2011) Structural basis for both pro- and anti-inflammatory response induced by mannose-specific legume lectin from *Cymbosema roseum*. Biochimie 93:806–816
114. Ducati RG, Basso LA, Santos DS, de Azevedo WF Jr (2010) Crystallographic and docking studies of purine nucleoside phosphorylase from *Mycobacterium tuberculosis*. Bioorg Med Chem 18:4769–4774

Chapter 8

Molecular Dynamics Simulations with NAMD2

Gabriela Bitencourt-Ferreira and Walter Filgueira de Azevedo Jr.

Abstract

X-ray diffraction crystallography is the primary technique to determine the three-dimensional structures of biomolecules. Although a robust method, X-ray crystallography is not able to access the dynamical behavior of macromolecules. To do so, we have to carry out molecular dynamics simulations taking as an initial system the three-dimensional structure obtained from experimental techniques or generated using homology modeling. In this chapter, we describe in detail a tutorial to carry out molecular dynamics simulations using the program NAMD2. We chose as a molecular system to simulate the structure of human cyclin-dependent kinase 2.

Key words Force fields, NAMD2, Molecular dynamics, Cyclin-dependent kinase 2, Drug design, Molecular recognition

1 Introduction

Molecular dynamics of biomolecular systems is an active area of research in the computational simulation of proteins and nucleic acids and complexes involving biological macromolecules. These computational simulations play a fundamental role in crystallographic [1–12] and nuclear magnetic resonance studies [13–21] of biological macromolecules as well as in theoretical approaches [22–28].

The basic idea of molecular dynamics simulations of biomolecules is the assessment of the flexibility of the macromolecular structures through a computer simulation over time. Typically, in the analysis of molecular dynamics simulations, the trajectory of the macromolecule through time is evaluated, which provides a molecular view of the flexibility of the system as well as a dynamical view of intermolecular interactions when the simulation focuses on complexes composed of two or more molecules. It is possible to carry out molecular dynamics simulations of protein-ligand [29], protein-protein [30], protein-membrane [31], and nucleic acid-protein [32], to mention a few among the most common systems.

Walter Filgueira de Azevedo Jr. (ed.), *Docking Screens for Drug Discovery*, Methods in Molecular Biology, vol. 2053,
https://doi.org/10.1007/978-1-4939-9752-7_8, © Springer Science+Business Media, LLC, part of Springer Nature 2019

All molecular dynamics simulations rely on two primary computational methodologies. First, it requires physical modeling to express the potential energy of the systems. This model involves one equation to evaluate the potential energy of the system and a set of parameters to define each intramolecular and intermolecular interaction. The combination of the equation to assess the potential energy and the set of parameters for the intramolecular and intermolecular interactions is named as molecular force field. In general, the expression to calculate the potential energy (V) of a biomolecular system has the following expression:

$$
\begin{aligned}
V = & \sum_{(i,j)\in B} K_{ij}^{b}\left(r_{ij}-\bar{r}_{ij}\right)^2 + \sum_{(i,j,k)\in A} K_{ijk}^{a}\left(\theta_{ijk}-\bar{\theta}_{ijk}\right)^2 \\
& + \sum_{(i,j,k,l)\in D} \frac{K_{ijkl}^{d}}{2}\left[1+\cos\left(n_{ijkl}\emptyset_{ijkl}-\gamma_{ijkl}\right)\right] \\
& + \sum_{j\in F}\sum_{[i<j]}\left[\frac{A_{ij}}{r_{ij}^{12}}-\frac{B_{ij}}{r_{ij}^{6}}\right] + K_c \sum_{j\in F}\sum_{[i<j]}\frac{q_i q_j}{\varepsilon_{ij} r_{ij}}
\end{aligned}
\quad (1)
$$

In the above equation, the first term shows the potential energy relative to deviation from the equilibrium distances ($\bar{r}_{ij}$) for covalently bonded atoms (i, j) with an interatomic distance of r_{ij}. The parameter K_{ij}^{b} is the bond stretch force constant applied when the atom (i) is covalently bonded to the atom (j). The first summation is taken over all pairs of bonded atoms (B). The second summation considers deviations from an ideal angle ($\bar{\theta}_{ijk}$) involving three atoms (i, j, and k) and the angles involving them, θ_{ijk}. The parameter K_{ijk}^{a} is the force constant applied for the bond angle formed by the atoms (i, j, and k). The constant A is the set of three atoms (i, j, and k) that form the angle θ_{ijk}. The third summation considers the contribution of dihedral angles ($\emptyset_{ijkl}$) formed by four consecutive bonded atoms (i, j, k, and l). The constant n_{ijkl} is the periodicity of the dihedral angle, and γ_{ijkl} is the phase offset. This third summation is taken over all elements of the set D, which is formed by quadruplets of consecutive atoms. The parameter K_{ijkl}^{d} is the constant force for the dihedral angle formed the quadruplets of consecutive atoms. Molecular dynamics programs use the constant force parameters determined from empirical observations of experimental molecular structures. These first three summations represent the energy of bonded atoms.

The last two terms of the above equation represent non-bonded interactions in biological systems. The fourth is the van der Waals term, given by the 12–6 potential, where r_{ij} is the distance between the atoms (i and j). The coefficients A_{ij} and B_{ij} are the Lennard-Jones parameters for the pair of atoms (i, j). We take this fourth summation for all non-bonded atoms (set F) without repetitions. The last summation considers the electrostatic potential energy between charges q_i and q_j with an interatomic distance of

r_{ij}. The constant K_c is a conversion factor needed to obtain energy in kcal/mol. Most of molecular dynamics programs use $K_c = 332$ (kmol/mol)(Å/esu^2), where esu means the electrostatic unit of charge and its value is 1 esu = $3.335640951982 \times 10^{-10}$ °C. The above equation illustrates the main features of any modern potential energy implemented in molecular dynamics programs. Nevertheless, there are variations in the force fields, either on the equation itself or in the set of parameters the programs use to perform energy calculation. The molecular dynamics programs have extensive tables to provide the values for these parameters.

There are several molecular force fields suited to the simulation of biomolecular systems including ECEPP (Empirical Conformational Energy Program for Peptides) [33, 34], AMBER (Assisted Model Building with Energy Refinement) [35, 36], CHARMM22 (Chemistry at Harvard Macromolecular Mechanics) [37], GROMOS (GROningen MOlecular Simulation) [38–40], CVFF (Consistent-Valence Force Field) [41], and OPLS (Optimized Potentials for Liquid Simulations) [42]. The differences among these forces fields are on the set of parameters and in the implementation of Eq. (1).

In summary, when we refer to a specific force field, we are dealing with a mathematical expression of the potential energy for the system that uses pre-defined parameters to estimate each type of interaction present in the potential energy function. It is clear that the precise evaluation of this potential energy could be reached through computational demanding quantum mechanics methods [43–52]. Nevertheless, reliable assessment of binding affinity can be achieved through fast methods based on semi-empirical force fields [33–42].

In this chapter, we describe a detailed tutorial explaining the use of molecular dynamics simulation of a protein system. Due to the easy use and the free availability of the program, we chose NAMD (Nanoscale Molecular Dynamics) software [53].

2 NAMD2

The molecular dynamics package NAMD is a parallel package developed for high-performance simulation of biological macromolecules. Based on CHARMM22 [37] parallel objects, NAMD can make use of hundreds of cores for usual molecular dynamics simulations and beyond 500,000 cores for the simulations of the largest biological systems. NAMD employs the Visual Molecular Dynamics (VMD) [54] program for initial setup and analysis of the results. Furthermore, NAMD is also file-compatible with AMBER [35, 36] and X-PLOR [55, 56].

3 Biological System

In this chapter, we show how to carry out molecular dynamics simulation of cyclin-dependent kinase 2 (CDK2) (EC 2.7.11.22) with NAMD2 [53]. Figure 1 shows the electrostatic molecular surface of the ATP-binding pocket with the structure of the inhibitor roscovitine bound to CDK2 crystallographic structure [57]. CDK2 is a target for the development of anticancer drugs [58–68].

The first high-resolution crystallographic structure of CDK2 was obtained in 1993 at the University of California, Berkeley [70]. Analysis of the CDK2 structure indicated a typical bilobal architecture of serine/threonine protein kinases (EC 2.7.11.1). Figure 2 shows the structure of CDK2 in complex with ATP (PDB access code: 1HCK) [71]. Analysis of the structure of CDK2 shows that the N-terminal domain is mainly built by a distorted beta-sheet and a short alpha helix. A helix bundle forms the C-terminal. The two lobes of the CDK2 structure allow the binding of the ATP molecule, as we can see in Fig. 2.

Fig. 1 Electrostatic surface for ATP-binding pocket of human CDK2 in complex with the inhibitor roscovitine. This figure was generated using Molegro Virtual Docker (MVD) [69]. PDB access code: 2A4L [57]

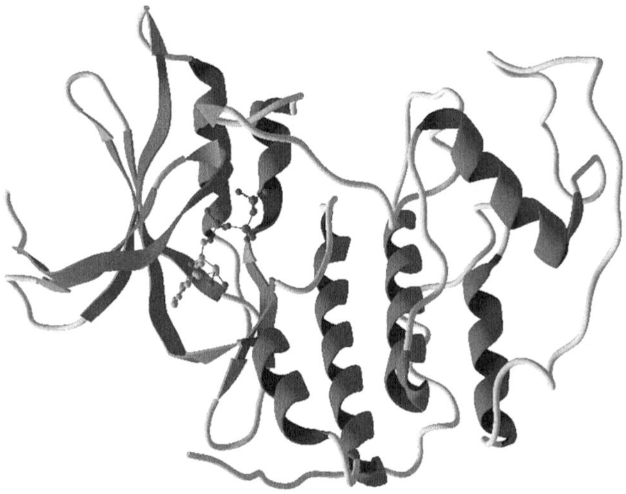

Fig. 2 Crystallographic structure of human CDK2 in complex with ATP. This figure was generated using Molegro Virtual Docker (MVD) [69]. PDB access code: 1HCK [71]

4 Graphical Tutorial

For this tutorial, it is necessary to have VMD [54] installed and running. We used this program to prepare the PDB (Protein Data Bank) and PSF (Protein Structure Format) files required to run the molecular dynamics simulation using NAMD2. To obtain the coordinates necessary for this tutorial, we may go to the Protein Data Bank (PDB) [72–74] (www.rcsb.org/pdb) and download the atomic coordinates for CDK2 in complex with roscovitine (PDB access code: 2A4L) [57]. Next, we must split the original PDB file into two files, one for roscovitine *(lig.pdb)* and the other for the CDK2 (*prot.pdb*). In this tutorial, we carried out molecular dynamics simulation of the protein only. Therefore, we need just *prot.pdb* file. Since there are missing residues in the structure 2A4L, we carried out a homology modeling to have the complete structure. We used the MODELLER program for homology modeling [75, 76].

Besides having the VMD and NAMD2 installed on our computer, we also need to have the following files in the same folder to run the molecular dynamics simulation with NAMD2.

```
prot.pdb
prot.pgn
top_all27_prot_lipid.inp
wat_box.tcl
par_all27_prot_lipid.inp
prot_wb_eq.conf
```

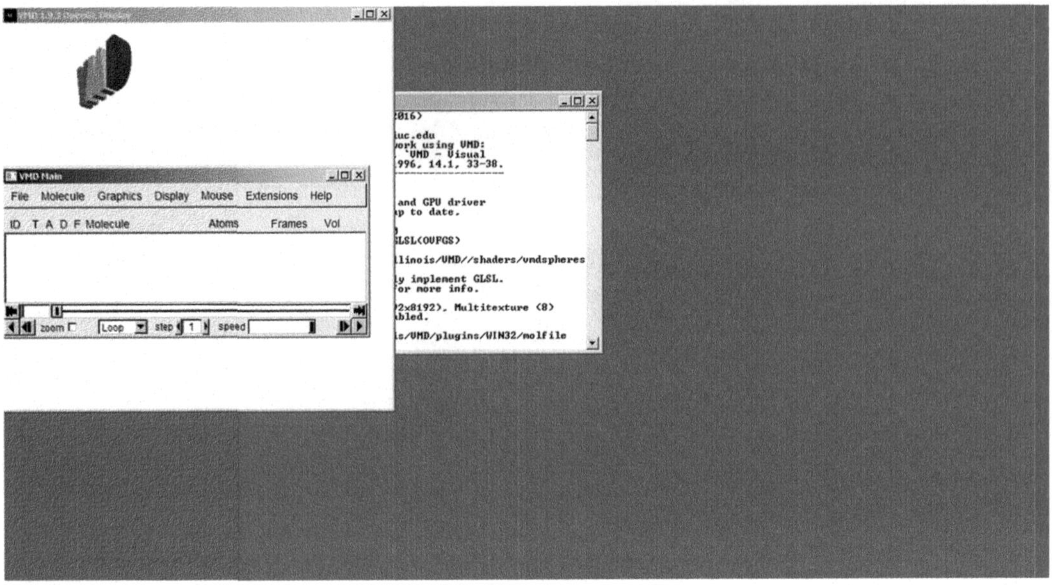

Fig. 3 VMD main graphics menu. From this menu, the user can access all tools to generate the input files necessary to run molecular dynamics simulations using NAMD2

Figure 3 shows the VMD main menu where we can access all available tasks necessary to prepare the files for molecular dynamics simulations. In this tutorial, we used the VMD version 1.9.3, but this tutorial should work for newer versions. We used VMD for Windows; it is mostly the same for the Linux and Mac OS X versions.

On the *VMD Main* menu, click on *File→New Molecule*. Then browse to the folder where the *prot.pdb* file is. Select the *prot.pdb* file and click on the *Open* button. Click on the *Load* button. Close the *Molecule File Browser* pop-up window. On the *OpenGL Display*, we have the CDK structure (Fig. 4) with the Lines representation.

On the *VMD Main* menu, click on *Extensions→Tk Console*. VMD calls the *Tk Console*. Make sure that we are on the folder where the *prot.pdb* file is. The *Tk Console* works as a Linux emulator. Type *pwd* to check the folder. It is possible to change the folder by typing *cd name_of_the_folder*. On the *Tk Console*, type the following commands:

```
set prot [atomselect top protein]
$prot writepdb protp.pdb
```

Now VMD has created the *protp.pdb* file. It is possible to check the folder content by typing *ls*. The *protp.pdb* file contains the atomic coordinates for the CDK2 structure without hydrogen atoms. On the Tk Console type the following command:

```
quit
```

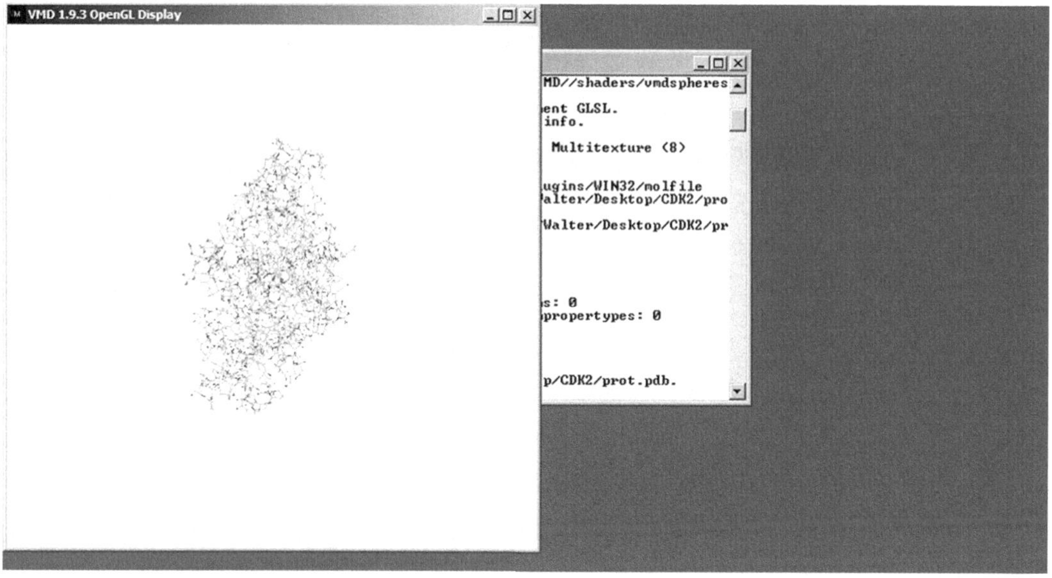

Fig. 4 Lines representation of the structure of CDK2 generated using the program VMD

To create the *prot.psf* file, we need to have the *prot.pgn* file. We used it as input for VMD. It should be in the same folder that has the *protp.pdb* file. In the command prompt (Terminal on Linux or Mac OS X), type *cd* to change to the folder where the files are. We may edit the *prot.pgn* file. Besides the *protp.pdb* file, we also need the topology information necessary to generate the psf file. The *top_all27_prot_lipid.inp* topology file should be in the same folder that has the *protp.pdb* file. The *prot.pgn* file has the following lines.

```
package require psfgen
topology top_all27_prot_lipid.inp
pdbalias residue HIS HSE
pdbalias atom ILE CD1 CD
segment U {pdb protp.pdb}
coordpdb protp.pdb U
guesscoord
writepdb prot.pdb
writepsf prot.psf
```

In the command prompt, type the following command:

```
vmd -dispdev text -e prot.pgn
```

If everything goes fine, we create a new *prot.psf* file.

Type the following command:

```
exit
```

To solvate the protein, type the following command:

```
vmd -dispdev text -e wat_box.tcl
```

If everything goes fine, VMD will create the *prot_wb.psf* and *prot_wb.pdb* files. These bring the protein structure centered inside a water box. After finishing generating the psf file for the CDK2 structure inside a water box, VMD writes down center information as follows:

```
CENTER OF MASS OF SPHERE IS: 101.54539489746094
88.6478271484375 84.6511459350586
```

Next, type the following command:

```
exit
```

Now we visualize the biological system (Fig. 5), with the CDK2 structure inside a water box. To start a new VMD session, on the command prompt, type the following command:

```
vmd
```

A new session for the VMD is initiated. To load the solvated structure, on the *VMD* Main menu, click on *File→New Molecule.* Click on the *Browse* button. Click on the *prot_wb.psf* file and then click on the *Open* button. Click on the *Load* button. To load the *prot_wb.pdb* file, click on the *Browse...* button. Click on the *prot_wb. pdb* file, then click on the *Open* button. Click on the *Load* button. Close the *Molecule File Browser* pop-up window. We see the solvated CDK2 structure on the *OpenGL Display* window (Fig. 5).

On the *VMD Main* menu, click on *Graphics→Representations.* On the *Graphical Representations* menu, click on the *Create Rep.* Button. VMD created two identical representations for the system. Leave one marked. On the *Drawing Method*, choose *New Cartoon.* Close *Graphical Representations* Menu. VMD shows a beautiful view of the CDK2 structure inside a water box (Fig. 6). On the *VMD Main* menu, click on *Extenstions→Tk Console.* VMD shows *Tk Console.* In the *Tk Console*, type the following commands:

```
set everyone [atomselect top all]
measure minmax $everyone
```

You will get the range for each coordinate axis.

```
{69.18399810791016 44.51900100708008 43.79399871826172}
{133.99899291992188 132.927001953125 125.56999969482422}
```

In the *Tk Console*, type the following commands:

```
measure center $everyone
```

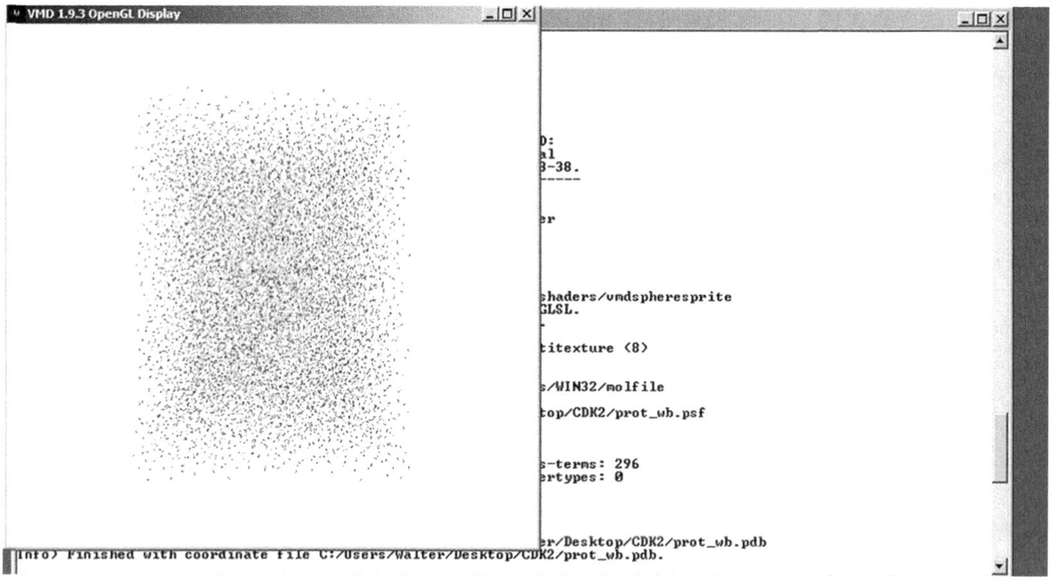

Fig. 5 Structure of CDK2 (lines representation) inserted in a water box

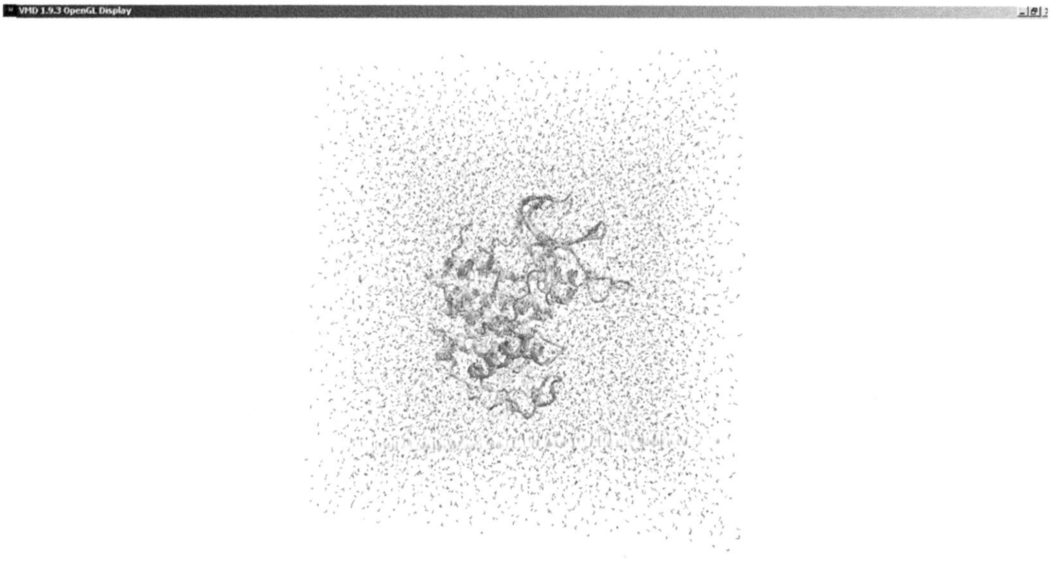

Fig. 6 CDK2 structure (secondary structure elements) inserted in a water box

You will get center for the system.

```
101.58609008789063 88.71961975097656 84.76802825927734
```

Type the following command:

```
quit
```

To generate the ionized forms for the molecular dynamics simulation, type the following command on the command prompt.

```
vmd
```

On the *VMD Main* menu, click on *File→New Molecule*. Click on the *Browse* button. Click on the *prot_wb.psf* file, and then click on the *Open* button. Click on the *Load* button. Click on the *Browse* button. Click on the *prot_wb.pdb* file, then click on the *Open* button. Click on the *Load* button. Close the *Molecule File Browser* pop-up window. On the *VMD Main* menu, click on *Extensions→Modeling→Add Ions*. On the *Autoionize* menu, click on the *Autoionize* button. Once finished, close the *Autoionize* menu. VMD has created *ionized.psf* and *ionized.pdb* files, which will be used later on for the molecular dynamics simulations. On the *VMD Main* menu, click on *File→Quit*.

You need the following files to run the molecular dynamics simulation with NAMD2:

```
ionized.pdb
ionized.psf
par_all27_prot_lipid.inp
prot_wb_eq.conf
```

In the command prompt, type the following command:

```
namd2 prot_ws_eq.conf > prot_ws_eq.log &
```

We have our molecular dynamics simulation running. To generate a plot for the energy terms, we used the Python script *plot_-namd_energy.py*. This script requires two input files, one the log file, *prot_ws_eq.log*. The second file is the *namd.in*, which brings information about how to generate the plot. The *namd.in* to generate a plot to the electrostatic term is shown below:

```
FILE_IN,"prot_wb_eq.log", # Namd energy log file
START_COLLECT,100, # Indicate the time step where to start to
collect
TERM,"ELECT", # Indicate which energy to plot
XMIN,100, # Minimum for x-axis
XMAX,2500, # Maximum for x-axis
XLABEL,"steps", # Label for x-axis
YLABEL,"Electrostatic Energy (kcal/mol)", # Label for y-axis
TITLE,"Molecular Dynamics Simulation of CDK2", # Plot title
YMIN,-150000, # Minimum for y-axis
YMAX,-147500, # Maximum for y-axis
LINE_COLOR,"black", # Line color
FILE_OUT,"ELECT.png", # Plot file name
```

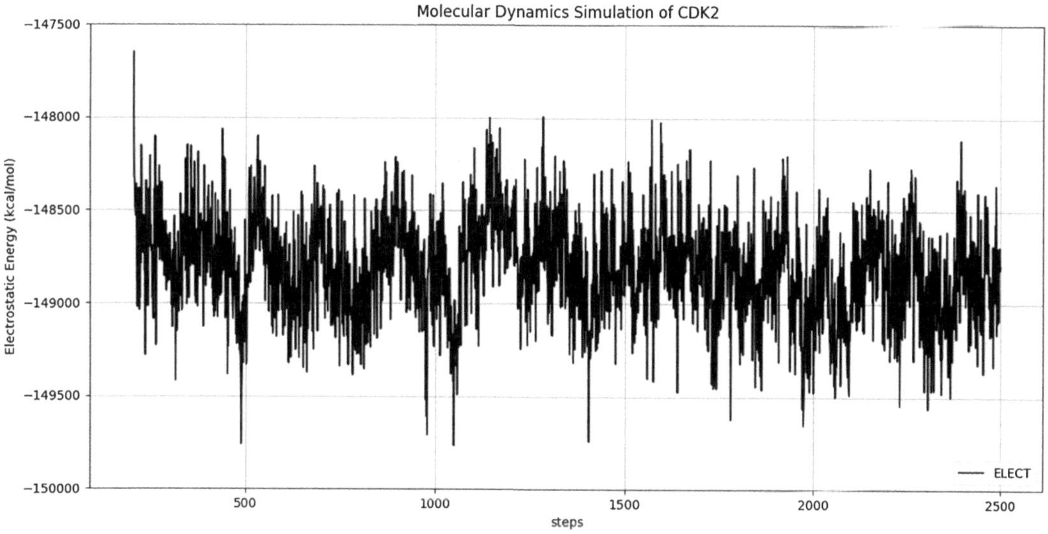

Fig. 7 Variation in the electrostatic potential energy of the structure of CDK2 during a molecular dynamics simulation of 5 ns. In the above plot, each step means 2 ft.

To run the Python script *plot_namd_energy.py*, we should open a command prompt (terminal on Linux and Mac OS X). Then go to the folder where the *namd.in* and *prot_ws_eq.log* files are. In this tutorial, all files are in the c:\users\Walter\Desktop\CDK2. Type the following command:

```
cd c:\Users\Walter\Desktop\CDK2
```

Considering that you have Python 3, Numpy and Matplotlib libraries installed on your computer, type the following command:

```
python plot_namd_energy.py
```

The Python script generated two files. The first is the *energy.csv* file, which brings the values for energy terms during the molecular dynamics simulation. The second is the *elect.png* file, shown in Fig. 7. From this plot, we may say that CDK2 structure is stable since its electrical potential energy does not increase during the simulation.

5 Availability

All files necessary to run this tutorial are available at https://azevedolab.net/resources/NAMD_CDK2.zip.

6 Colophon

We used the program Molegro Virtual Docker [69] to generate Figs. 1 and 2. We created Figs. 3–6 using the program VMD [54]. We used the python script *plot_namd_energy*.py to make Fig. 7. We performed molecular dynamics simulations described in this chapter using a Desktop PC with 4 GB memory, a 1 TB hard disk, and an Intel® Core® i3-2120 @ 3.30 GHz processor running Windows 8.1.

7 Final Remarks

Molecular dynamics simulations of biological systems open the possibility to access the flexibility of organic molecules in a solvent environment. Such simulations can assess the dynamical behavior of the molecule, which allows us to investigate the biomolecule in the physical situation closer to the biological environment where we expect to find proteins, nucleic acids, and membranes. The use of the program NAMD2 to simulate the dynamical features biomolecules has been successfully applied to a wide range of biological systems [77–96], which further validates the importance of this program in the simulation of such complex systems.

Acknowledgments

This work was supported by grants from CNPq (Brazil) (308883/2014-4). This study was financed in part by the Coordenação de Aperfeiçoamento de Pessoal de Nivel Superior—Brasil (CAPES)—Finance Code 001. GB-F acknowledges support from PUCRS/BPA fellowship. WFA is a senior researcher for CNPq (Brazil) (Process Numbers: 308883/2014-4 and 309029/2018-0).

References

1. Depristo MA, de Bakker PI, Johnson RJ, Blundell TL (2005) Crystallographic refinement by knowledge-based exploration of complex energy landscapes. Structure 13:1311–1319
2. Adams PD, Pannu NS, Read RJ, Brünger AT (1997) Cross-validated maximum likelihood enhances crystallographic simulated annealing refinement. Proc Natl Acad Sci U S A 94:5018–5023
3. Rice LM, Brünger AT (1994) Torsion angle dynamics: reduced variable conformational sampling enhances crystallographic structure refinement. Proteins 19:277–290
4. Clarage JB, Phillips GN Jr (1994) Cross-validation tests of time-averaged molecular dynamics refinements for determination of protein structures by X-ray crystallography. Acta Crystallogr D Biol Crystallogr 50:24–36
5. Gros P, Betzel C, Dauter Z, Wilson KS, Hol WG (1989) Molecular dynamics refinement of a thermitase-eglin-c complex at 1.98 A resolution and comparison of two crystal forms that

differ in calcium content. J Mol Biol 210:347–367

6. Kuriyan J, Petsko GA, Levy RM, Karplus M (1986) Effect of anisotropy and anharmonicity on protein crystallographic refinement. An evaluation by molecular dynamics. J Mol Biol 190:227–254
7. Westhof E, Chevrier B, Gallion SL, Weiner PK, Levy RM (1986) Temperature-dependent molecular dynamics and restrained X-ray refinement simulations of a Z-DNA hexamer. J Mol Biol 191:699–712
8. Wendoloski JJ, Wasserman ZR, Salemme FR (1988) Computer simulation of biological interactions and reactivity. J Comput Aided Mol Des 1:313–322
9. Ichiye T, Karplus M (1988) Anisotropy and anharmonicity of atomic fluctuations in proteins: implications for X-ray analysis. Biochemistry 27:3487–3497
10. Postma JP, Parker MW, Tsernoglou D (1989) Application of molecular dynamics in the crystallographic refinement of colicin A. Acta Crystallogr A 45:471–477
11. Gros P, Fujinaga M, Dijkstra BW, Kalk KH, Hol WG (1989) Crystallographic refinement by incorporation of molecular dynamics: thermostable serine protease thermitase complexed with eglin c. Acta Crystallogr B 45:488–499
12. Canduri F, de Azevedo WF (2008) Protein crystallography in drug discovery. Curr Drug Targets 9:1048–1053
13. Campagne S, Krepl M, Sponer J, Allain FH (2019) Combining NMR spectroscopy and molecular dynamic simulations to solve and analyze the structure of protein-RNA complexes. Methods Enzymol 614:393–422
14. Kämpf K, Izmailov SA, Rabdano SO, Groves AT, Podkorytov IS, Skrynnikov NR (2018) What drives 15N spin relaxation in disordered proteins? combined NMR/MD study of the H4 histone tail. Biophys J 115:2348–2367
15. Bochicchio A, Krepl M, Yang F, Varani G, Sponer J, Carloni P (2018) Molecular basis for the increased affinity of an RNA recognition motif with re-engineered specificity: a molecular dynamics and enhanced sampling simulations study. PLoS Comput Biol 14:e1006642
16. Purslow JA, Nguyen TT, Egner TK, Dotas RR, Khatiwada B, Venditti V (2018) Active site breathing of human Alkbh5 revealed by solution NMR and accelerated molecular dynamics. Biophys J 115:1895–1905
17. Quinn CM, Wang M, Fritz MP, Runge B, Ahn J, Xu C et al (2018) Dynamic regulation of HIV-1 capsid interaction with the restriction factor TRIM5α identified by magic-angle spinning NMR and molecular dynamics simulations. Proc Natl Acad Sci U S A 115:11519–11524
18. Cousin SF, Kadeřávek P, Bolik-Coulon N, Gu Y, Charlier C, Carlier L (2018) Time-resolved protein side-chain motions unraveled by high-resolution relaxometry and molecular dynamics simulations. J Am Chem Soc 140:13456–13465
19. Papaleo E, Camilloni C, Teilum K, Vendruscolo M, Lindorff-Larsen K (2018) Molecular dynamics ensemble refinement of the heterogeneous native state of NCBD using chemical shifts and NOEs. PeerJ 6:e5125
20. Sforça ML, Oyama S Jr, Canduri F, Lorenzi CC, Pertinhez TA, Konno K et al (2004) How C-terminal carboxyamidation alters the biological activity of peptides from the venom of the eumenine solitary wasp. Biochemistry 43:5608–5617
21. Fadel V, Bettendorff P, Herrmann T, de Azevedo WF Jr, Oliveira EB, Yamane T et al (2005) Automated NMR structure determination and disulfide bond identification of the myotoxin crotamine from *Crotalus durissus terrificus*. Toxicon 46:759–767
22. de Azevedo WF Jr (2011) Molecular dynamics simulations of protein targets identified in *Mycobacterium tuberculosis*. Curr Med Chem 18:1353–1366
23. Ganai SA (2018) Designing isoform-selective inhibitors against classical HDACs for effective anticancer therapy: insight and perspectives from in silico. Curr Drug Targets 19:815–824
24. Abdolmaleki A, Ghasemi JB, Ghasemi F (2017) Computer aided drug design for multi-target drug design: SAR/QSAR, molecular docking and pharmacophore methods. Curr Drug Targets 18:556–575
25. Kontoyianni M, Lacy B (2018) Toward computational understanding of molecular recognition in the human metabolizing cytochrome P450s. Curr Med Chem 25:3353–3373
26. Gentile L, Uccella NA, Sivakumar G (2017) Oleuropein: molecular dynamics and computation. Curr Med Chem 24:4315–4328
27. Hernández-Rodríguez M, Rosales-Hernández MC, Mendieta-Wejebe JE, Martínez-Archundia M, Basurto JC (2016) Current tools and methods in molecular dynamics (MD) simulations for drug design. Curr Med Chem 23:3909–3924
28. Tamay-Cach F, Villa-Tanaca ML, Trujillo-Ferrara JG, Alemán-González-Duhart D, Quintana-Pérez JC, González-Ramírez IA et al (2016) In silico studies most employed

in the discovery of new antimicrobial agents. Curr Med Chem 23:3360–3373

29. Perricone U, Gulotta MR, Lombino J, Parrino B, Cascioferro S, Diana P et al (2018) An overview of recent molecular dynamics applications as medicinal chemistry tools for the undruggable site challenge. Medchemcomm 9:920–936
30. Wang W, Donini O, Reyes CM, Kollman PA (2001) Biomolecular simulations: recent developments in force fields, simulations of enzyme catalysis, protein-ligand, protein-protein, and protein-nucleic acid noncovalent interactions. Annu Rev Biophys Biomol Struct 30:211–243
31. Ray A, Jatana N, Thukral L (2017) Lipidated proteins: Spotlight on protein-membrane binding interfaces. Prog Biophys Mol Biol 128:74–84
32. Mackerell AD Jr, Nilsson L (2008) Molecular dynamics simulations of nucleic acid-protein complexes. Curr Opin Struct Biol 18:194–199
33. Arnautova YA, Jagielska A, Scheraga HÁ (2006) A new force field (ECEPP-05) for peptides, proteins, and organic molecules. J Phys Chem B 110:5025–5044
34. Arnautova YA, Vorobjev YN, Vila JA, Scheraga HÁ (2009) Identifying native-like protein structures with scoring functions based on all-atom ECEPP force fields, implicit solvent models and structure relaxation. Proteins 77:38–51
35. Cornell WD, Cieplak P, Bayly CI, Gould IR, Merz KM, Ferguson DM et al (1995) A second generation force field for the simulation of proteins, nucleic acids, and organic molecules. J Am Chem Soc 117:5179–5197
36. Duan Y, Wu C, Chowdhury S, Lee MC, Xiong G, Zhang W et al (2003) A point-charge force field for molecular mechanics simulations of proteins based on condensed-phase quantum mechanical calculations. Comput Chem 24:1999–2002
37. AD MK Jr, Bashford D, Bellott M, Dunbrack RL Jr, Evanseck J, Field MJ et al (1998) All-atom empirical potential for molecular modeling and dynamics studies of proteins. Phys Chem B 102:3586–3616
38. Oostenbrink C, Soares TA, van der Vegt NF, van Gunsteren WF (2005) Validation of the 53A6 GROMOS force field. Eur Biophys J 34:273–384
39. Soares TA, Hünenberger PH, Kastenholz MA, Kräutler V, Lenz T, Lins RD et al (2005) An improved nucleic acid parameter set for the GROMOS force field. J Comput Chem 26:725–737
40. Lin Z, van Gunsteren WF (2013) Refinement of the application of the GROMOS 54A7 force field to β-peptides. J Comput Chem 34:2796–2805
41. Ewig CS, Berry R, Dinur U, Hill J-R, Hwang M-J, Li H et al (2001) Derivation of class II force fields. VIII. Derivation of a general quantum mechanical force field for organic compounds. J Comput Chem 22:1782–1800
42. Kaminski GA, Friesner RA, Tirado-Rives J, Jorgensen WL (2001) Evaluation and reparametrization of the OPLS-AA force field for proteins via comparison with accurate quantum chemical calculations on peptides. J Phys Chem B 105:6474–6487
43. Adeniyi AA, Soliman MES (2017) Implementing QM in docking calculations: is it a waste of computational time? Drug Discov Today 22:1216–1223
44. Crespo A, Rodriguez-Granillo A, Lim VT (2017) Quantum-mechanics methodologies in drug discovery: applications of docking and scoring in lead optimization. Curr Top Med Chem 17:2663–2680
45. Yilmazer ND, Korth M (2016) Recent progress in treating protein-ligand interactions with quantum-mechanical methods. Int J Mol Sci 17:742
46. Cavasotto CN, Adler NS, Aucar MG (2018) Quantum chemical approaches in structure-based virtual screening and lead optimization. Front Chem 6:188
47. Hitzenberger M, Schuster D, Hofer TS (2017) The binding mode of the sonic hedgehog inhibitor Robotnikinin, a combined docking and QM/MM MD study. Front Chem 5:76
48. Ekhteiari Salmas R, Serhat Is Y, Durdagi S, Stein M, Yurtsever M (2018) A QM protein-ligand investigation of antipsychotic drugs with the dopamine D2 receptor (D2R). J Biomol Struct Dyn 36:2668–2677
49. Phipps MJ, Fox T, Tautermann CS, Skylaris CK (2017) Intuitive density functional theory-based energy decomposition analysis for protein-ligand interactions. J Chem Theory Comput 13:1837–1850
50. Hylsová M, Carbain B, Fanfrlík J, Musilová L, Haldar S, Köprülüoğlu C et al (2017) Explicit treatment of active-site waters enhances quantum mechanical/implicit solvent scoring: Inhibition of CDK2 by new pyrazolo[1,5-a] pyrimidines. Eur J Med Chem 126:1118–1128
51. Pecina A, Meier R, Fanfrlík J, Lepšík M, Řezáč J, Hobza P et al (2016) The SQM/COSMO filter: reliable native pose identification based on the quantum-mechanical description of protein-ligand

interactions and implicit COSMO solvation. Chem Commun (Camb) 52:3312–3315
52. Yang Z, Liu Y, Chen Z, Xu Z, Shi J, Chen K et al (2015) A quantum mechanics-based halogen bonding scoring function for protein-ligand interactions. J Mol Model 21:138
53. Phillips JC, Braun R, Wang W, Gumbart J, Tajkhorshid E, Villa E et al (2005) Scalable molecular dynamics with NAMD. J Comput Chem 26:1781–1802
54. Humphrey W, Dalke A, Schulten K (1996) VMD—visual molecular dynamics. J Mol Graph 14:33–38
55. Brünger AT, Kuriyan J, Karplus M (1987) Crystallographic R factor refinement by molecular dynamics. Science 235:458–460
56. de Azevedo WF Jr, Canduri F, Fadel V, Teodoro LG, Hial V, Gomes RA (2001) Molecular model for the binary complex of uropepsin and pepstatin. Biochem Biophys Res Commun 287:277–281
57. De Azevedo WF, Leclerc S, Meijer L, Havlicek L, Strnad M, Kim SH (1997) Inhibition of cyclin-dependent kinases by purine analogues: crystal structure of human cdk2 complexed with roscovitine. Eur J Biochem 243:518–526
58. Morgan DO (1995) Principles of CDK regulation. Nature 374:131–134
59. Murray AW (1994) Cyclin-dependent kinases: regulators of the cell cycle and more. Chem Biol 1:191–195
60. Kim SH, Schulze-Gahmen U, Brandsen J, de Azevedo Junior WF (1996) Structural basis for chemical inhibition of CDK2. Prog Cell Cycle Res 2:137–145
61. De Azevedo WF Jr, Mueller-Dieckmann HJ, Schulze-Gahmen U, Worland PJ, Sausville E, Kim SH (1996) Structural basis for specificity and potency of a flavonoid inhibitor of human CDK2, a cell cycle kinase. Proc Natl Acad Sci U S A 93:2735–2740
62. Canduri F, de Azevedo WF Jr (2005) Structural basis for interaction of inhibitors with cyclin-dependent kinase 2. Curr Comput Aided Drug Des 1:53–64
63. Krystof V, Cankar P, Frysová I, Slouka J, Kontopidis G, Dzubák P (2006) 4-arylazo-3,5-diamino-1H-pyrazole CDK inhibitors: SAR study, crystal structure in complex with CDK2, selectivity, and cellular effects. J Med Chem 49:6500–6509
64. de Azevedo WF Jr (2016) Opinion paper: targeting multiple cyclin-dependent kinases (CDKs): a new strategy for molecular docking studies. Curr Drug Targets 17:2
65. Levin NM, Pintro VO, de Ávila MB, de Mattos BB, De Azevedo WF Jr (2017) Understanding the structural basis for inhibition of cyclin-dependent kinases. New pieces in the molecular puzzle. Curr Drug Targets 18:1104–1111
66. de Ávila MB, Xavier MM, Pintro VO, de Azevedo WF (2017) Supervised machine learning techniques to predict binding affinity. A study for cyclin-dependent kinase 2. Biochem Biophys Res Commun 494:305–310
67. Levin NMB, Pintro VO, Bitencourt-Ferreira G, Mattos BB, Silvério AC, de Azevedo WF Jr (2018) Development of CDK-targeted scoring functions for prediction of binding affinity. Biophys Chem 235:1–8
68. Volkart PA, Bitencourt-Ferreira G, Souto AA, de Azevedo WF (2019) Cyclin-dependent kinase 2 in cellular senescence and cancer. A structural and functional review. Curr Drug Targets 20(7):716–726. https://doi.org/10.2174/1389450120666181204165344
69. Thomsen R, Christensen MH (2006) MolDock: a new technique for high-accuracy molecular docking. J Med Chem 49:3315–3321
70. De Bondt HL, Rosenblatt J, Jancarik J, Jones HD, Morgan DO, Kim SH (1993) Crystal structure of cyclin-dependent kinase 2. Nature 363:595–602
71. Schulze-Gahmen U, De Bondt HL, Kim SH (1996) High-resolution crystal structures of human cyclin-dependent kinase 2 with and without ATP: bound waters and natural ligand as guides for inhibitor design. J Med Chem 39:4540–4546
72. Berman HM, Westbrook J, Feng Z, Gilliland G, Bhat TN, Weissig H et al (2000) The Protein Data Bank. Nucleic Acids Res 28:235–242
73. Berman HM, Battistuz T, Bhat TN, Bluhm WF, Bourne PE, Burkhardt K et al (2002) The Protein Data Bank. Acta Crystallogr D Biol Crystallogr 58:899–907
74. Westbrook J, Feng Z, Chen L, Yang H, Berman HM (2003) The Protein Data Bank and structural genomics. Nucleic Acids Res 31:489–491
75. Sali A, Blundell TL (1993) Comparative protein modelling by satisfaction of spatial restraints. J Mol Biol 234:779–815
76. Uchôa HB, Jorge GE, Freitas Da Silveira NJ, Camera JC Jr, Canduri F, De Azevedo WF Jr (2004) Parmodel: a web server for automated comparative modeling of proteins. Biochem Biophys Res Commun 325:1481–1486
77. Daniyan MO, Ojo OT (2019) In silico identification and evaluation of potential interaction

of *Azadirachta indica* phytochemicals with *Plasmodium falciparum* heat shock protein 90. J Mol Graph Model 87:144–164

78. Chandra N, Biswas S, Rout J, Basu G, Tripathy U (2018) Stability of β-turn in LaR2C-N7 peptide for its translation-inhibitory activity against hepatitis C viral infection: A molecular dynamics study. Spectrochim Acta A Mol Biomol Spectrosc 211:26–33
79. Uba AI, Yelekçi K (2018) Pharmacophore-based virtual screening for identification of potential selective inhibitors of human histone deacetylase 6. Comput Biol Chem 77:318–330
80. Miao Y, Bhattarai A, Nguyen ATN, Christopoulos A, May LT (2018) Structural basis for binding of allosteric drug leads in the adenosine A1 receptor. Sci Rep 8:16836
81. Liamas E, Kubiak-Ossowska K, Black RA, Thomas ORT, Zhang ZJ, Mulheran PA (2018) Adsorption of fibronectin fragment on surfaces using fully atomistic molecular dynamics simulations. Int J Mol Sci 19:3321
82. Rezapour N, Rasekh B, Mofradnia SR, Yazdian F, Rashedi H, Tavakoli Z (2019) Molecular dynamics studies of polysaccharide carrier based on starch in dental cavities. Int J Biol Macromol 121:616–624
83. Jiang W, Thirman J, Jo S, Roux B (2018) Reduced free energy perturbation/hamiltonian replica exchange molecular dynamics method with unbiased alchemical thermodynamic axis. J Phys Chem B 122:9435–9442
84. Zhang R, Zhang L, Zheng Q, Gao P, Zhao J, Yang J (2018) Direct Z-scheme water splitting photocatalyst based on two-dimensional Van Der Waals heterostructures. J Phys Chem Lett 9:5419–5424
85. Kulke M, Geist N, Möller D, Langel W (2018) Replica-based protein structure sampling methods: compromising between explicit and implicit solvents. J Phys Chem B 122:7295–7307
86. Sarkar R, Habib M, Pal S, Prezhdo OV (2018) Ultrafast, asymmetric charge transfer and slow charge recombination in porphyrin/CNT composites demonstrated by time-domain atomistic simulation. Nanoscale 10:12683–12694
87. Chen H, Fu H, Shao X, Chipot C, Cai W (2018) ELF: an extended-lagrangian free energy calculation module for multiple molecular dynamics engines. J Chem Inf Model 58:1315–1318
88. Childers MC, Daggett V (2018) Validating molecular dynamics simulations against experimental observables in light of underlying conformational ensembles. J Phys Chem B 122:6673–6689
89. Uba AI, Yelekçi K (2018) Carboxylic acid derivatives display potential selectivity for human histone deacetylase 6: Structure-based virtual screening, molecular docking and dynamics simulation studies. Comput Biol Chem 75:131–142
90. Mishra V, Pathak C (2018) Structural insights into pharmacophore-assisted *in silico* identification of protein-protein interaction inhibitors for inhibition of human toll-like receptor 4 - myeloid differentiation factor-2 (hTLR4-MD-2) complex. J Biomol Struct Dyn 29:1–24
91. Serçinoglu O, Ozbek P (2018) gRINN: a tool for calculation of residue interaction energies and protein energy network analysis of molecular dynamics simulations. Nucleic Acids Res 46:554–562
92. Banu H, Joseph MC, Nisar MN (2018) *In-silico* approach to investigate death domains associated with nano-particle-mediated cellular responses. Comput Biol Chem 75:11–23
93. Mena-Ulecia K, MacLeod-Carey D (2018) Interactions of 2-phenyl-benzotriazole xenobiotic compounds with human Cytochrome P450-CYP1A1 by means of docking, molecular dynamics simulations and MM-GBSA calculations. Comput Biol Chem 74:253–262
94. Kurniawan F, Kartasasmita RE, Yoshioka N, Mutalib A, Tjahjono DH (2018) Computational study of imidazolylporphyrin derivatives as a radiopharmaceutical ligand for melanoma. Curr Comput Aided Drug Des 14:191–199
95. Khezri A, Karimi A, Yazdian F, Jokar M, Mofradnia SR, Rashedi H et al (2018) Molecular dynamic of curcumin/chitosan interaction using a computational molecular approach: emphasis on biofilm reduction. Int J Biol Macromol 114:972–978
96. Subasri S, Chaudhary SK, Sekar K, Kesherwani M, Velmurugan D (2017) Molecular docking and molecular dynamics simulations of fumarate hydratase and its mutant H235N complexed with pyromellitic acid and citrate. J Bioinforma Comput Biol 15:1750026

Chapter 9

Docking with AutoDock4

Gabriela Bitencourt-Ferreira, Val Oliveira Pintro, and Walter Filgueira de Azevedo Jr.

Abstract

AutoDock is one of the most popular receptor-ligand docking simulation programs. It was first released in the early 1990s and is in continuous development and adapted to specific protein targets. AutoDock has been applied to a wide range of biological systems. It has been used not only for protein-ligand docking simulation but also for the prediction of binding affinity with good correlation with experimental binding affinity for several protein systems. The latest version makes use of a semi-empirical force field to evaluate protein-ligand binding affinity and for selecting the lowest energy pose in docking simulation. AutoDock4.2.6 has an arsenal of four search algorithms to carry out docking simulation including simulated annealing, genetic algorithm, and Lamarckian algorithm. In this chapter, we describe a tutorial about how to perform docking with AutoDock4. We focus our simulations on the protein target cyclin-dependent kinase 2.

Key words AutoDock, Molecular docking, Cyclin-dependent kinase 2, Drug design, Protein-ligand interactions

1 Introduction

The development of molecular docking methods began in the early 1980s [1]. As soon as these programs became available, in silico methodologies were effectively used to discover many approved drugs including HIV-1 protease inhibitors [2–7]. We may say that drug development has progressed significantly from the use of in silico methodologies, which currently is the first approach in drug discovery and development [8, 9].

We can envisage the molecular docking problem as an optimization problem, where we attempt to locate the optimal position for an organic molecule ligand into the protein structure. As to computer-aided drug design, molecular docking methodology is the most common approach that has been extensively used to drug development ever since the early 1980s, and the rise of the

Walter Filgueira de Azevedo Jr. (ed.), *Docking Screens for Drug Discovery*, Methods in Molecular Biology, vol. 2053,
https://doi.org/10.1007/978-1-4939-9752-7_9,

computational capacity and the availability of protein structures have been the main factors for the progress of the field [10–24].

It is usual with a simple workstation to perform docking of thousands of ligands against a protein target—furthermore, the availability of modern open source protein-ligand docking programs such as AutoDock [25–28], AutoDock Vina [29], and GemDock [30, 31] to mention a few, made possible to research laboratories even with a modest budget to perform robust protein-ligand docking projects [32–48]. Also, the integration of the docking programs in a workflow makes it possible to carry out docking simulations in an integrated way that facilitates the simulations and the analysis of the docking results [49].

Studies using docking simulation were able to find binders to wide-spectrum druggable targets [50–60]. Along with the development of protein-ligand docking programs, we have also seen an increase in the number of protein–drug complexes available in the Protein Data Bank (PDB) [61–63]. Furthermore, the availability of experimental information about inhibition constant (K_i), dissociation constant (K_d), half maximal inhibitory concentration (IC_{50}), and Gibbs free energy of binding (ΔG) offers a solid framework of structural and binding affinity data that permits us to explore the structural basis for inhibition of protein targets. Experimental binding affinity data are available at MOAD [64], BindingDB [65], and PDBbind [66].

Among the most used protein-ligand docking programs, we would like to highlight here the AutoDock. The program AutoDock provides an integrated computational environment for docking simulations and calculation of protein-ligand binding affinities. There are 1160 studies about the application of AutoDock to docking simulations (search carried out on October 26, 2018, using the keyword "autodock" in the PubMed). Integration of AutoDock4 in the program SAnDReS [49] makes it possible to perform protein-ligand docking simulations in a well-designed and fast computational tool.

We have successfully employed SAnDReS to study the coagulation factor Xa [49], cyclin-dependent kinases [36, 39, 41, 67], HIV-1 protease [38], estrogen receptor [35], cannabinoid receptor 1 [34], 3-dehydroquinate dehydratase [33], and enoyl-[acyl carrier protein] reductase (InhA) from *Mycobacterium tuberculosis* [68]. Also, we used SAnDReS to develop a machine-learning model to predict the Gibbs free energy of binding for protein-ligand complexes [32]. In the next sections, we describe a tutorial for the application of the AutoDock4 to carry out docking simulations against the structure cyclin-dependent kinase 2 and highlight the main integrated tools available for protein-ligand docking simulations and analysis of the predictive performance of this in silico methodology.

2 Biological System

In this tutorial, we show how to perform protein-ligand docking simulations of cyclin-dependent kinase 2 (CDK2) (EC 2.7.11.22) with AutoDock4 [28]. Figure 1 shows the intermolecular hydrogen bond interaction of the ATP-binding pocket with the structure of the inhibitor roscovitine bound to CDK2 crystallographic structure [69]. This vital protein kinase has been intensively studied as a target for the development of anticancer drugs [70–75].

The first high-resolution crystallographic structure of CDK2 was determined in 1993 at the University of California, Berkeley [77]. Analysis of the CDK2 structure indicated a typical bilobal architecture of serine/threonine protein kinases (EC 2.7.11.1). Figure 2 shows the structure of CDK2 in complex with ATP (PDB access code: 1HCK) [78]. Analysis of the structure of CDK2 shows that the N-terminal domain is mainly built by a distorted beta-sheet and a short alpha helix. A helix bundle forms the C-terminal. The two lobes of the CDK2 structure allow the binding of the ATP molecule, as we can *see* in Fig. 2.

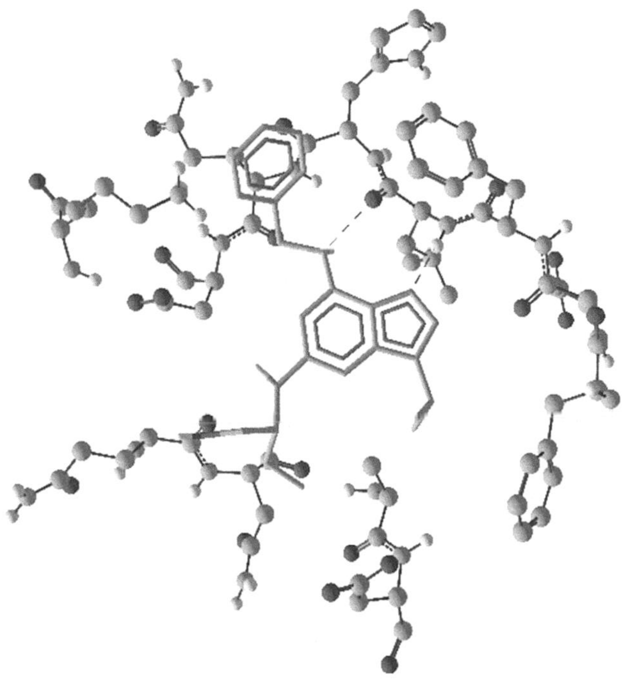

Fig. 1 Intermolecular hydrogen bonds of human CDK2 in complex with the inhibitor roscovitine. This figure was generated using Molegro Virtual Docker (MVD) [76]. PDB access code: 2A4L [69]. MVD indicates hydrogen bonds as dashed lines. MVD used stick representation for ligand and ball-and-stick representation for the amino acid structures

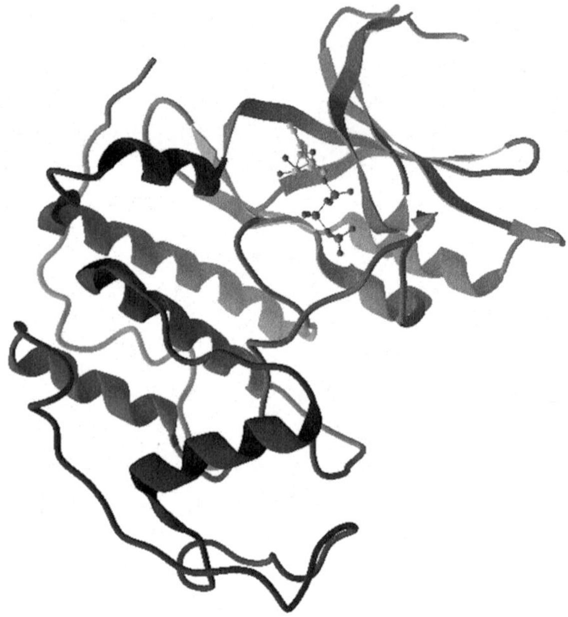

Fig. 2 Crystallographic structure of human CDK2 in complex with ATP. This figure was generated using Molegro Virtual Docker (MVD) [76]. PDB access code: 1HCK [78]

3 Graphical Tutorial

We consider that you have AutoDockTools4 (ADT) and AutoDock4 installed on your computer (Fig. 3). We used the version 1.5.6 for Windows; it is mostly the same for Mac OS X and Linux versions. Below you have files we will need for this tutorial: a PDB file for a protein without a ligand (2A4L), a PDB file for a ligand (RRC_300), and executables files for AutoGrid and Autodock.

autodock4.exe, autogrid4.exe, 2a4l.pdb, and RRC_300.pdb

To facilitate our work, we will change the directory. Start AutoDockTools4 (ADT) and click *File→Preferences→Set* as shown in Fig. 4. Go to the directory where we have the files for this tutorial. Copy and paste it on the startup directory as shown in Fig. 4. Then click *Make Default→Dismiss.*

In the next step, we will need the protein PDB file with no ligands. Click *on File→Read Molecule.* Select your PDB file and open it. In your screen will appear the protein structure and water molecules. Choose the color scheme by atom type for better visualization (Fig. 5). Now our structure has a pattern for color according to the atom. In the next step, we will need to delete crystallographic water molecules. Click on *Select→Select From String* (Fig. 6).

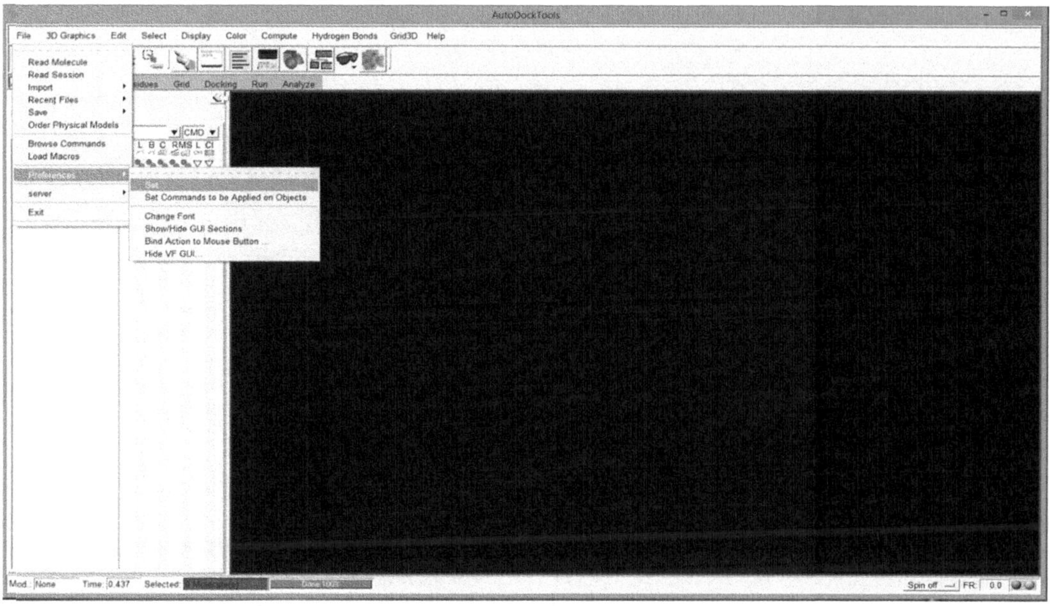

Fig. 3 The main window of AutoDockTools4 [28]

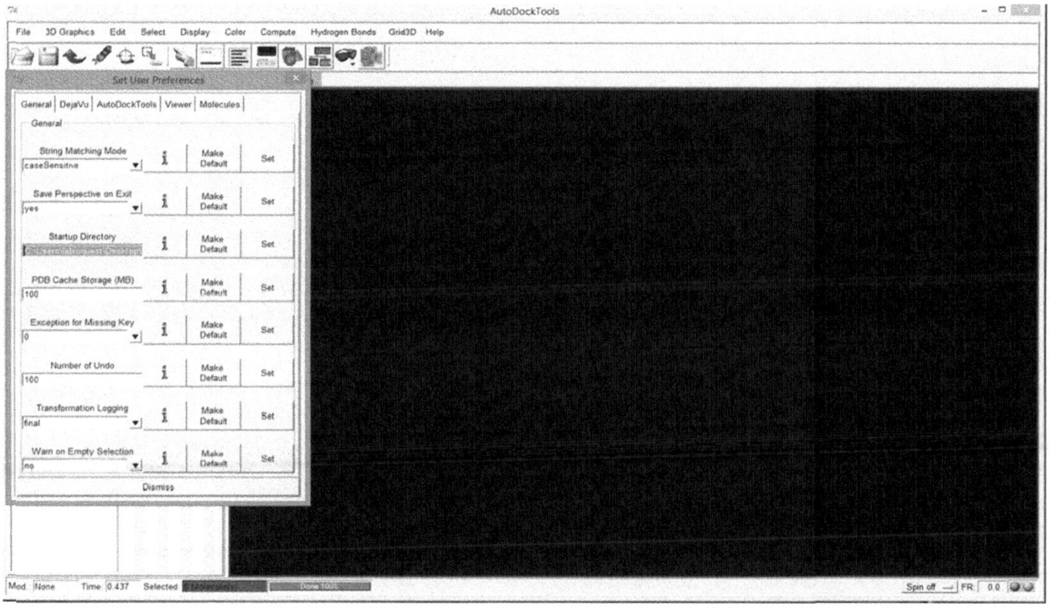

Fig. 4 The main window of AutoDockTools4 [28] showing how to set up a working directory

In the Residue field, type ***HOH****∗* and only ∗ for the Atom field (Fig. 7). After selecting the molecules, we will need to delete them. Click on *Edit→Delete→Delete Selected Atoms* (Fig. 8). A warning set will appear to confirm this action. Once we go on, this command cannot the undone. After deleting water molecules, we will

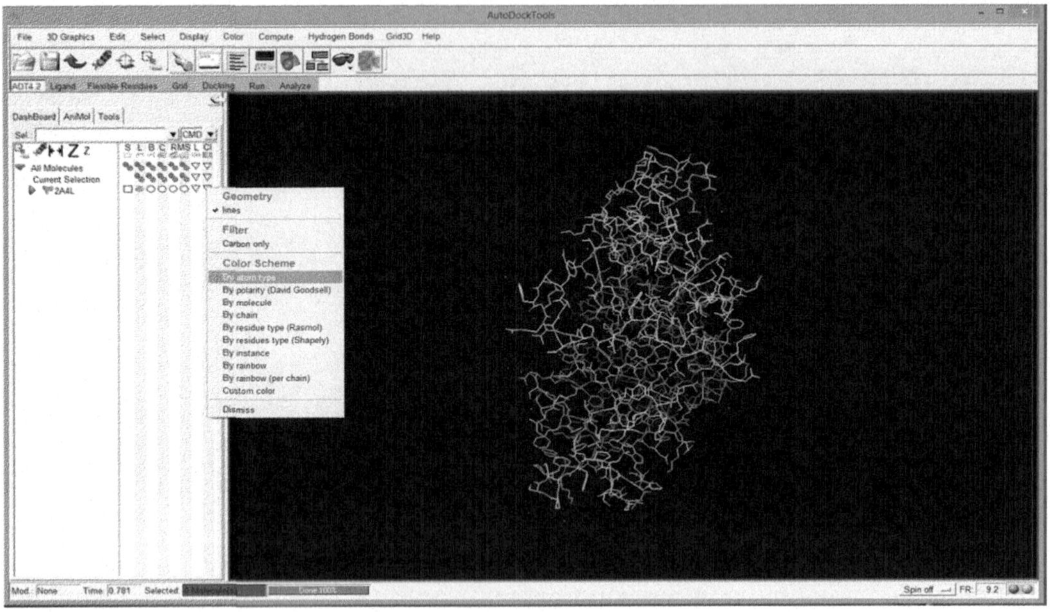

Fig. 5 The main window of AutoDockTools4 [28] showing how to set up the color scheme

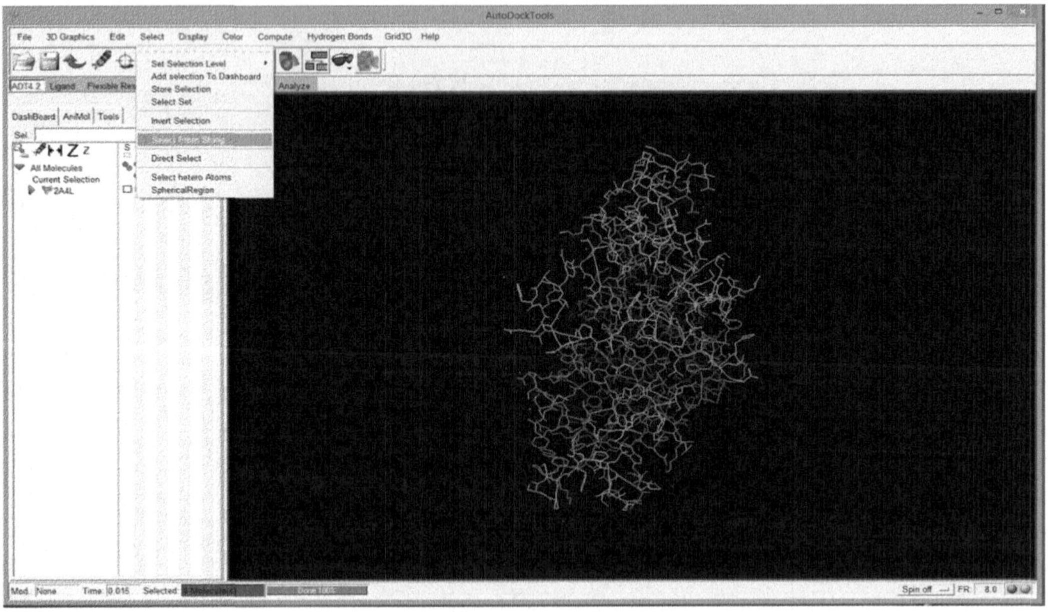

Fig. 6 The main window of AutoDockTools4 [28] showing how to select a string

need to add hydrogens. Click *Edit→Hydrogens→Add*. A window will appear to choose the chemical parts on adding hydrogens. For this tutorial, we will use the default, as shown below. Click on the OK button. We recommend saving this file. Click on *File→Save-*

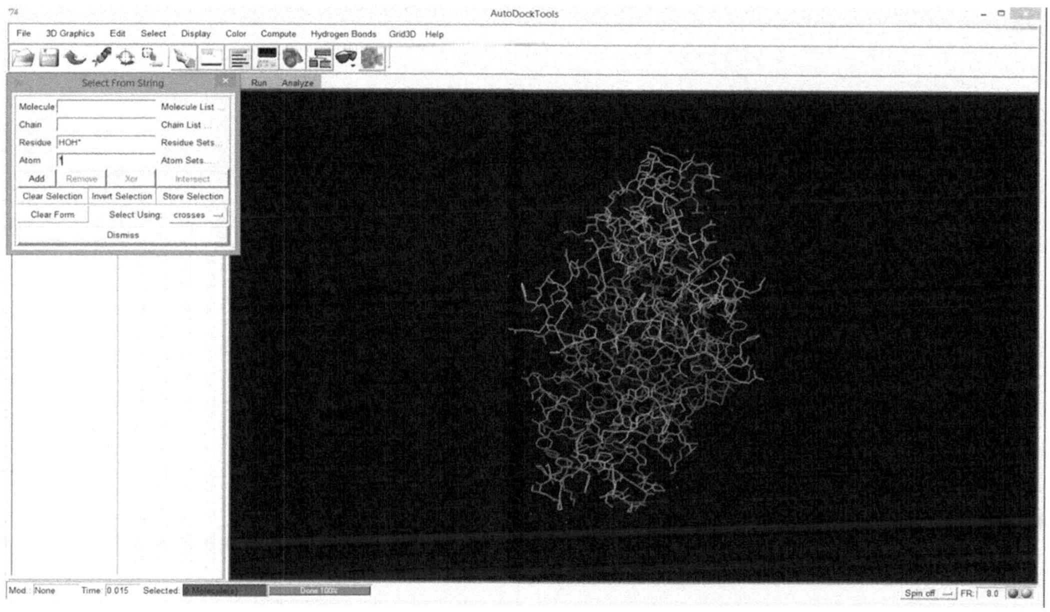

Fig. 7 The main window of AutoDockTools4 [28] showing how to select residue and type of atom

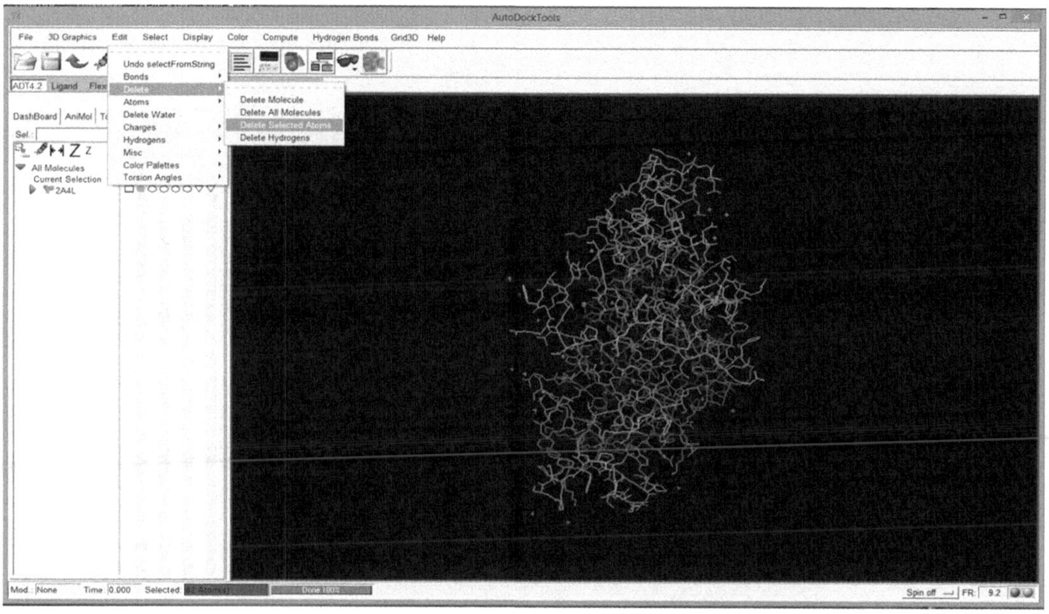

Fig. 8 The main window of AutoDockTools4 [28] showing how to delete selected atoms

Write PDB. Choose the option Atom on PDB Records to be saved. Click Ok, and now we have a protein file prepared (Fig. 9).

The program asks if we want to overwrite the file. Click Yes. Next step is the preparation of the ligand PDB file. Click *Ligand→Input→Open* (Fig. 10). If the format is PDBQT, change

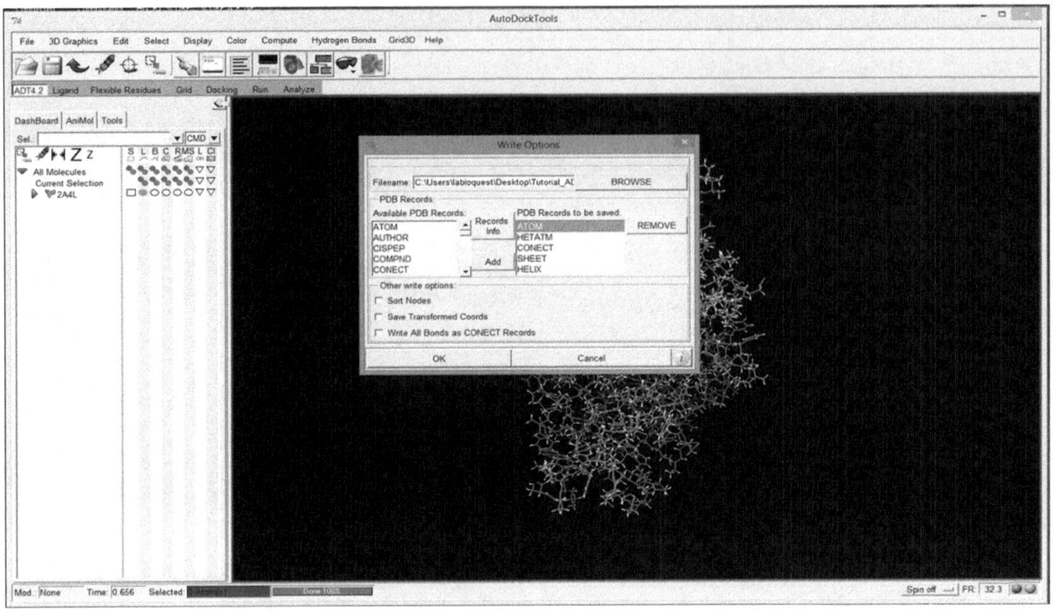

Fig. 9 The main window of AutoDockTools4 [28] showing how to set up the type of atom will be included in the output file

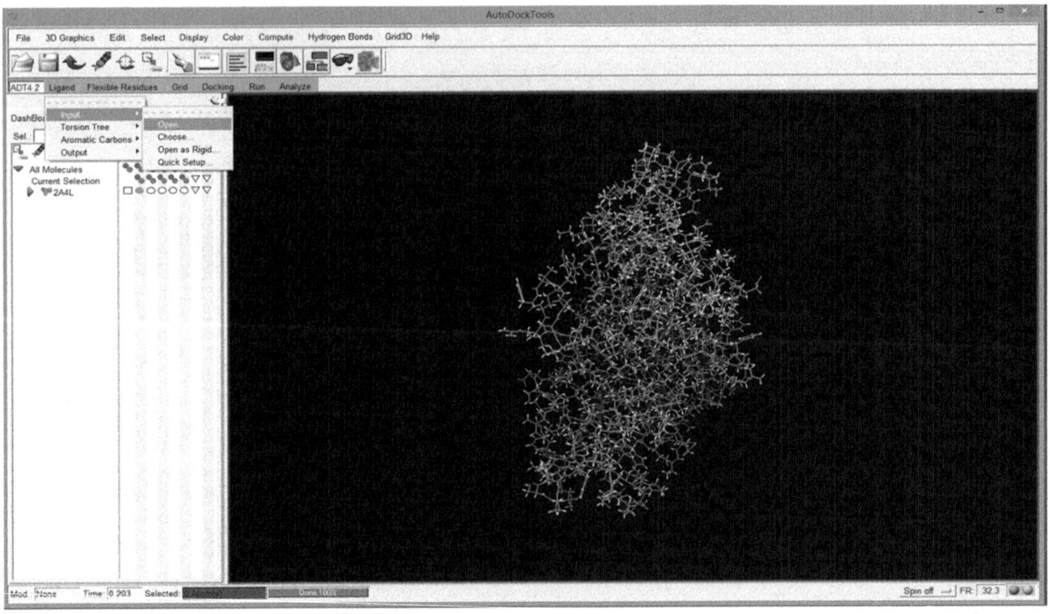

Fig. 10 The main window of AutoDockTools4 [28] showing how to load a ligand file

the option to PDB and select the ligand. Save the file. Once selected, a pop-up window shows information about ligand as the rotatable bonds (Fig. 11). Click on the OK button. The ligand will appear in the center of the window. Note that we hide the protein

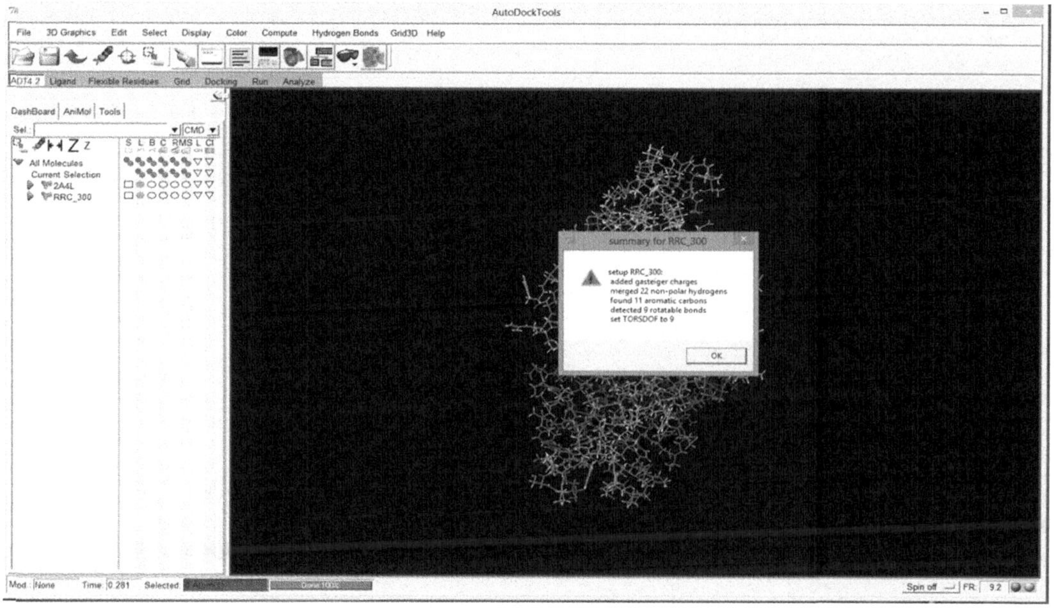

Fig. 11 The main window of AutoDockTools4 [28] showing overall information about a ligand

clicking on the gray rectangle in the dashboard, so, just the ligand is shown on the screen (Fig. 12).

In the next step, the AutoDockTools4 detects the central atom and uses it as the root. The result is a green sphere in the center of ligand. Click on *Ligand→Torsion Tree→Detect Root* (Fig. 13). Now we display the numbers of currently active bounds. Click on *Ligand→Torsion Tree→Choose Torsions.* In the new pop-up window appears the number of rotatable bonds in the ligand, and the maximum allowed by AutoDockTools4 is 32 (Fig. 14). We can also select which rotatable bounds will be considered. For this tutorial, we maintain the default. Click on the Done button.

In the next step, we select the number of torsions for the ligand. Click *on Ligand→Torsion Tree→Set Number of Torsions.* The default of AutoDockTools4 is the fewest atoms, and for the ligand, RRC_300 has 9 active torsions. We keep the default again. Click on the Dismiss button. Now the ligand file is ready, and we must save it in the pdbqt format. Click on *Ligand→Output→Save as PDBQT.* Use the name of the ligand for the pdbqt format file (RRC_300.pdbqt). The protein and ligand files are ready. Note that we unselected the ligand and kept the protein file. First, we will open the protein file and save it in the gpf format. Click on *Grid→Macromolecule→Choose* (Fig. 15).

In the sequence, click on *Select molecule→Dismiss* (Fig. 16). A warning pop-up window appears with some information about hydrogens on the protein. Click on the OK button. We must save protein file as *.pdbqt* too and use the protein's name for the file. In

Fig. 12 The main window of AutoDockTools4 [28] showing the ligand structure

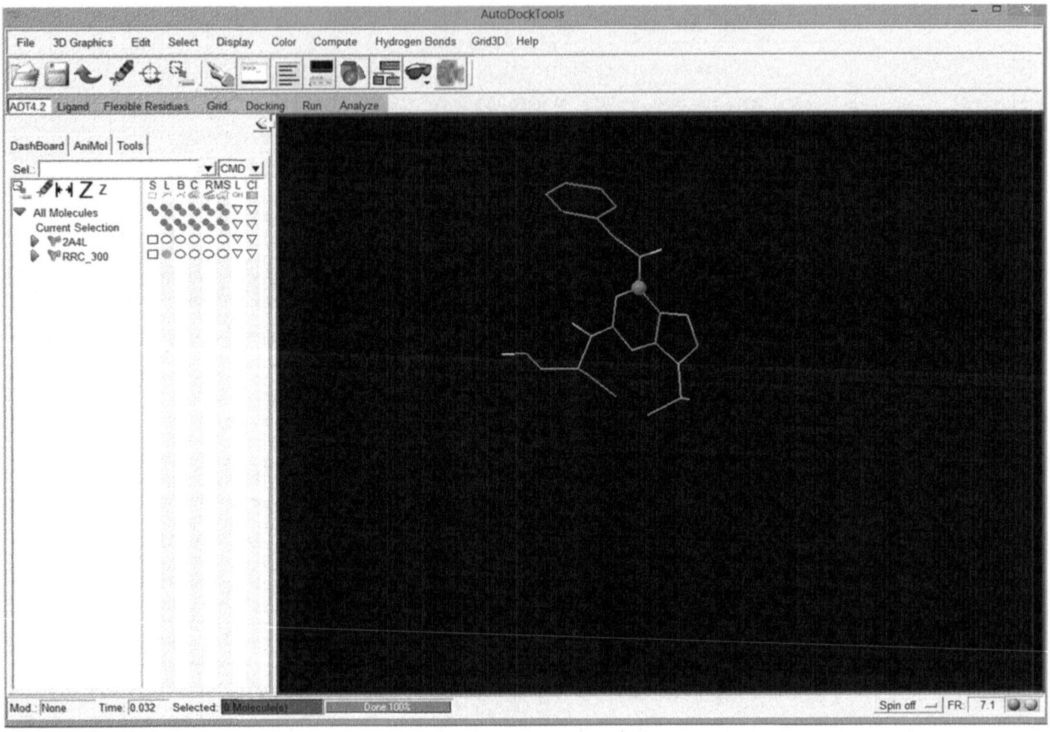

Fig. 13 The main window of AutoDockTools4 [28] showing the ligand structure and the root to identify torsion angles

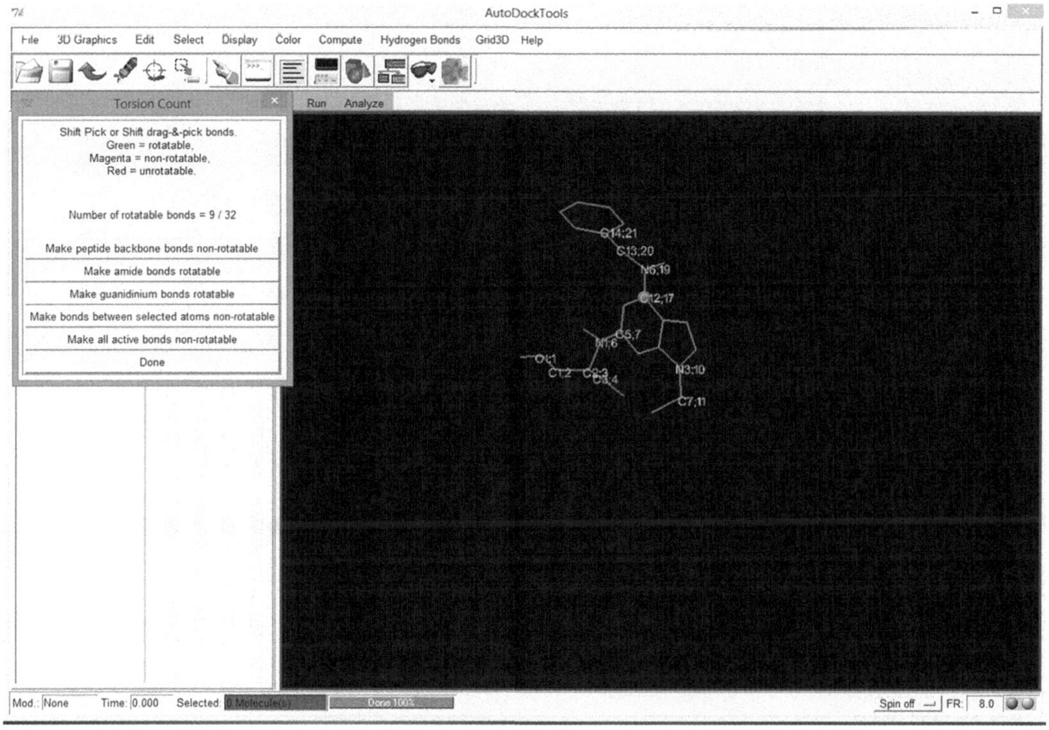

Fig. 14 The main window of AutoDockTools4 [28] showing the torsion angles in the ligand structure

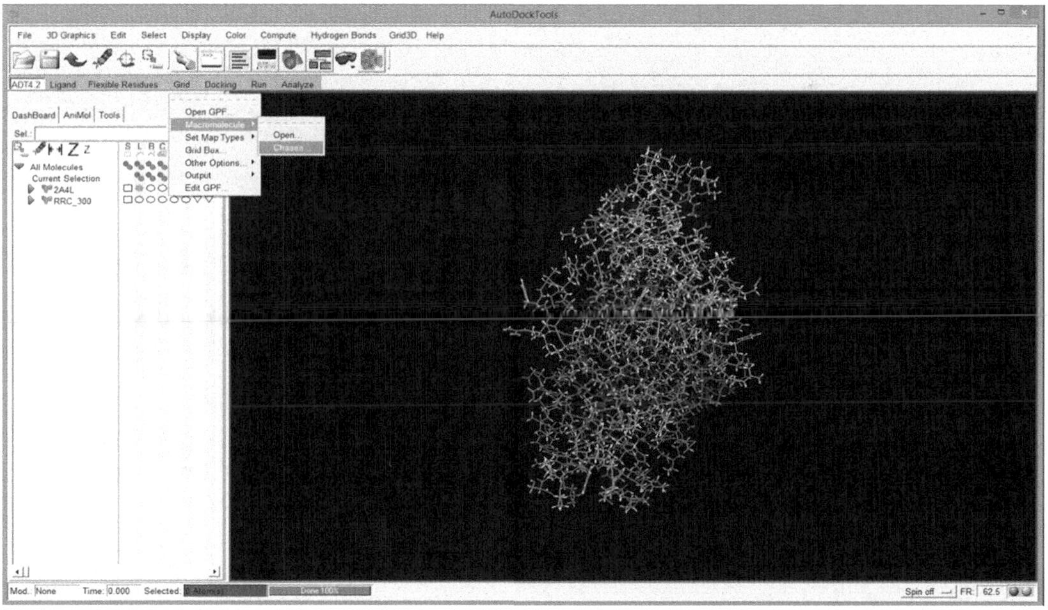

Fig. 15 The main window of AutoDockTools4 [28] showing the protein structure

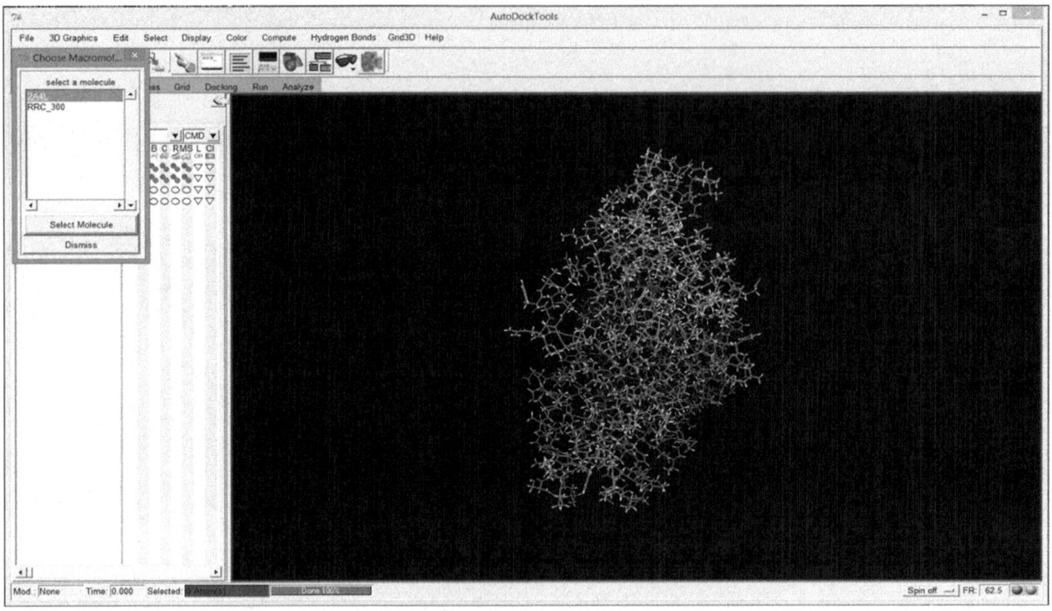

Fig. 16 The main window of AutoDockTools4 [28] showing a pop-up window for selection of the protein structure (2A4L)

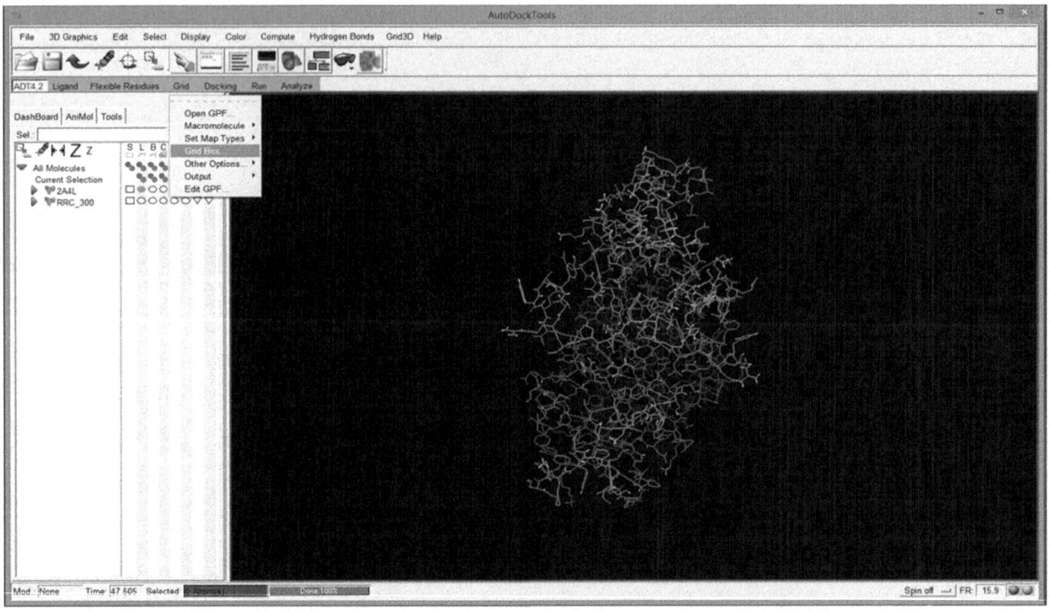

Fig. 17 The main window of AutoDockTools4 [28] showing to set up the grid box

the next step, we choose the location and define a grid box where the docking simulation will take place. Click on *Grid→Grid Box* (Fig. 17).

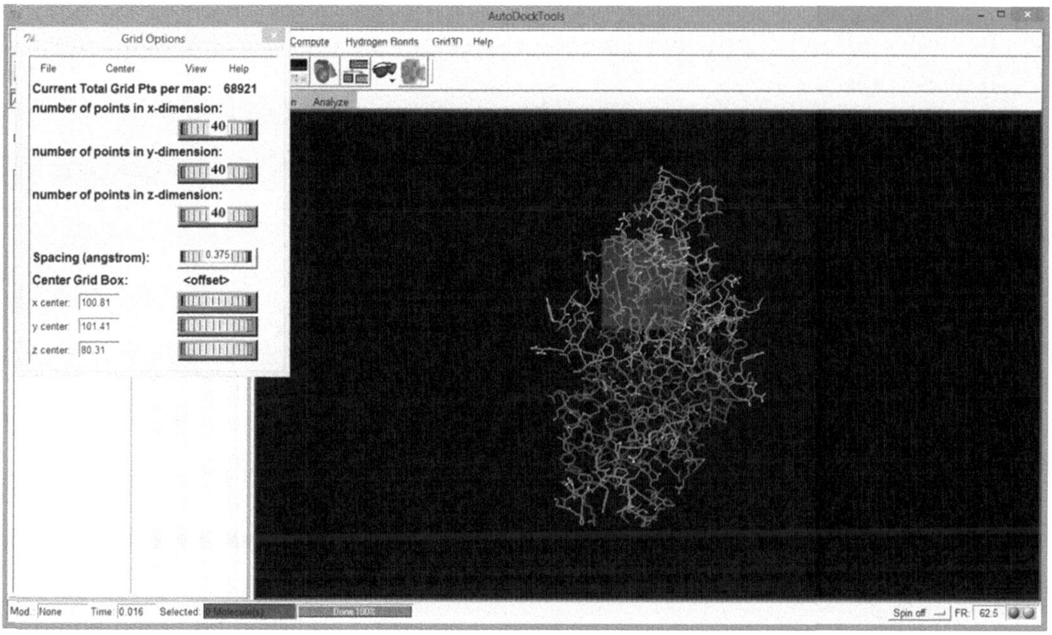

Fig. 18 The main window of AutoDockTools4 [28] showing how to change the grid box center

We keep the default for the number of points in *X*, *Y*, and *Z*, and spacing. Change the center of *X*, *Y*, and *Z* of the ligand (Fig. 18). On the Grid Option pop-up window, click on *File→Close saving current*. AutoDock4 uses a pre-calculated map for docking simulations. We must select the ligand. Click on *Grid→Set map types→Choose Ligand*. Select the ligand (Fig. 19). Click *on Select Ligand→Dismiss*. We must save the protein file as *.gpf*. Click on *Grid→Output→Save GPF*. We keep the name of protein file just taking care of format that must be *.gpf*. Click on the OK button.

To carry out docking simulations, we need to prepare the parameter file for docking. First, we select the protein and ligand files. Click on *Docking→Macromolecule→Set Rigid Filename* (Fig. 20). Then, we select the protein file previously saved as *.pdbqt* and click on the Open button. In the sequence, we click on *Docking→Ligand→Choose*. We choose the ligand (Fig. 21). Click on the Dismiss button after selecting the ligand (RRC_300). Now, we set up the docking parameters for the ligand. We keep the default values (Fig. 22). Click on the Accept button.

In the next step, we define the search algorithm. Click on *Docking→Search Parameters→Genetic Algorithm*, as shown in Fig. 23. We change the Maximum Number of evals to short. For the rest of the fields, we keep the default values and click on the Accept button. In the following, we set up the docking parameter. We click on *Docking→Docking Parameters*. We keep the default values and click on the Accept button.

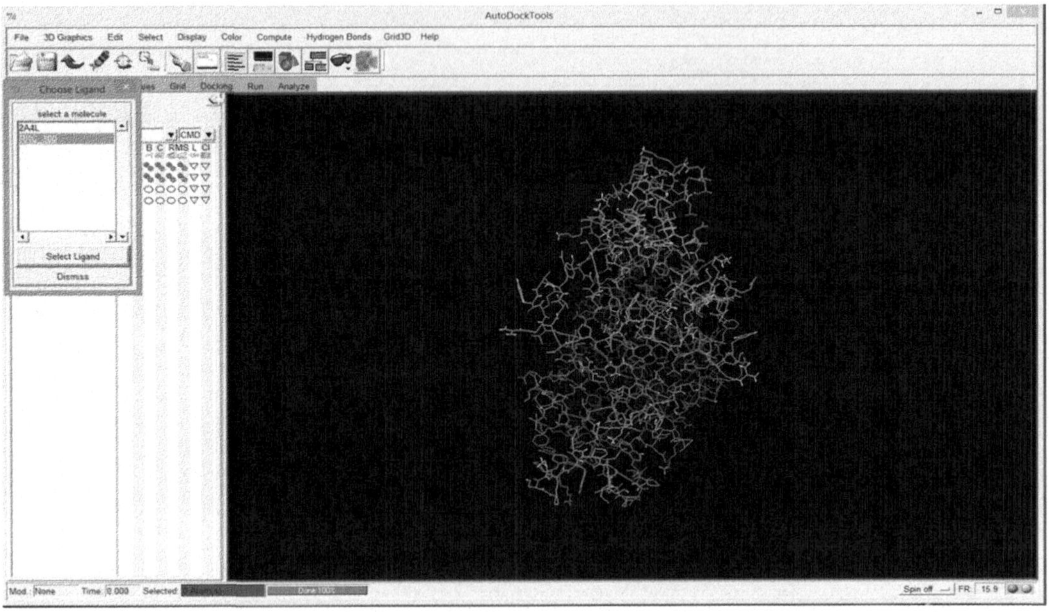

Fig. 19 The main window of AutoDockTools4 [28] showing how to choose the ligand

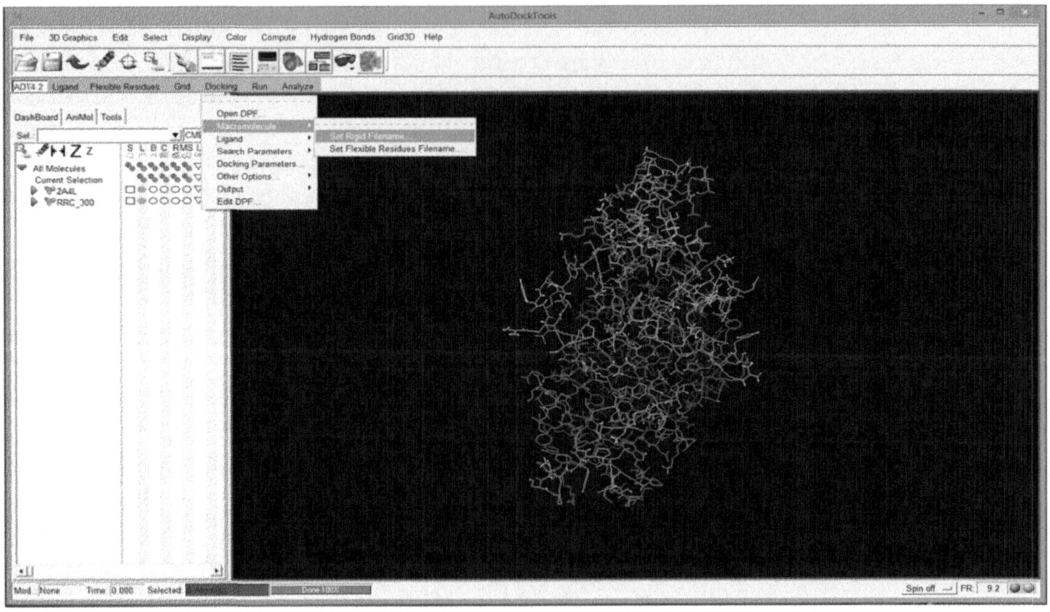

Fig. 20 The main window of AutoDockTools4 [28] showing how to select the protein target

Next, we define the file with the docking parameters and instructions for the Lamarckian genetic algorithm. Click on *Docking→Output→Lamarckian GA (4.2)*. Save the fine as docking.dpf. To run docking simulations with AutoDock4, first we need to run AutoGrid. Click on *Run→AutoGrid* (Fig. 24). In the new pop-up

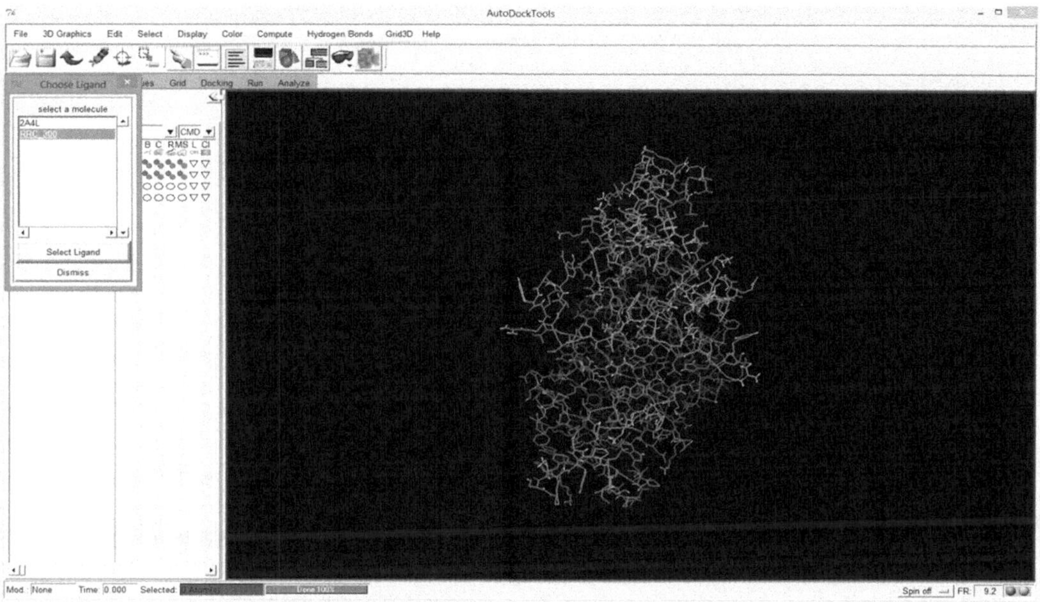

Fig. 21 The main window of AutoDockTools4 [28] showing how to select the ligand

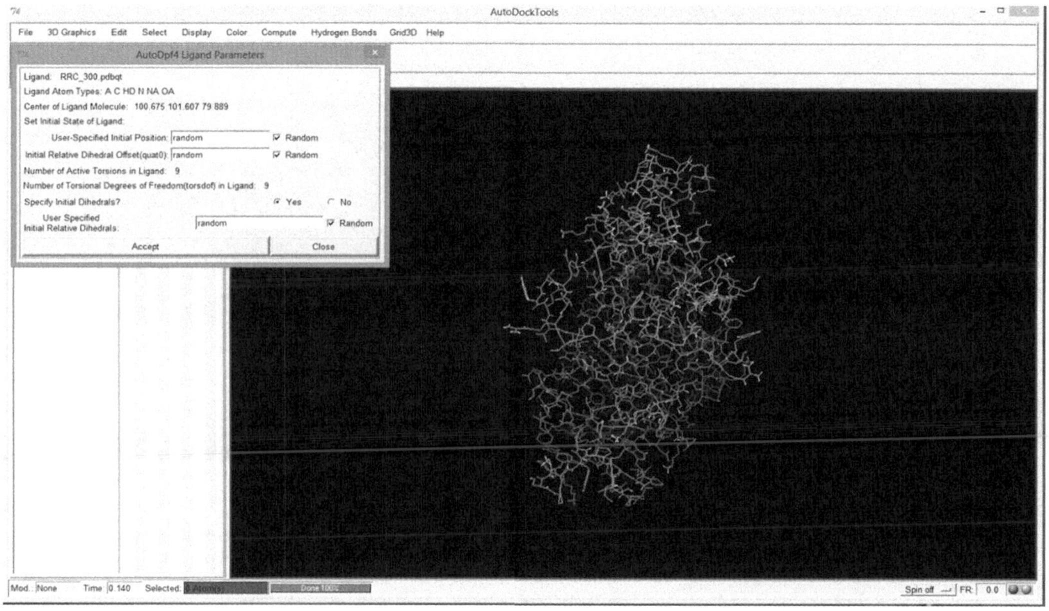

Fig. 22 The main window of AutoDockTools4 [28] showing how to set up ligand parameters

window (Fig. 25), check the Working Directory. If it is not correct, click on the Browse button and locate the tutorial directory. Click on the Launch button to start AutoGrid. The pop-up window below (Fig. 26) appears during the AutoGrid execution and will be closed when it is done. To run AutoDock4, click on

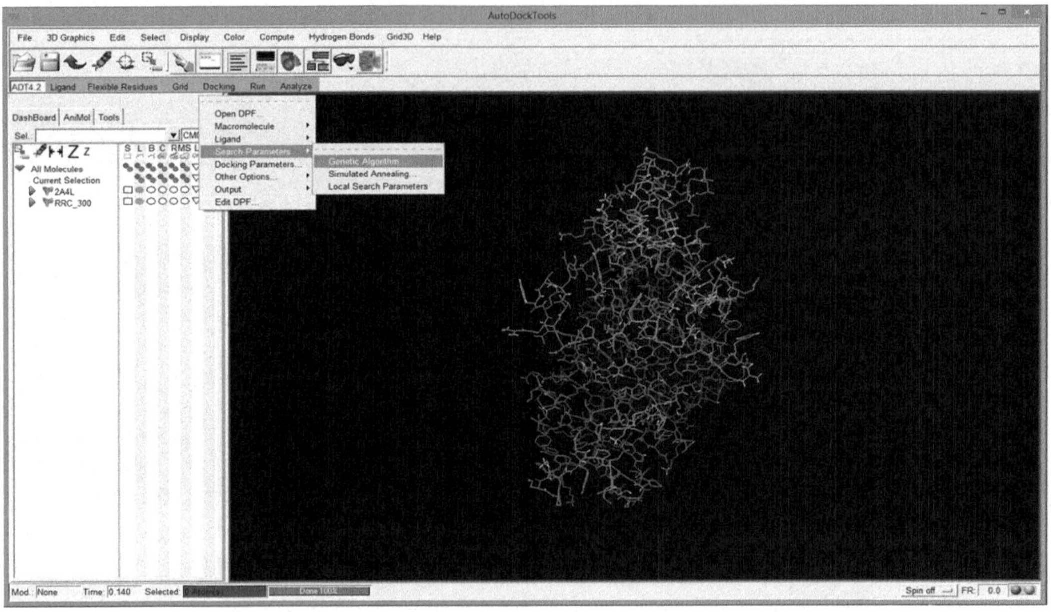

Fig. 23 The main window of AutoDockTools4 [28] showing how to set up the search parameters

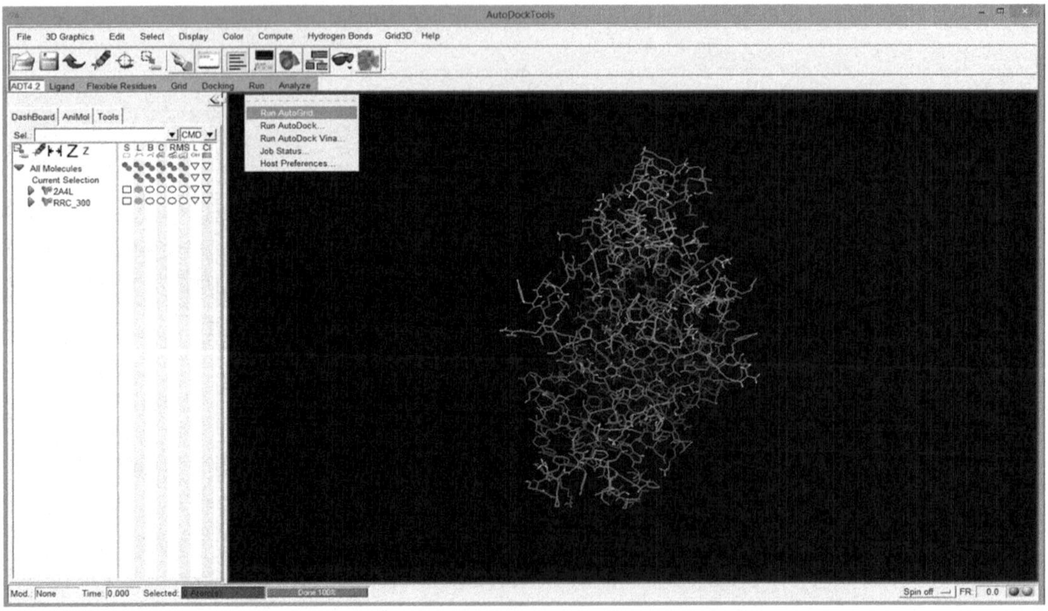

Fig. 24 The main window of AutoDockTools4 [28] showing how to run AutoGrid

Run→AutoDock. In the new pop-up window (Fig. 27), check the Working Directory. For docking simulations on the directory *C:/Users/labioquest/Desktop/Tutorial_ADT/* the Run AutoDock window will be as shown in Fig. 28. To run AutoDock4, click on the Launch button.

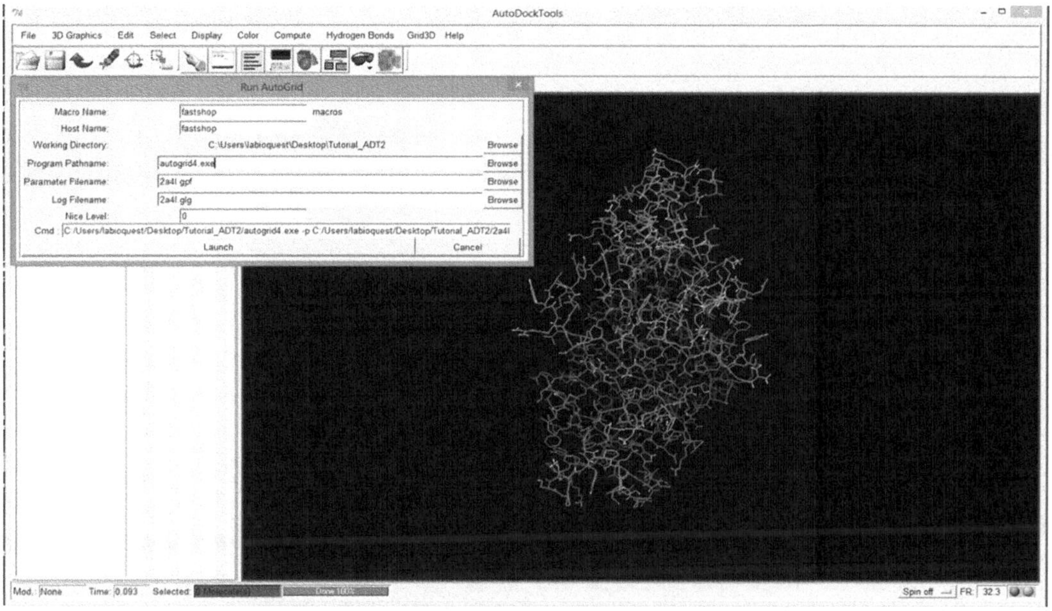

Fig. 25 The main window of AutoDockTools4 [28] showing how to set up AutoGrid parameters

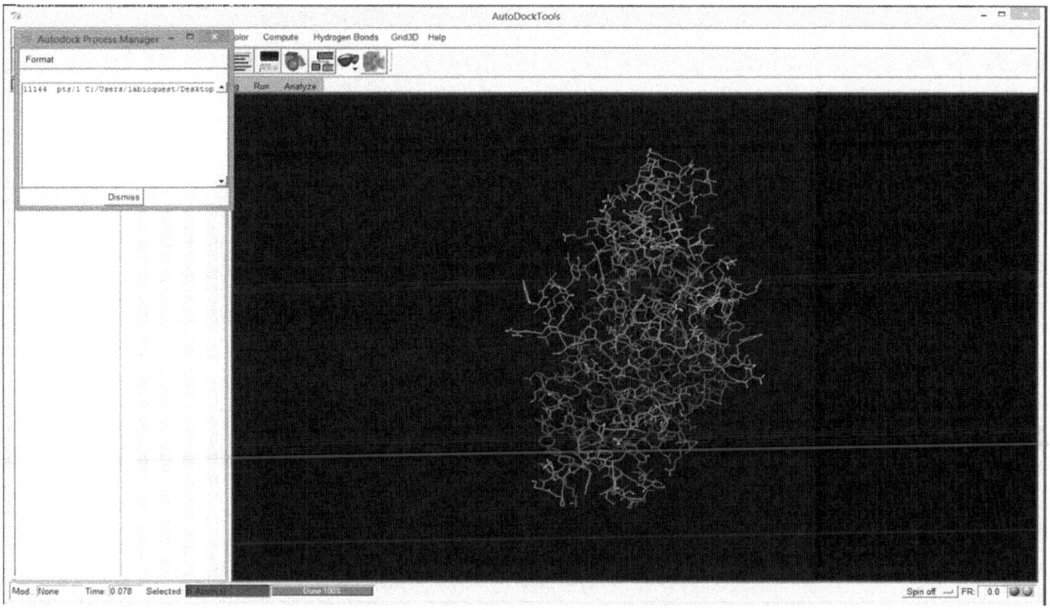

Fig. 26 The main window of AutoDockTools4 [28] showing that AutoGrid is running

We have a new pop-up window (Fig. 29) indicating that AutoDock is running. It can take a few minutes. Once the protein-ligand docking simulation is finished, we can carry out the analysis of the results. Click on *Analyze→Docking→Open*. Then we must choose

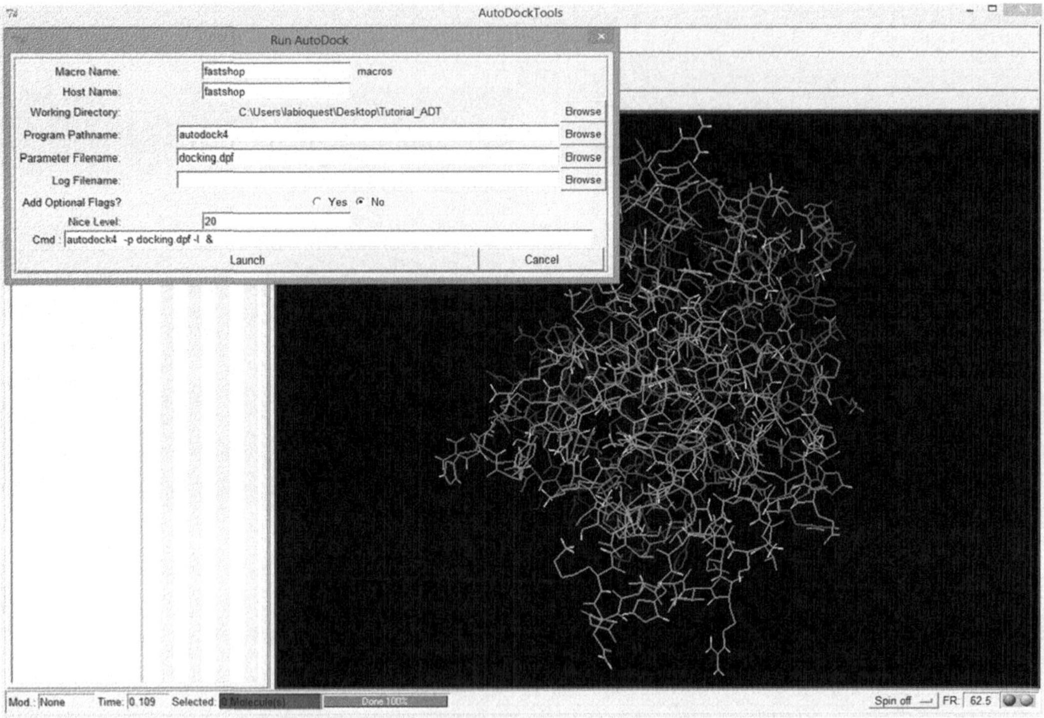

Fig. 27 The main window of AutoDockTools4 [28] showing how to set up AutoDock parameters

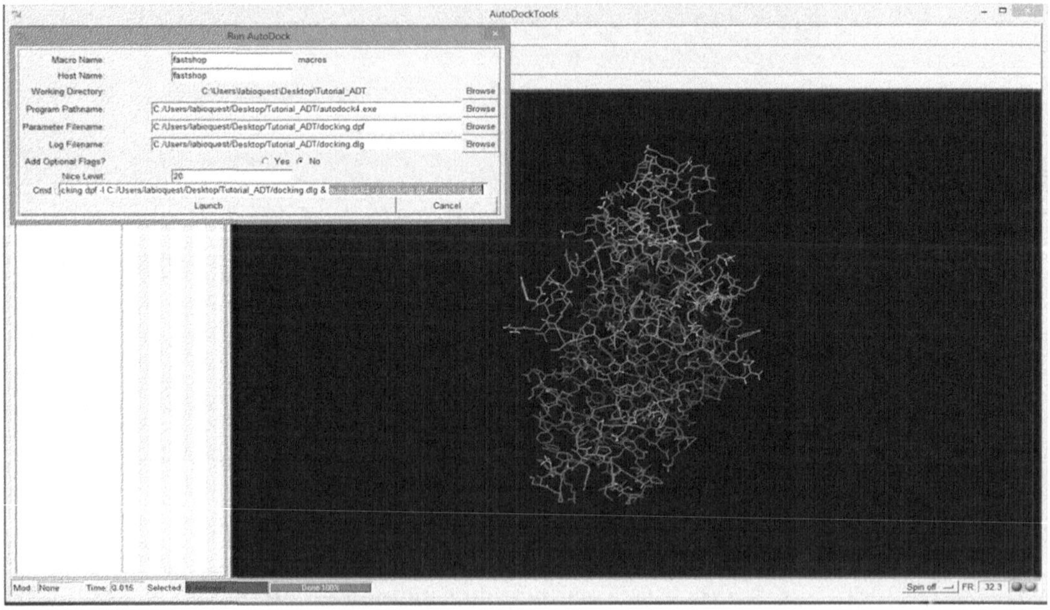

Fig. 28 The main window of AutoDockTools4 [28] showing how to run AutoDock

the *docking.dlg* file. Click on the Open button. We have a new pop-up window, as shown in Fig. 30. Click on the OK button.

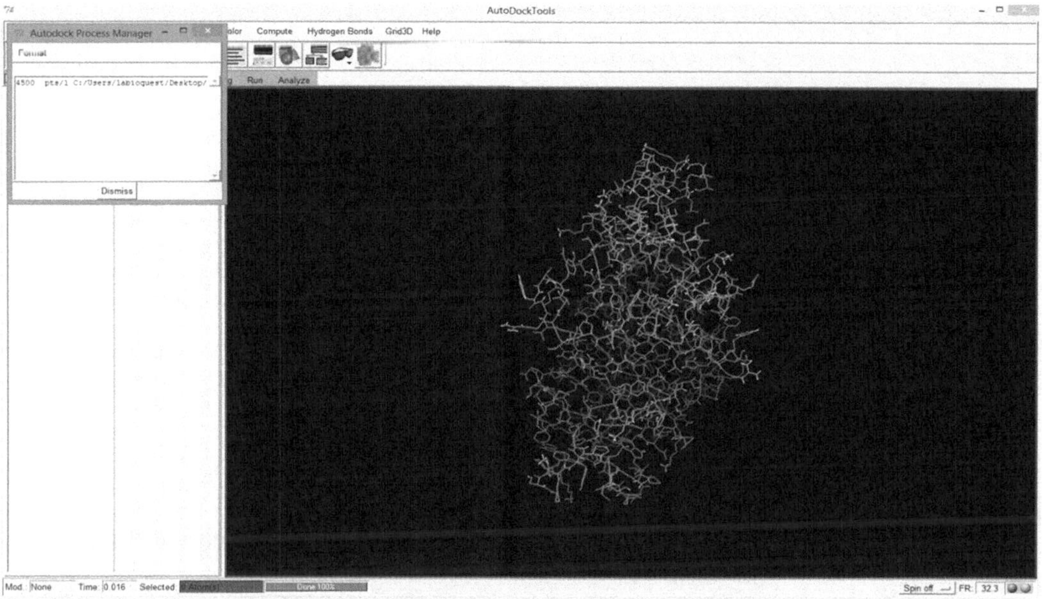

Fig. 29 The main window of AutoDockTools4 [28] showing that AutoDock is running

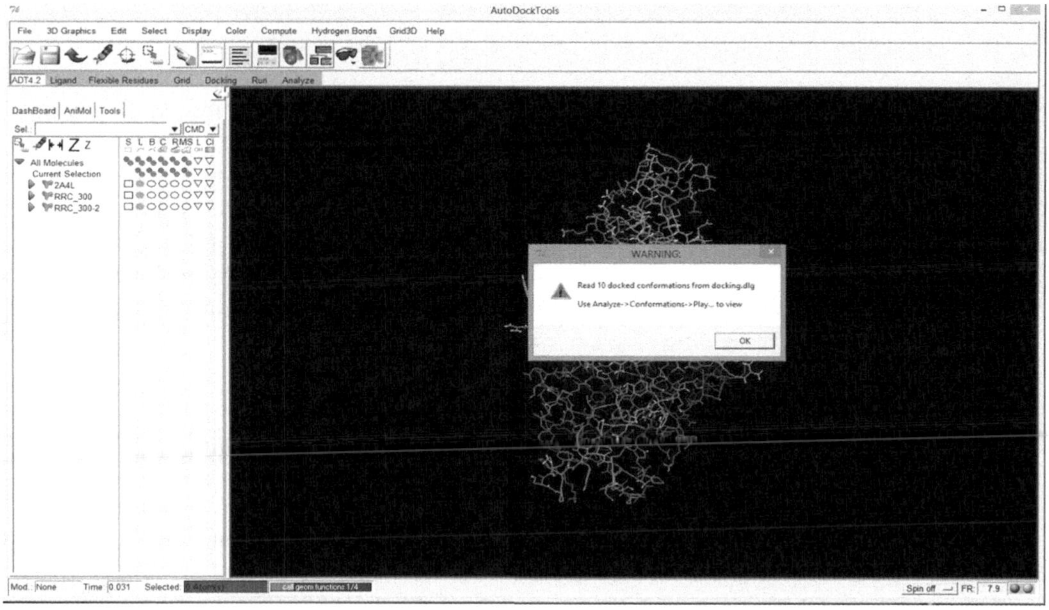

Fig. 30 The main window of AutoDockTools4 [28] indicating that ten poses were generated

Now we have the crystallographic position of the ligand (RRC_300) and the ten poses for this ligand generated during the docking simulation (RRC_300-2). To analyze the poses, we click on *Analyze→Conformation-Load*. Then, we have access to all conformations, as shown in Fig. 31. Choose RRC_300-2 1_1 that

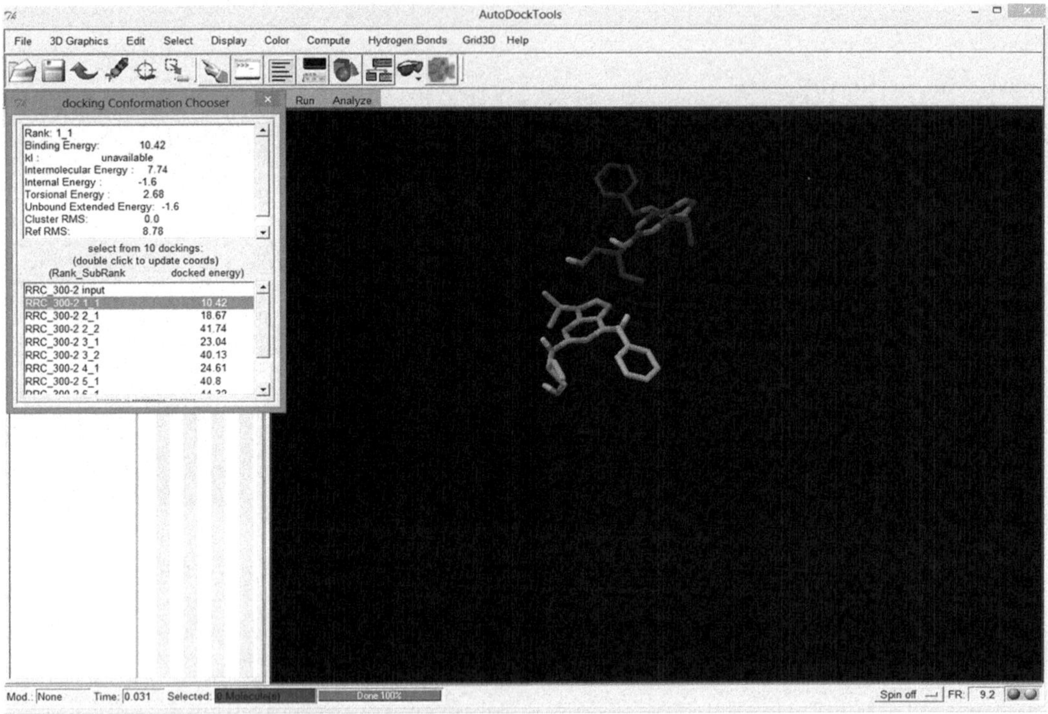

Fig. 31 The main window of AutoDockTools4 [28] showing how to select different poses

has the lowest docked energy. As we can in the docking Conformation Chooser window, this pose has docking root mean squared deviation from the crystallographic position of ligand higher than 2.0 Å. There are several ways that we may use to improve docking results. We may change the docking search algorithm. Besides the Lamarckian algorithm, AutoDock4 has options for search algorithm, the local search, the genetic algorithm, and simulated annealing. We can easily set up new docking simulations as shown in Fig. 23.

We call this type of docking simulation redocking since we recover the crystallographic position of the ligand. Our main goal here is to validate the docking protocol; once checked, we may apply this docking protocol to investigate the binding of small organic molecules to the binding site of the protein; we call this method virtual screen.

4 Availability

All files necessary to run this tutorial are available at https://azevedolab.net/resources/2A4L_AutoDock4_Tutorial.zip.

5 Colophon

We used the program Molegro Virtual Docker [76] to generate Figs. 1 and 2. We created Figs. 3–31 using the program AutoDock-Tools4 [28]. We performed molecular docking simulations described in this chapter using a Desktop PC with 4 GB memory, a 1 TB hard disk, and an Intel® Core® i3-2120 @ 3.30 GHz processor running Windows 8.1.

6 Final Remarks

Protein-ligand docking simulations of biological systems open the possibility to identify the new ligand for a protein target. The program AutoDock4 can play with four search algorithms and. AutoDockTools4 allow us to perform docking simulations using a graphical interface that integrates simulations and analysis of the results in one computational tool. Programs such as SAnDReS can run directly AutoDock4 in an integrated computational environment and analyze the docking results, generating a statistical analysis of docking results such as docking RMSD and docking accuracy. Furthermore, SAnDReS can make use of the concept of scoring function space [40] and generate a scoring function targeted to the biological system of interest, which may improve docking accuracy and create a scoring function with superior predictive performance.

Acknowledgments

This work was supported by grants from CNPq (Brazil) (308883/2014-4) and CAPES. GB-F acknowledges support from PUCRS/BPA fellowship. WFA is a senior researcher for CNPq (Brazil) (Process Number: 308883/2014-4).

References

1. Kuntz ID, Blaney JM, Oatley SJ, Langridge R, Ferrin TE (1982) A geometric approach to macromolecule-ligand interactions. J Mol Biol 161:269–288
2. DesJarlais RL, Dixon JS (1994) A shape- and chemistry-based docking method and its use in the design of HIV-1 protease inhibitors. J Comput Aided Mol Des 8:231–242
3. Lunney EA, Hagen SE, Domagala JM, Humblet C, Kosinski J, Tait BD et al (1994) A novel nonpeptide HIV-1 protease inhibitor: elucidation of the binding mode and its application in the design of related analogs. J Med Chem 37:2664–2677
4. Vaillancourt M, Cohen E, Sauvé G (1995) Characterization of dynamic state inhibitors of HIV-1 protease. J Enzyme Inhib 9:217–233
5. Gehlhaar DK, Verkhivker GM, Rejto PA, Sherman CJ, Fogel DB, Fogel LJ et al (1995) Molecular recognition of the inhibitor AG-1343 by HIV-1 protease: conformationally flexible docking by evolutionary programming. Chem Biol 2:317–324
6. King BL, Vajda S, DeLisi C (1996) Empirical free energy as a target function in docking and

design: application to HIV-1 protease inhibitors. FEBS Lett 384:87–91
7. Wang S, Milne GW, Yan X, Posey IJ, Nicklaus MC, Graham L et al (1996) Discovery of novel, non-peptide HIV-1 protease inhibitors by pharmacophore searching. J Med Chem 39:2047–2054
8. Muegge I, Bergner A, Kriegl JM (2017) Computer-aided drug design at Boehringer Ingelheim. J Comput Aided Mol Des 31:275–285
9. Hillisch A, Heinrich N, Wild H (2015) Computational chemistry in the pharmaceutical industry: from childhood to adolescence. ChemMedChem 10:1958–1962
10. Potemkin V, Grishina M (2018) Grid-based technologies for in silico screening and drug design. Curr Med Chem 25:3526–3537
11. Elmessaoudi-Idrissi M, Blondel A, Kettani A, Windisch MP, Benjelloun S, Ezzikouri S (2018) Virtual screening in hepatitis B virus drug discovery: current Stateof- the-art and future perspectives. Curr Med Chem 25:2709–2721
12. Vilar S, Sobarzo-Sanchez E, Santana L, Uriarte E (2017) Molecular docking and drug discovery in β-adrenergic receptors. Curr Med Chem 24:4340–4359
13. Krüger J, Thiel P, Merelli I, Grunzke R, Gesing S (2016) Portals and web-based resources for virtual screening. Curr Drug Targets 17:1649–1660
14. Abdolmaleki A, Ghasemi JB, Ghasemi F (2017) Computer aided drug design for multi-target drug design: SAR/QSAR, molecular docking and pharmacophore methods. Curr Drug Targets 18:556–575
15. de Azevedo WF (2016) Opinion paper: targeting multiple Cyclin-dependent kinases (CDKs): a new strategy for molecular docking studies. Curr Drug Targets 17:2
16. Scotti L, Mendonca Junior FJ, Ishiki HM, Ribeiro FF, Singla RK, Barbosa Filho JM et al (2017) Docking studies for multi-target drugs. Curr Drug Targets 18:592–604
17. Azevedo LS, Moraes FP, Xavier MM, Pantoja EO, Villavicencio B, Finck JA et al (2012) Recent Progress of molecular docking simulations applied to development of drugs. Curr Bioinf 7:352–365
18. Heberlé G, de Azevedo WF Jr (2011) Bio-inspired algorithms applied to molecular docking simulations. Curr Med Chem 18:1339–1352
19. De Azevedo WF Jr (2010) Structure-based virtual screening. Curr Drug Targets 11:261–263
20. De Azevedo WF Jr (2010) MolDock applied to structure-based virtual screening. Curr Drug Targets 11:327–334
21. Dias R, de Azevedo WF Jr (2008) Molecular docking algorithms. Curr Drug Targets 9:1040–1047
22. de Azevedo WF Jr, Dias R (2008) Computational methods for calculation of ligand-binding affinity. Curr Drug Targets 9:1031–1039
23. de Azevedo WF Jr (2008) Protein-drug interactions. Curr Drug Targets 9:1030
24. Breda A, Basso LA, Santos DS, de Azevedo WF Jr (2008) Virtual screening of drugs: score functions, docking, and drug design. Curr Comput Aided Drug Des 4:265–272
25. Goodsell DS, Olson AJ (1990) Automated docking of substrates to proteins by simulated annealing. Proteins 8:195–202
26. Morris GM, Goodsell DS, Huey R, Olson AJ (1996) Distributed automated docking of flexible ligands to proteins: parallel applications of AutoDock 2.4. J Comput Aided Mol Des 10:293–304
27. Morris GM, Goodsell DS, Halliday RS, Huey R, Hart WE, Belew RK et al (1998) Automated docking using a Lamarckian genetic algorithm and empirical binding free energy function. J Comput Chem 19:1639–1662
28. Morris GM, Huey R, Lindstrom W, Sanner MF, Belew RK, Goodsell DS et al (2009) AutoDock4 and AutoDockTools4: automated docking with selective receptor flexibility. J Comput Chem 30:2785–2791
29. Trott O, Olson AJ (2010) AutoDock Vina: improving the speed and accuracy of docking with a new scoring function, efficient optimization, and multithreading. J Comput Chem 31:455–461
30. Yang JM, Chen CC (2004) GEMDOCK: a generic evolutionary method for molecular docking. Proteins 55:288–304
31. Yang JM, Shen TW (2005) A pharmacophore-based evolutionary approach for screening selective estrogen receptor modulators. Proteins 59:205–220
32. Bitencourt-Ferreira G, de Azevedo WF Jr (2018) Development of a machine-learning model to predict Gibbs free energy of binding for protein-ligand complexes. Biophys Chem 240:63–69
33. de Ávila MB, de Azevedo WF Jr (2018) Development of machine learning models to predict inhibition of 3-dehydroquinate dehydratase. Chem Biol Drug Des 92:1468–1474

34. Russo S, de Azevedo WF (2018) Advances in the understanding of the cannabinoid receptor 1—focusing on the inverse agonists interactions. Curr Med Chem. https://doi.org/10.2174/0929867325666180417165247
35. Amaral MEA, Nery LR, Leite CE, de Azevedo Junior WF, Campos MM (2018) Pre-clinical effects of metformin and aspirin on the cell lines of different breast cancer subtypes. Investig New Drugs 36:782–796
36. Levin NMB, Pintro VO, Bitencourt-Ferreira G, Mattos BB, Silvério AC, de Azevedo WF Jr (2018) Development of CDK-targeted scoring functions for prediction of binding affinity. Biophys Chem 235:1–8
37. Freitas PG, Elias TC, Pinto IA, Costa LT, de Carvalho PVSD, Omote DQ et al (2018) Computational approach to the discovery of phytochemical molecules with therapeutic potential targets to the PKCZ protein. Lett Drug Des Discovery 15:488–499
38. Pintro VO, Azevedo WF (2017) Optimized virtual screening workflow. Towards target-based polynomial scoring functions for HIV-1 protease. Comb Chem High Throughput Screen 20:820–827
39. de Ávila MB, Xavier MM, Pintro VO, de Azevedo WF (2017) Supervised machine learning techniques to predict binding affinity. A study for cyclin-dependent kinase 2. Biochem Biophys Res Commun 494:305–310
40. Heck GS, Pintro VO, Pereira RR, de Ávila MB, Levin NMB, de Azevedo WF (2017) Supervised machine learning methods applied to predict ligand-binding affinity. Curr Med Chem 24:2459–2470
41. Levin NM, Pintro VO, de Ávila MB, de Mattos BB, De Azevedo WF Jr (2017) Understanding the structural basis for inhibition of Cyclin-dependent kinases. New pieces in the molecular puzzle. Curr Drug Targets 18:1104–1111
42. Teles CB, Moreira-Dill LS, Silva Ade A, Facundo VA, de Azevedo WF Jr, da Silva LH et al (2015) A Lupane-triterpene isolated from *Combretum leprosum* Mart. Fruit extracts that interferes with the intracellular development of *Leishmania* (L.) *amazonensis in vitro*. BMC Complement Altern Med 15:165
43. Coracini JD, de Azevedo WF Jr (2014) Shikimate kinase, a protein target for drug design. Curr Med Chem 21:592–604
44. Moraes FP, de Azevedo WF Jr (2012) Targeting imidazoline site on monoamine oxidase B through molecular docking simulations. J Mol Model 18:3877–3886
45. Soares MB, Silva CV, Bastos TM, Guimarães ET, Figueira CP, Smirlis D et al (2012) Anti-*Trypanosoma cruzi* activity of nicotinamide. Acta Trop 12:224–229
46. Vianna CP, de Azevedo WF Jr (2012) Identification of new potential *Mycobacterium tuberculosis* shikimate kinase inhibitors through molecular docking simulations. J Mol Model 18:755–764
47. Sá MS, de Menezes MN, Krettli AU, Ribeiro IM, Tomassini TC, Ribeiro dos Santos R et al (2011) Antimalarial activity of physalins B, D, F, and G. J Nat Prod 74:2269–2272
48. Canduri F, Perez PC, Caceres RA, de Azevedo WF Jr (2008) CDK9 a potential target for drug development. Med Chem 4:210–218
49. Xavier MM, Heck GS, de Avila MB, Levin NM, Pintro VO, Carvalho NL et al (2016) SAnDReS a computational tool for statistical analysis of docking results and development of scoring functions. Comb Chem High Throughput Screen 19:801–812
50. Kuntz ID (1992) Structure-based strategies for drug design and discovery. Science 257:1078–1082
51. Shoichet BK, Stroud RM, Santi DV, Kuntz ID, Perry KM (1993) Structure-based discovery of inhibitors of thymidylate synthase. Science 259:1445–1450
52. Rutenber E, Fauman EB, Keenan RJ, Fong S, Furth PS, Ortiz de Montellano PR et al (1993) Structure of a non-peptide inhibitor complexed with HIV-1 protease. Developing a cycle of structure-based drug design. J Biol Chem 268:15343–15346
53. Zheng Q, Kyle DJ (1996) Computational screening of combinatorial libraries. Bioorg Med Chem 4:631–638
54. Gschwend DA, Good AC, Kuntz ID (1996) Molecular docking towards drug discovery. J Mol Recognit 9:175–186
55. Finn PW (1996) Computer-based screening of compound databases for the identification of novel leads. Drug Discov Today 1:363–370
56. Horvath D (1997) A virtual screening approach applied to the search for trypanothione reductase inhibitors. J Med Chem 40:2412–2423
57. Toyoda T, Brobey RKB, Sano G, Horii T, Tomioka N, Itai A (1997) Lead discovery of inhibitors of the dihydrofolate reductase domain of *Plasmodium falciparum* dihydrofolate reductase-thymidylate synthase. Biochem Biophys Res Commun 235:515–519

58. Olson AJ, Goodsell DS (1998) Automated docking and the search for HIV protease inhibitors. SAR QSAR Environ Res 8:273–285
59. Walters WP, Stahl MT, Murcko MA (1998) Virtual screening—an overview. Drug Discov Today 3:160–178
60. Toney JH, Fitzgerald PMD, Groversharma N, Olson SH, May WJ, Sundelof JG et al (1998) Antibiotic sensitization using biphenyl Tetrazoles as potent inhibitors of *Bacteroides fragilis* Metallo-BetaLactamase. Chem Biol 5:185–196
61. Berman HM, Westbrook J, Feng Z, Gilliland G, Bhat TN, Weissig H et al (2000) The Protein Data Bank. Nucleic Acids Res 28:235–242
62. Berman HM, Battistuz T, Bhat TN, Bluhm WF, Bourne PE, Burkhardt K et al (2002) The Protein Data Bank. Acta Crystallogr D Biol Crystallogr 58:899–907
63. Westbrook J, Fen Z, Chen L, Yang H, Berman HM (2003) The protein data Bank and structural genomics. Nucleic Acids Res 31:489–491
64. Hu L, Benson ML, Smith RD, Lerner MG, Carlson HA (2005) Binding MOAD (mother of all databases). Proteins 60:333–340
65. Liu T, Lin Y, Wen X, Jorrisen RN, Gilson MK (2007) BindingDB: a web-accessible database of experimentally determined protein-ligand binding affinities. Nucleic Acids Res 35:198–201
66. Wang R, Fang X, Lu Y, Wang S (2004) The PDBbind database: collection of binding affinities for protein-ligand complexes with known three-dimensional structures. J Med Chem 47:2977–2980
67. Volkart PA, Bitencourt-Ferreira G, Souto AA, de Azevedo WF (2019) Cyclin-dependent kinase 2 in cellular senescence and cancer. A structural and functional review. Curr Drug Targets 20(7):716–726. https://doi.org/10.2174/1389450120666181204165344
68. de Ávila MB, Bitencourt-Ferreira G, de Azevedo WF Jr (2018) Structural basis for inhibition of Enoyl-[acyl carrier protein] Reductase (InhA) from *Mycobacterium tuberculosis*. Curr Med Chem. https://doi.org/10.2174/0929867326666181203125229
69. De Azevedo WF, Leclerc S, Meijer L, Havlicek L, Strnad M, Kim SH (1997) Inhibition of cyclin-dependent kinases by purine analogues: crystal structure of human cdk2 complexed with roscovitine. Eur J Biochem 243:518–526
70. Morgan DO (1995) Principles of CDK regulation. Nature 374:131–134
71. Murray AW (1994) Cyclin-dependent kinases: regulators of the cell cycle and more. Chem Biol 1:191–195
72. Kim SH, Schulze-Gahmen U, Brandsen J, de Azevedo Júnior WF (1996) Structural basis for chemical inhibition of CDK2. Prog Cell Cycle Res 2:137–145
73. De Azevedo WF Jr, Mueller-Dieckmann HJ, Schulze-Gahmen U, Worland PJ, Sausville E, Kim SH (1996) Structural basis for specificity and potency of a flavonoid inhibitor of human CDK2, a cell cycle kinase. Proc Natl Acad Sci U S A 93:2735–2740
74. Canduri F, de Azevedo WF Jr (2005) Structural basis for interaction of inhibitors with Cyclin-dependent kinase 2. Curr Comput Aided Drug Des 1:53–64
75. Krystof V, Cankar P, Frysová I, Slouka J, Kontopidis G, Dzubák P et al (2006) 4-arylazo-3,5-diamino-1H-pyrazole CDK inhibitors: SAR study, crystal structure in complex with CDK2, selectivity, and cellular effects. J Med Chem 49:6500–6509
76. Thomsen R, Christensen MH (2006) MolDock: a new technique for high-accuracy molecular docking. J Med Chem 49:3315–3321
77. De Bondt HL, Rosenblatt J, Jancarik J, Jones HD, Morgan DO, Kim SH (1993) Crystal structure of cyclin-dependent kinase 2. Nature 363:595–602
78. Schulze-Gahmen U, De Bondt HL, Kim SH (1996) High-resolution crystal structures of human cyclin-dependent kinase 2 with and without ATP: bound waters and natural ligand as guides for inhibitor design. J Med Chem 39:4540–4546

Chapter 10

Molegro Virtual Docker for Docking

Gabriela Bitencourt-Ferreira and Walter Filgueira de Azevedo Jr.

Abstract

Molegro Virtual Docker is a protein-ligand docking simulation program that allows us to carry out docking simulations in a fully integrated computational package. MVD has been successfully applied to hundreds of different proteins, with docking performance similar to other docking programs such as AutoDock4 and AutoDock Vina. The program MVD has four search algorithms and four native scoring functions. Considering that we may have water molecules or not in the docking simulations, we have a total of 32 docking protocols. The integration of the programs SAnDReS (https://github.com/azevedolab/sandres) and MVD opens the possibility to carry out a detailed statistical analysis of docking results, which adds to the native capabilities of the program MVD. In this chapter, we describe a tutorial to carry out docking simulations with MVD and how to perform a statistical analysis of the docking results with the program SAnDReS. To illustrate the integration of both programs, we describe the redocking simulation focused the cyclin-dependent kinase 2 in complex with a competitive inhibitor.

Key words Molegro Virtual Docker, MolDock, Molecular docking, Cyclin-dependent kinase 2, Drug design, Protein-ligand interactions

1 Introduction

Computational determination of the position of a potential drug in the binding site of a protein target is of pivotal importance for computer-aided drug design [1–10]. Such computational approaches have two significant benefits. First, they are cheaper than in vitro tests of binding affinity of a ligand for a protein target [11–15]. Through computational simulations, we may test the interaction of a potential ligand and assess its ligand binding affinity, generating an outcome that indicates whether or not a new molecule can interact with a protein target [16–20]. Such relative easiness in the assessment of the interaction of a potential ligand with a target allows us to simulate thousands or even millions of molecules available in free databases such as ZINC [21, 22] and DrugBank [23, 24].

Second, the computational approaches for the assessment of protein-ligand interaction add plasticity to the analysis of this

Walter Filgueira de Azevedo Jr. (ed.), *Docking Screens for Drug Discovery*, Methods in Molecular Biology, vol. 2053,
https://doi.org/10.1007/978-1-4939-9752-7_10, © Springer Science+Business Media, LLC, part of Springer Nature 2019

biological system since we may vary the sophistication of the computational representation of the biological model relevant for the drug discovery. For instance, let us consider a library of small molecules extracted from natural products [21, 22]. We can perform computational tests of the interaction of these molecules with a protein target through protein-ligand docking simulations [25–27]. We could carry out these simulations without taking into account the participation of water molecules and seeing the protein system as a rigid body, not allowing flexibility to the macromolecule structure. Such simplification is somehow unrealistic since we know that proteins are flexible entities [28] and that water molecules would be present in the biological environment.

Nevertheless, such distance from the real biological system is acceptable, since it speeds up the computer simulations and therefore might generate reliable results. Considering the best results obtained in virtual screenings [29], we could carry out most demanding computational simulations on the top-ranked ligands. It is common to combine protein-ligand docking with molecular dynamics simulations [30].

In summary, the computational approach of docking has increasing participation on drug design and development, being relatively fast when compared with molecular dynamics simulations [31]; in the first try, the molecular docking identifies new potential ligands for a given protein target.

It is customary with a desktop computer to perform protein-ligand docking simulations of thousands of potential ligands against a protein target. Also, the availability of modern docking programs such as AutoDock [32–35], AutoDock Vina [36], GemDock [37, 38], and Molegro Virtual Docker (MVD) [39–41], to mention a few, made possible to research laboratories even with a modest budget to perform robust protein-ligand docking projects [42–51].

Our goal in this chapter is to describe a detailed tutorial to carry protein-ligand docking simulations with the program MVD. The first version of MVD was released in 2006, and it has been applied to a wide range of protein systems [39–41]. MVD has a graphical user interface that allows the users to perform all tasks related to docking simulations from this window. In this tutorial, we initially describe how to run molecular docking simulations with MVD. We focus our discussion on the docking against cyclin-dependent kinase 2. We chose CDK2 due to its importance for the development of anticancer drugs and the abundance of experimental data for this protein target.

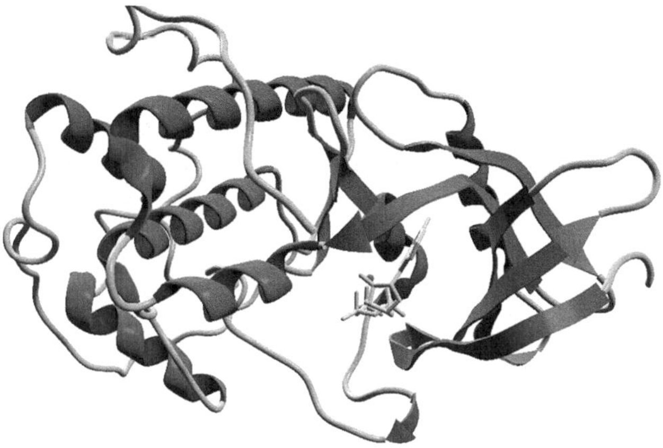

Fig. 1 Crystallographic structure of human CDK2 in complex with ATP. This figure was generated using Molegro Virtual Docker (MVD) [39]. PDB access code: 1HCK [63]

2 Biological System

In this tutorial, we show how to perform protein-ligand docking simulations of cyclin-dependent kinase 2 (CDK2) (EC 2.7.11.22) with MVD [39]. This critical protein kinase has been intensively studied as a target for the development of anticancer drugs [52–61]. The first crystallographic structure of CDK2 was determined in 1993 at the University of California, Berkeley [62]. Analysis of the CDK2 structure indicated a typical bilobal architecture of serine/threonine protein kinases (EC 2.7.11.1). Figure 1 shows the structure of CDK2 in complex with ATP (PDB access code: 1HCK) [63]. Analysis of the structure of CDK2 shows that the N-terminal domain comprises a distorted beta sheet and a short alpha helix. A helix bundle forms the C-terminal. The two lobes of the CDK2 structure allow the binding of the ATP molecule, as shown in Fig. 1.

3 Overview

The MVD version 6 brings the possibility of applying four search algorithms: MolDock Optimizer (MDO) (based on differential evolution [64]), MolDock Simplex Evolution (MDSE) (a modified algorithm based on Nelder-Mead local search algorithm [65]), Iterated Simplex (IS) (based on Nelder-Mead algorithm), and iterated simplex with ant colony optimization (ISACO) [66]. Also, it is possible to choose four scoring functions in each search algorithm. Furthermore, it is possible to consider the presence of water molecules in the system. In summary, we may say that we have 32 combinations of the search algorithms, scoring

Table 1
Combinations of search algorithms and scoring functions available in MVD [39]

Search algorithm	Scoring function	Presence of water
Iterated Simplex (Ant Colony Optimization)	MolDock Score	Yes/no
Iterated Simplex (Ant Colony Optimization)	MolDock Score [GRID]	Yes/no
Iterated Simplex (Ant Colony Optimization)	Plants Score	Yes/no
Iterated Simplex (Ant Colony Optimization)	Plants Score [GRID]	Yes/no
Iterated Simplex	MolDock Score	Yes/no
Iterated Simplex	MolDock Score [GRID]	Yes/no
Iterated Simplex	Plants Score	Yes/no
Iterated Simplex	Plants Score [GRID]	Yes/no
MolDock (Simplex Evolution) (SE)	MolDock Score	Yes/no
MolDock (Simplex Evolution) (SE)	MolDock Score [GRID]	Yes/no
MolDock (Simplex Evolution) (SE)	Plants Score	Yes/no
MolDock (Simplex Evolution) (SE)	Plants Score [GRID]	Yes/no
MolDock Optimizer	MolDock Score	Yes/no
MolDock Optimizer	MolDock Score [GRID]	Yes/no
MolDock Optimizer	Plants Score	Yes/no
MolDock Optimizer	Plants Score [GRID]	Yes/no

functions, and the presence of water molecules in the simulation, as highlighted in Table 1.

We consider that we have MVD installed on your computer. We used version 6.0 for Windows; it is mostly the same for Mac OS X and Linux versions. Here, we will recover the atomic coordinates of the ligand roscovitine bound to the structure of CDK2 (PDB access code: 2A4L) [67]. This simulation is of pivotal importance to validate a docking protocol. Our goal is to recover the ligand position and to assess the quality of the simulation.

4 Tutorial for Redocking

In the flowchart below (Fig. 2), we see the main steps to redock a ligand in the structure of a protein using MVD [39]. Initially, we need to have a PDB file of a protein complexed with a ligand. The MVD can read this file, and we have to identify the active ligand, which is the ligand we submit to the docking simulation. For instance, if we are interested in an enzyme–inhibitor complex, the inhibitor is our active ligand. Keep in mind that we are concerned

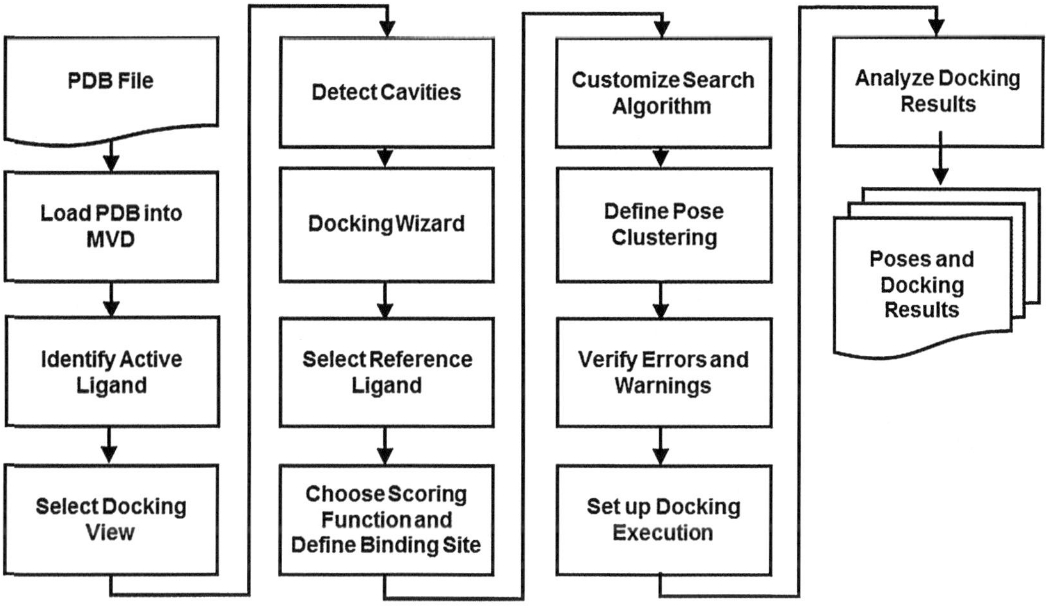

Fig. 2 Flowchart showing the main steps to carry out redocking with MVD [39]

here with protein-ligand docking simulations, where the ligand is a small molecule. In the sequence, we select the option Docking View, to have a difference in the color scheme between the crystallographic position and the computer-generated position (pose). MVD allows the user to identify the cavities present the structure of the protein, where we expect to find our ligand bound. In preparation for docking simulations, we indicate the active ligand for the MVD and define the scoring function and define the binding site. Following, we choose the search algorithm. Then we may start the docking simulation. In the end, we can evaluate the docking results, using the docking root mean square deviation as a criterion for the quality of the simulation. We expect that pose is close to the crystallographic position of the ligand, with RMSD < 2.0 Å. In the MVD, we have on the left side of the screen Workspace Explorer (Fig. 3), where all atomic coordinates files are highlighted once loaded. On the right, we have the graphical screen (black background).

To load a PDB file, click on *File→Import Molecule*. We could also drag and drop the PDB file on the graphical screen. Go to the directory, where we have the PDB file (2A4L) [67]. Click on the PDB file and open it. We have a pop-up window that shows the PDB file content. Click on the Preparation button. Change Assign All Below to Always. Then, click on the Import button. We have the molecule on the graphical screen (Fig. 4). On the left, we have the Workspace Explorer, click on the checkboxes on the left of each content to turn on and off the visualization of the specific part of

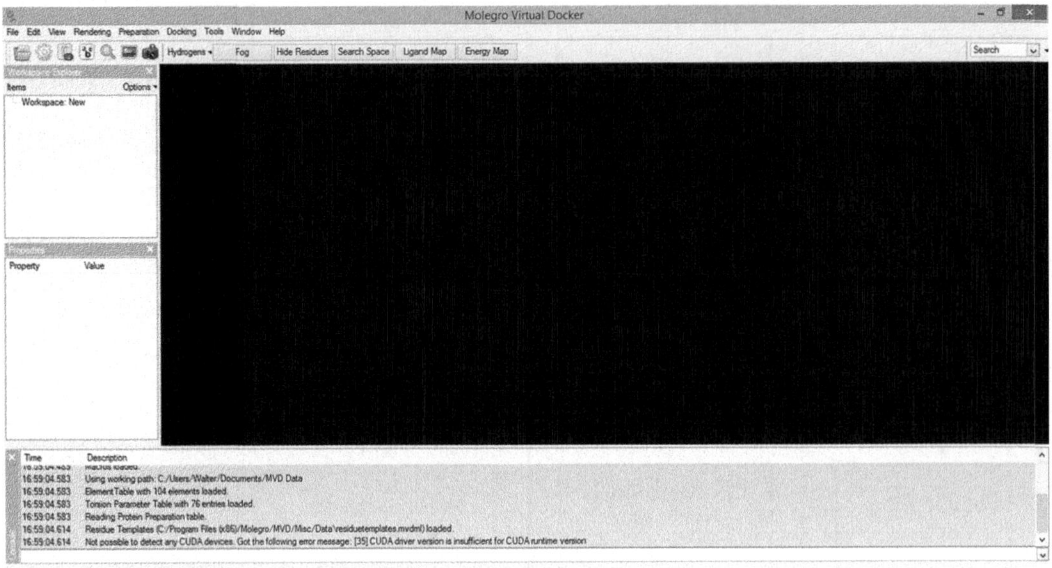

Fig. 3 The main windows of the MVD [39]

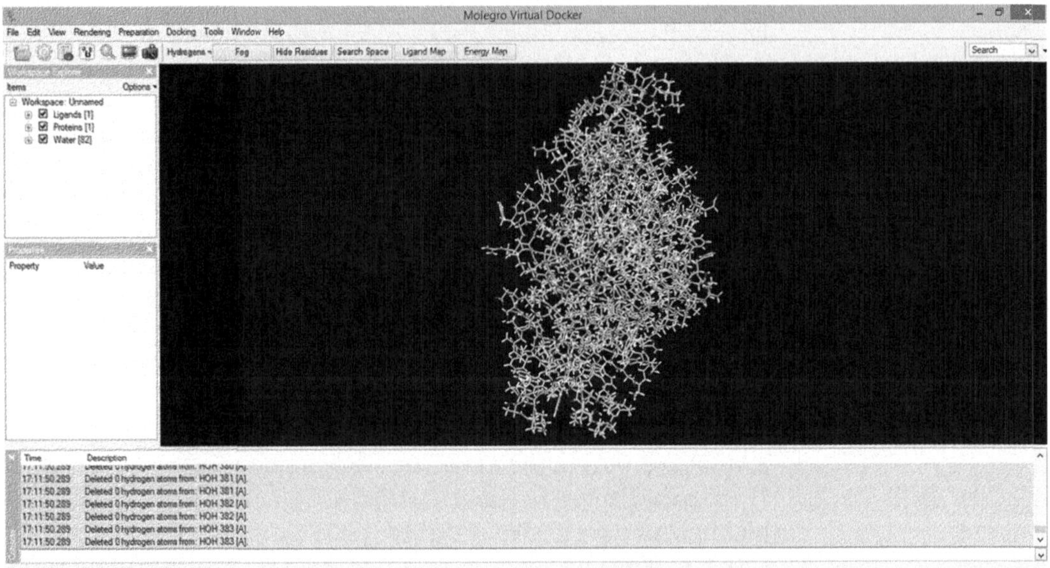

Fig. 4 Protein structure in the graphical screen represented with lines

the molecule. We click on the "+" signal to expand the tree. For instance, click on "+" signal on the left of Water. Now we have the 82 water molecules, which were found in the crystallographic structure 2A4L. Click on the "−" to return to the previous situation. Let us check the ligands. Click on "+" on the left of the Ligand. We have as an active ligand RRC_300[A], which is the code for roscovitine. We were lucky—MVD does not always find the right active ligand. It is better to check at the PDB site for information about the ligands. The active ligand is the one to be redocked. Keep in mind this information (RRC_300[A]), we will need it later to specify the reference ligand on the MVD Docking Wizard.

We could have hundreds of ligands, but only one is the active ligand. If we are not sure which ligand to choose to redock, we should get additional information about the molecular system we are about to simulate. For instance, for enzymes, most likely the active ligand is the inhibitor, with binding-affinity information. Now, we set up MVD to represent a pose and crystallographic position of the ligand with different colors. Click on *View→Docking View*. MVD change the representation of the ligand, from Ball and Stick to Stick. Uncheck the boxes for water and protein that we will have a clear view of the ligand. Figure 5 shows the crystallographic position of the ligand. Bring back protein and water molecules.

Following, we have to click on *Preparation→Detect Cavities* to detect potential binding sites in the protein structure. Click on the OK button. Then, we have potential binding pockets shown in the graphical screen (Fig. 6). The active ligand should be at least partially inserted in the predicted cavity. We are ready to redock. Click on *Docking→Docking Wizard*. On the new pop-up window, we have to choose the reference ligand (RRC_300[A]). Now we have to select the scoring function. MVD has four options, shown here (Fig. 7). We choose MolDock Score and click all options for

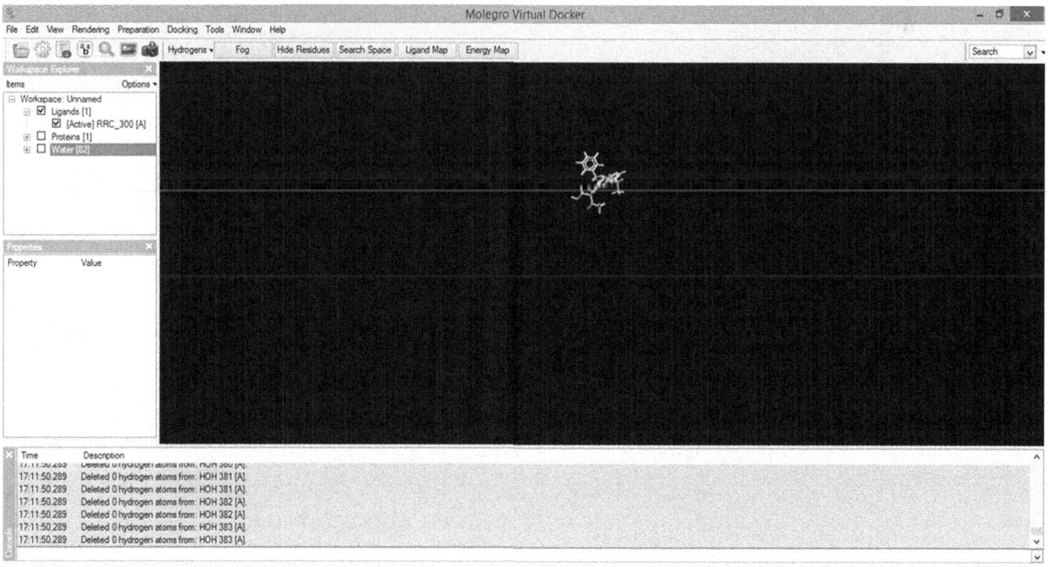

Fig. 5 Crystallographic position of the active ligand in the graphical screen

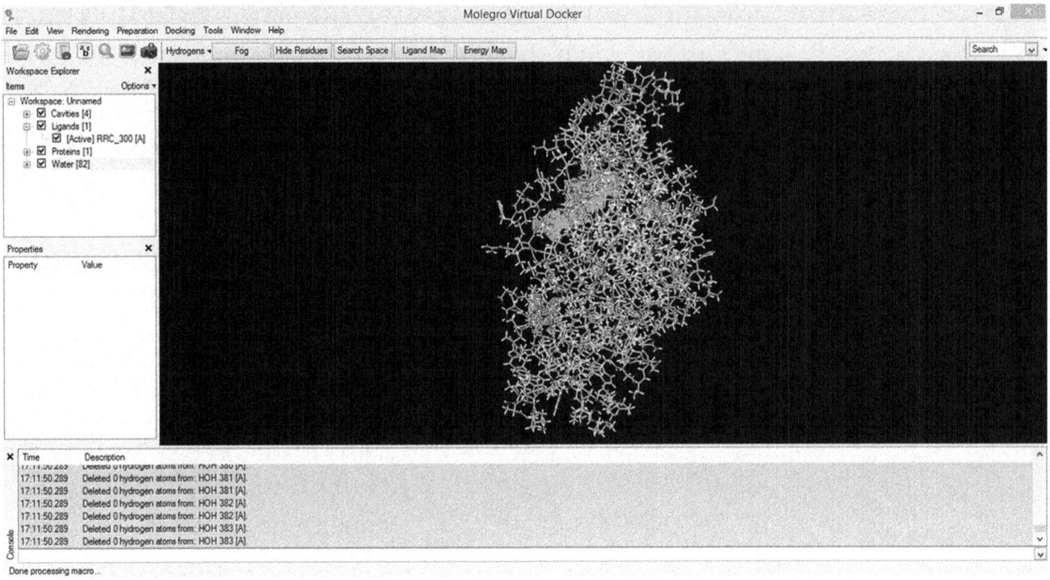

Fig. 6 Cavities found in the protein structure

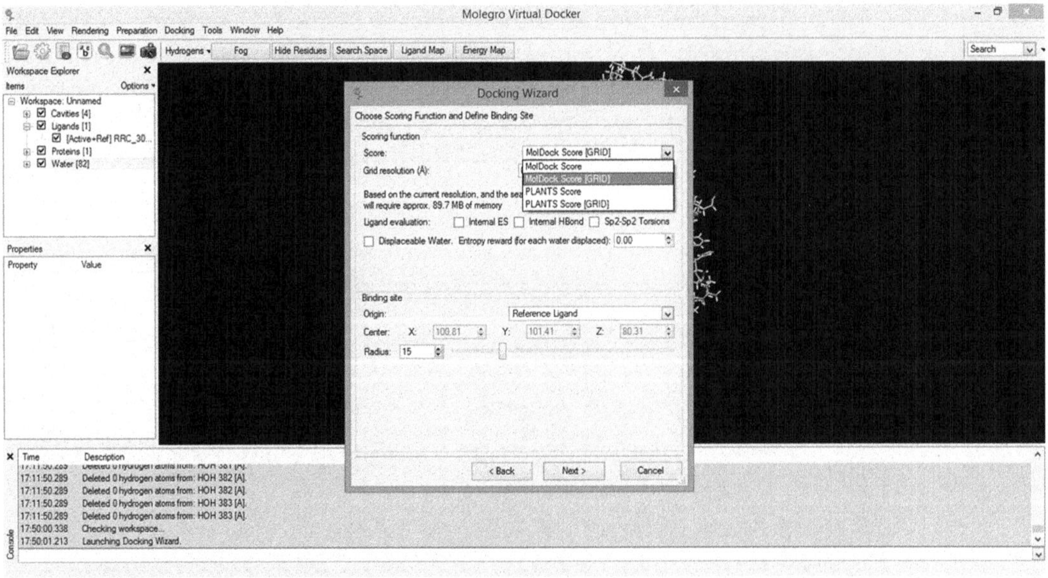

Fig. 7 Pop-up window with the scoring functions available in the MVD

ligand evaluation, except for Displaceable Water, as shown below. Then we click on the Next button.

Now we have to choose the search algorithm, we also have four options, as shown in Fig. 8. We choose MolDock Optimizer, which is an implementation of the differential evolution algorithm [64]. We change the number of runs to 20; the rest we leave the

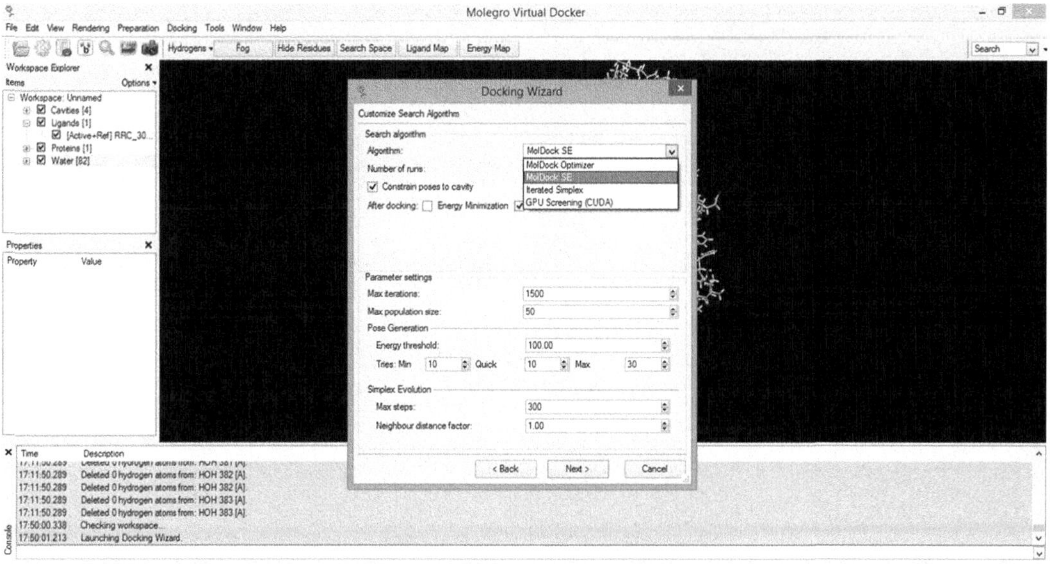

Fig. 8 Pop-up window with the search algorithms available in the MVD

default values. Then we click on the Next button. Then, we change the Max number of poses returned to 50 and modify the other parameters as shown below. Click on the Next button. We get the message "No Errors and Warnings. Click on the Next button. Now we have to choose where the results will be stored. Choose the same folder where we have our PDB file. Click on... to select the directory. Move to the directory where we want to leave the docking results and click on the OK button. Then, change pose format to mol2. We are ready to dock. Click on the Start button (Fig. 9). The docking simulations should start, and we can follow the docking process as illustrated in Fig. 10.

Once finished, we get the message "Finished". The docking results are in the *DockingResults.mvdresults* file. We can check this file dragging the Icon Results (Fig. 11) to the graphical screen. In the pop-up window with the docking results, we can sort the poses by the scoring function values, as shown in Fig. 12. As we can see, ranking poses with MolDock Score generated poses with RMSD < 2.0 Å. We could visualize clicking on the box on the left and then clicking on the OK button. The result is shown in Fig. 13. As we can see, we have an excellent superposition of the crystallographic position and the pose. Click on *File→Exit* to finish the execution of the MVD program.

To analyze docking results generated using Molegro Virtual Docker [39], we may use free software SAnDReS [68]. SAnDReS is an integrated computational environment for statistical analysis of docking simulations and application of machine-learning techniques to predict ligand binding affinity. In Fig. 14, we have the main GUI window of SAnDReS 1.0.2.

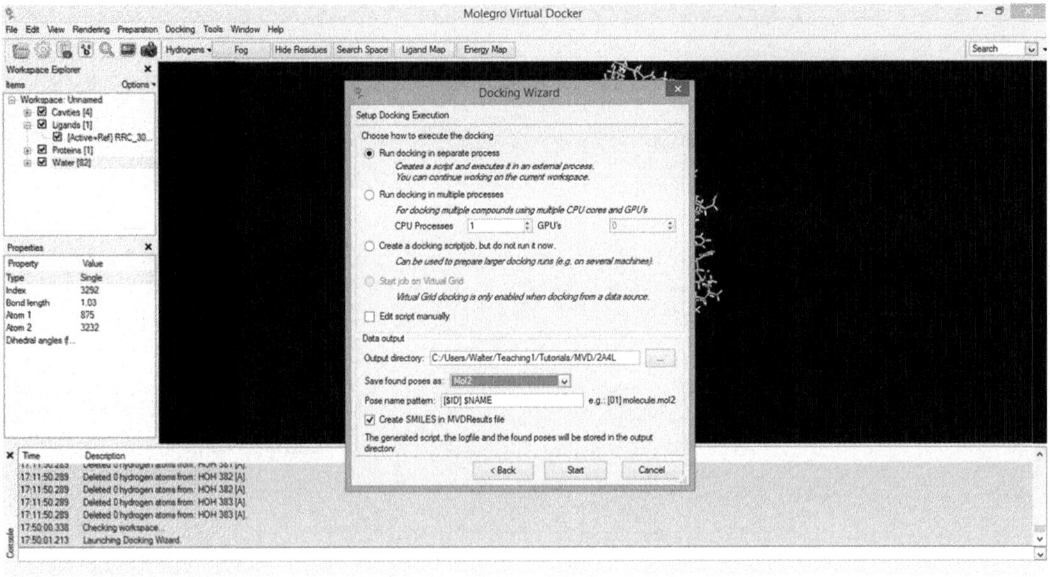

Fig. 9 Pop-up window to launch the docking simulation

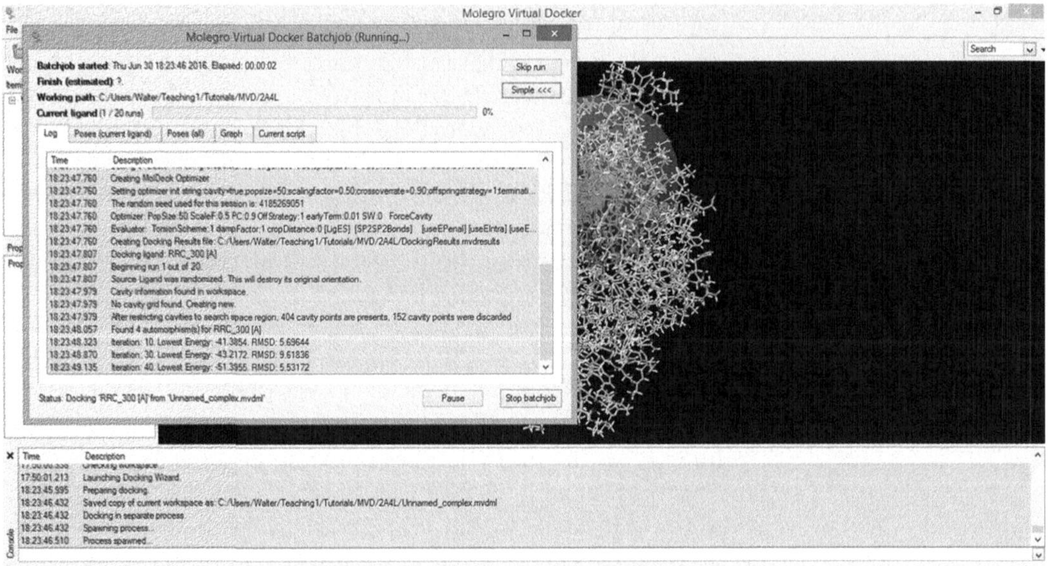

Fig. 10 Pop-up window showing the evolution of the docking simulation

To use SAnDReS to analyze docking results generated by Molegro Virtual Docker, we should have DockingResults.mvdresults (result of the docking simulation) file in the Project Directory. This directory should be updated in the first time we use SAnDReS to carry out analysis of the docking results. On the main GUI window, click on Find button and browse to the directory where

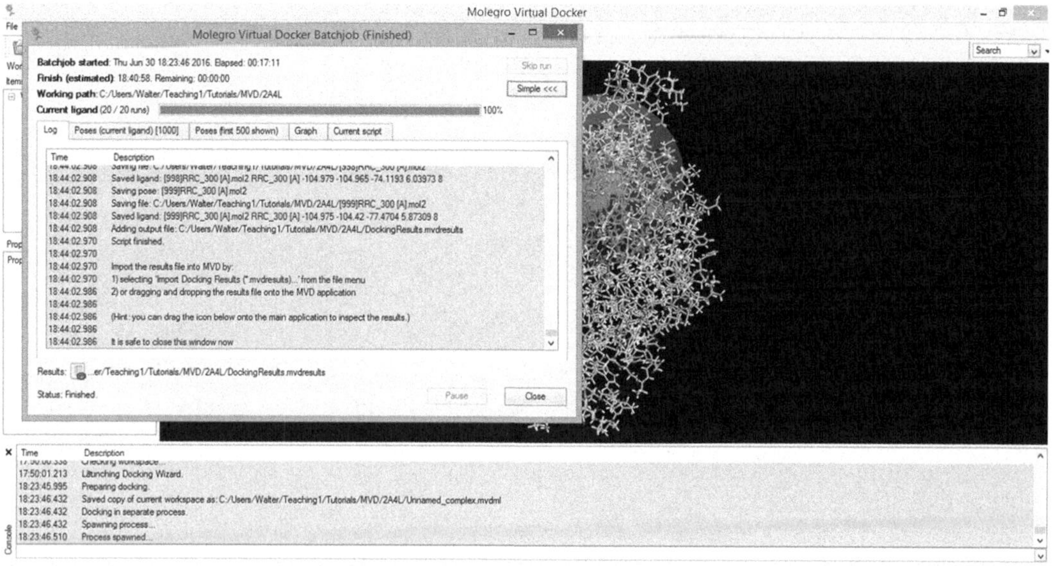

Fig. 11 Pop-up window showing that the simulation is finished. We may drag the Results icon to the black screen to have access to the docking results

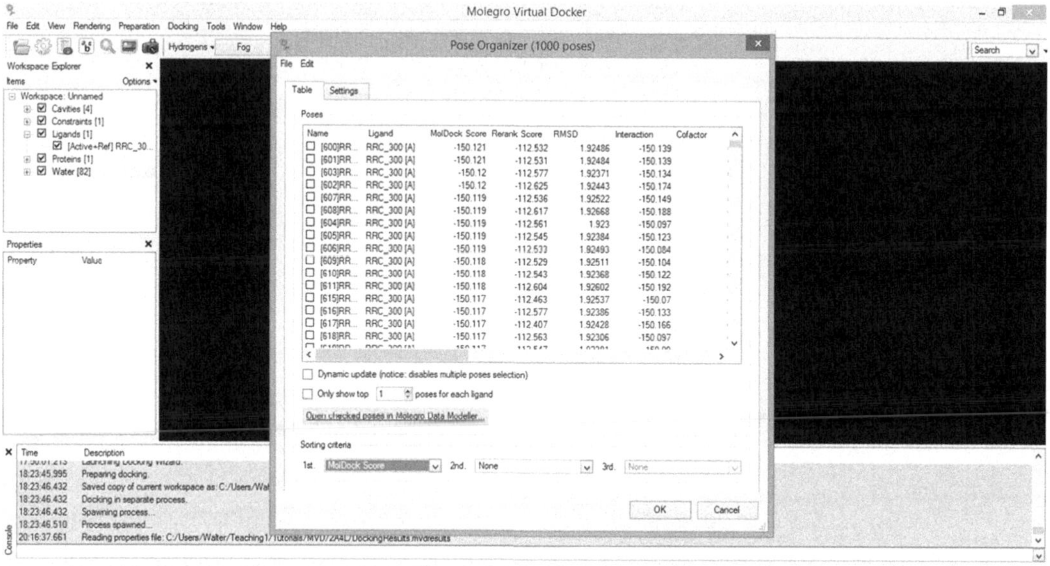

Fig. 12 Pop-up window showing the docking results sorted by MolDock score

the docking results are. Select the folder. Then, on the main GUI window, click on *Docking Hub→Import Docking Results*. We have a new pop-up window, where we can select the source of docking results (Fig. 15). Click on the Molegro Virtual Docker button.

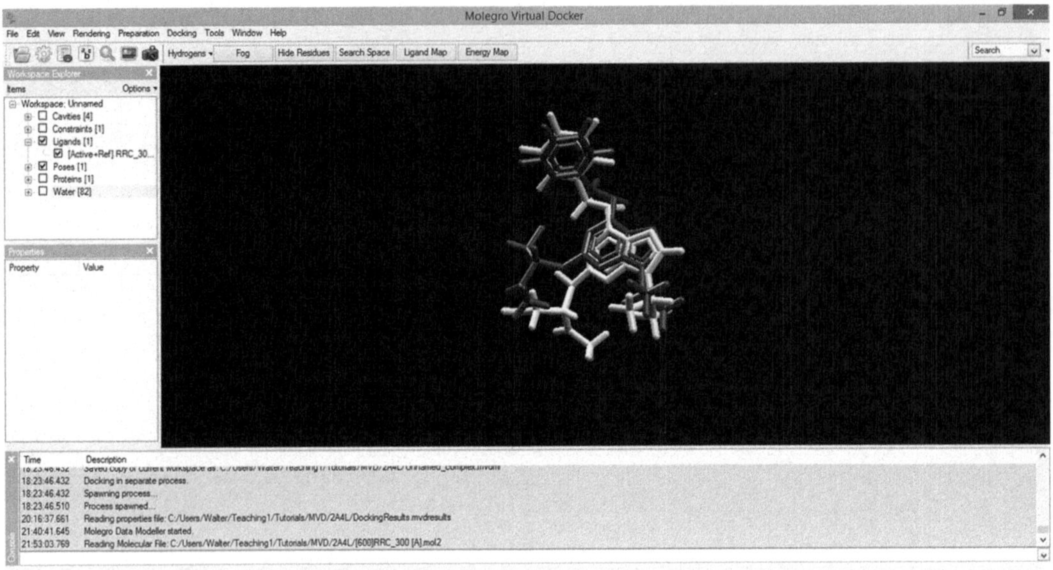

Fig. 13 Crystallographic position for the ligand (white sticks) and the pose (gray sticks)

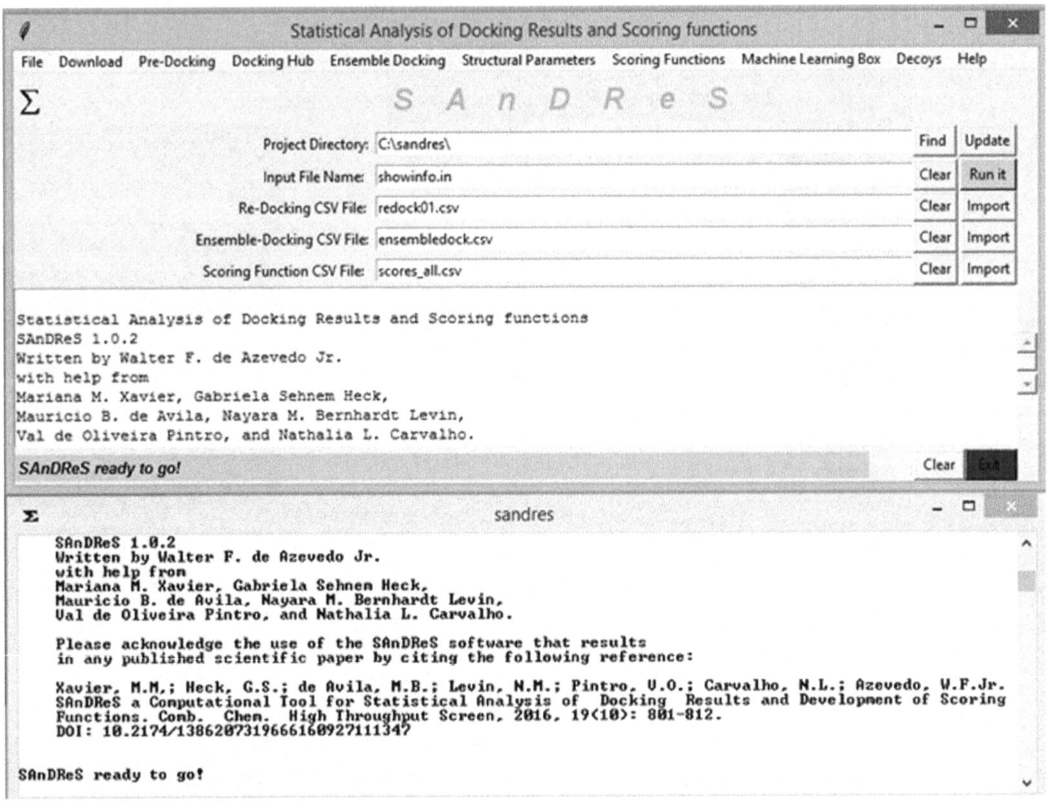

Fig. 14 SAnDReS main GUI

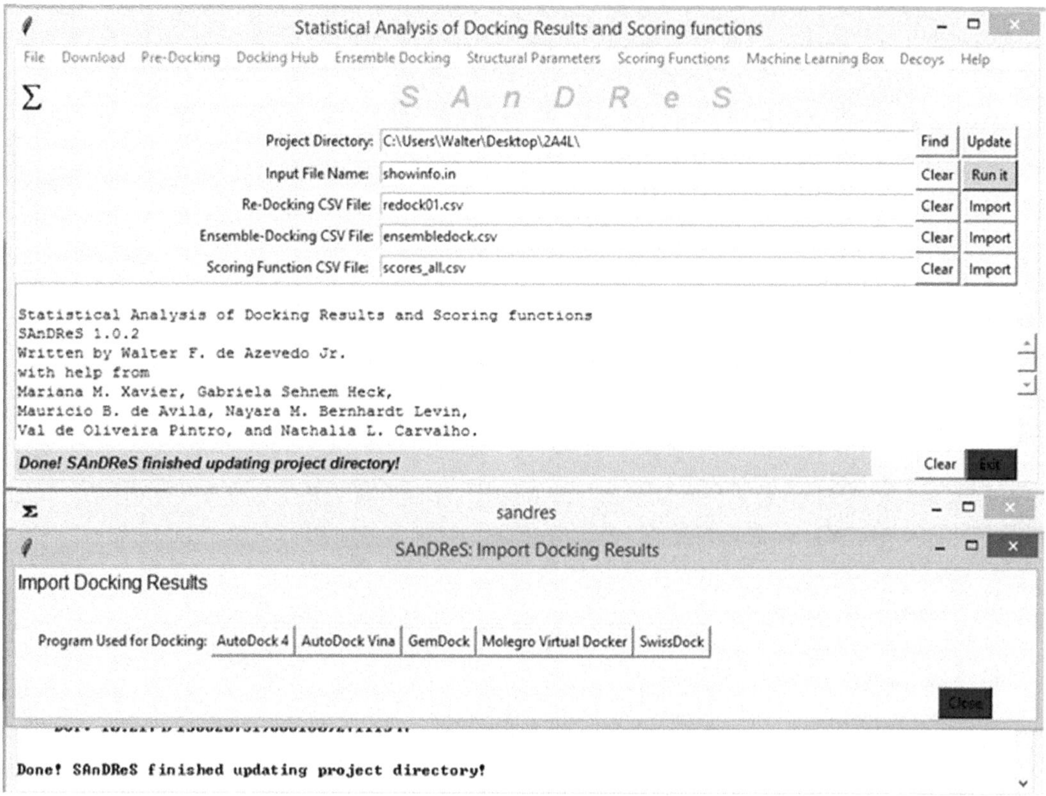

Fig. 15 Pop-up window to select the source of our docking results

On this new pop-up window (Fig. 16), we have all information necessary to convert Molegro Virtual Docker results (*DockingResults.mvdresults*) to a CSV format file (redock01.csv), which can be used by SAnDReS to carry out statistical analysis of docking results, such as root mean square deviation (RMSD), docking accuracy (DA1 and DA2), and correlation coefficients. Click on the Generate CSV File button; then we click on the Close button.

If everything goes fine, we will get the following message on the text window: "New CSV file has been written with RMSD data: redock01.csv", which means that we can proceed to carry out a statistical analysis of our docking results. Click on the Close button. To analyze the docking results, click on *Docking Hub→Statistical Analysis of Scoring Functions* vs. *RMSD*. Then, we click on the Yes button. SAnDReS generates a CSV file with the statistical analysis (*strmsd.csv*) and shows the partial results on the main GUI window. SAnDReS also creates individual CSV files for each scoring function, as shown in the column in the black rectangle (Fig. 17).

To generate plots, click on *Docking Hub→Prepare Files to Plot Redock Results*. On the new pop-up window, select the plot

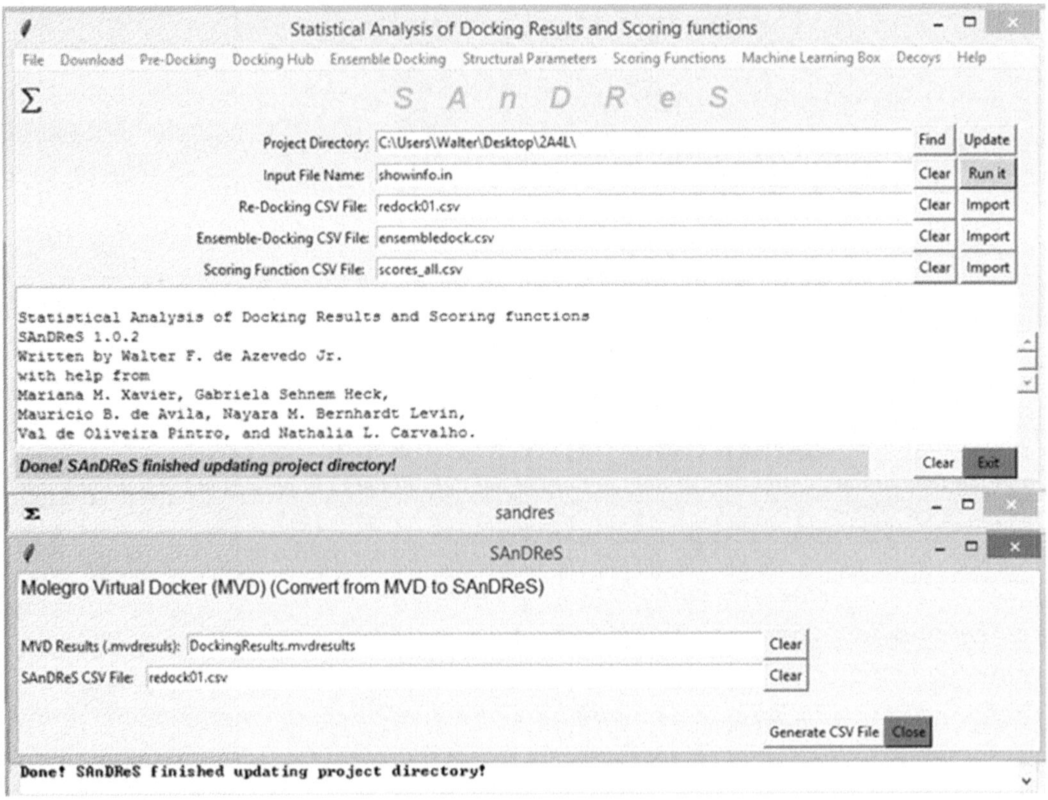

Fig. 16 Pop-up window to define the source of the docking results

parameters, click on the Generate Files button and then click on the Close button. Then, we click on the Plot Redock Results (Scatter Plot) button. On the new pop-up window, click on the Plot pltcsv File button. SAnDReS shows the generated plot file on the screen (Fig. 18). All generated files are on the Project Directory. We may click Exit button to finish SAnDReS. As we can see from Fig. 18, the lowest energy pose shows docking RMSD below 2.0 Å, which validates our docking protocol.

To carry out virtual screening simulation with MVD, we delete the ligand from the workspace explorer and follow all previously described steps. The only difference is when we choose the reference ligand. In this step, we get an error message since we do not have a reference ligand. To overcome this problem, we load an sdf file with all ligands that we want to dock against our protein target. The rest of the procedure is the same previously described. We select the best ligand using the scoring function value as criteria.

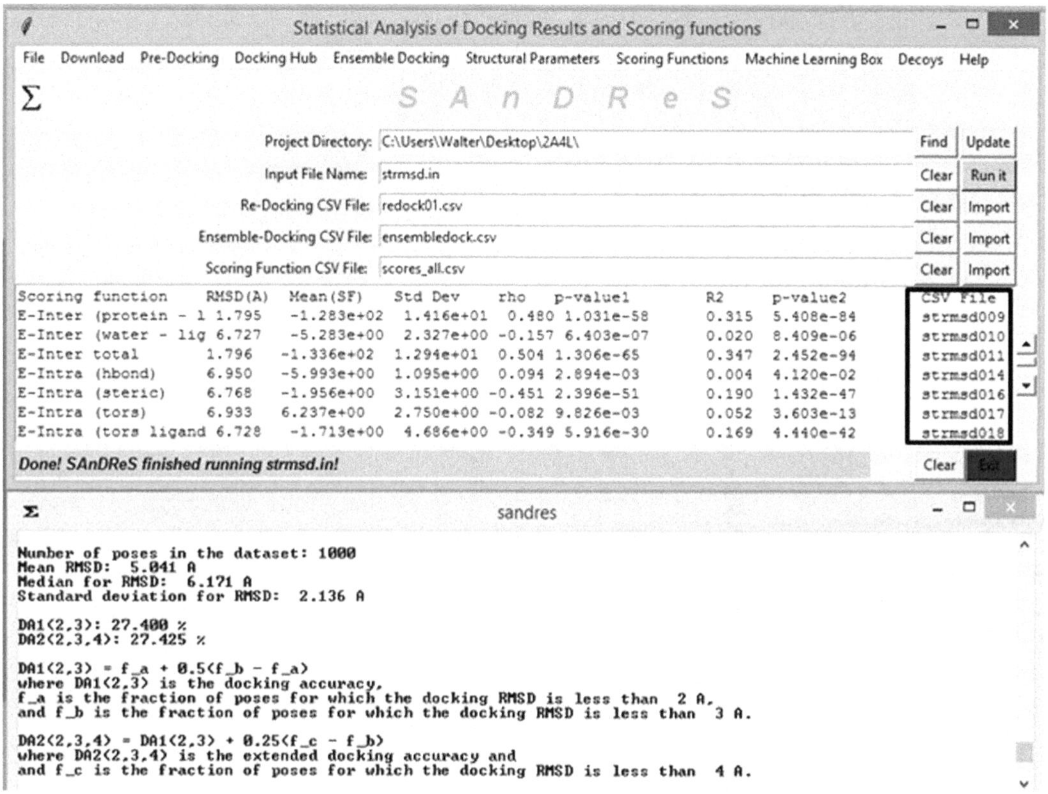

Fig. 17 CSV files generated with the docking results for each energy term available in the scoring functions of the program MVD are highlighted. The program SAnDReS also shows docking accuracy and RMSD

5 Availability

All files necessary to run this tutorial are available at https://azevedolab.net/resources/2A4L.zip. The program SAnDReS is available to download at https://github.com/azevedolab/sandres.

6 Colophon

We used the program Molegro Virtual Docker [35] to generate Figs. 1, 3–13. We created Fig. 2 using Microsoft PowerPoint 2016. We employed the program SAnDReS [68] to make Figs. 14–18. We performed molecular docking simulations described in this chapter using a Desktop PC with 4 GB memory, a 1 TB hard disk, and an Intel® Core® i3-2120 @ 3.30 GHz processor running Windows 8.1.

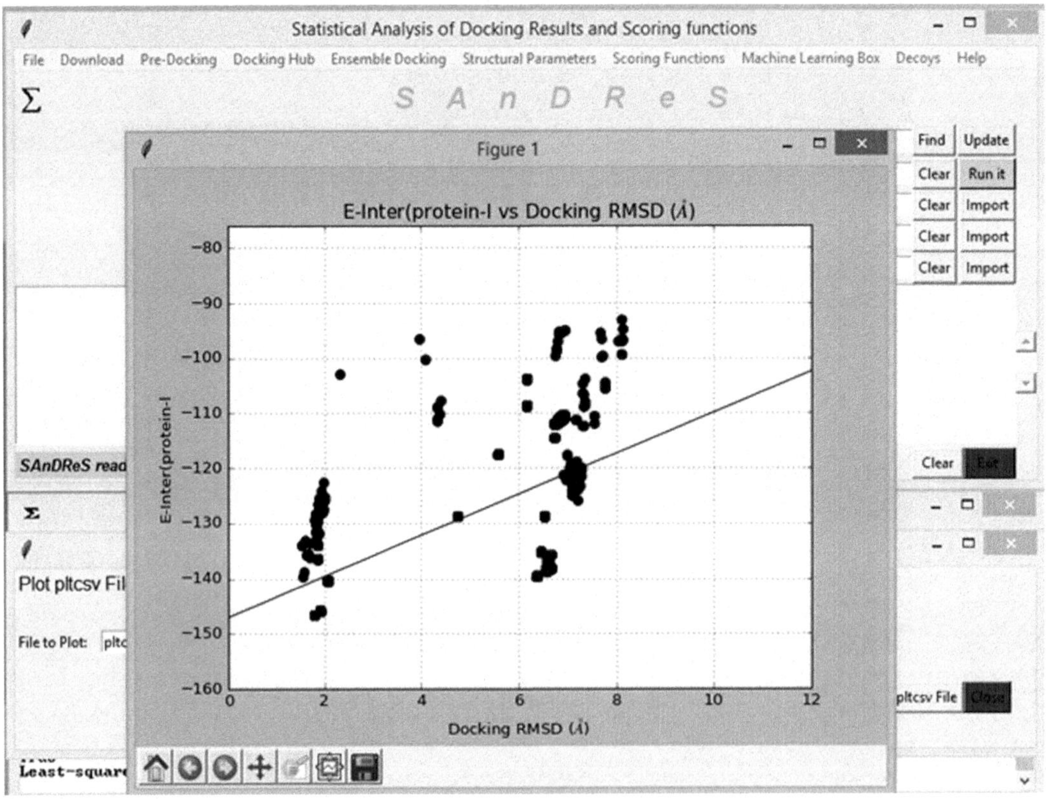

Fig. 18 Scattering plot generated with the program SAnDReS

7 Final Remarks

The program MVD allows us to carry out docking simulations in an integrated and intuitive platform. As we described here, MVD can handle all steps of the docking simulations with graphical capabilities, which made possible to generate high-quality figures of the docking results. Integration of MVD-SAnDReS opens the possibility to assess the docking accuracy and to create scatter plots between the docking RMSD and scoring functions. SAnDReS also makes a full statistical analysis of the docking results.

Acknowledgments

This work was supported by grants from CNPq (Brazil) (308883/2014-4). This study was financed in part by the Coordenação de Aperfeiçoamento de Pessoal de Nivel Superior—Brasil (CAPES)—Finance Code 001. GB-F acknowledges support from PUCRS/BPA fellowship. WFA is a researcher for CNPq (Brazil) (Process Numbers: 308883/2014-4 and 309029/2018-0).

References

1. Aarthy M, Singh SK (2018) Discovery of potent inhibitors for the inhibition of dengue envelope protein: an in silico approach. Curr Top Med Chem 18:1585–1602
2. Sehgal SA, Hammad MA, Tahir RA, Akram HN, Ahmad F (2018) Current therapeutic molecules and targets in neurodegenerative diseases based on in silico drug design. Curr Neuropharmacol 16:649–663
3. Zloh M, Kirton SB (2018) The benefits of *in silico* modeling to identify possible small-molecule drugs and their off-target interactions. Future Med Chem 10:423–432
4. Ishiki HM, Filho JMB, da Silva MS, Scotti MT, Scotti L (2018) Computer-aided drug design applied to Parkinson targets. Curr Neuropharmacol 16:865–880
5. Baig MH, Ahmad K, Rabbani G, Danishuddin M, Choi I (2018) Computer aided drug design and its application to the development of potential drugs for neurodegenerative disorders. Curr Neuropharmacol 16:740–748
6. Crespo A, Rodriguez-Granillo A, Lim VT (2017) Quantum-mechanics methodologies in drug discovery: applications of docking and scoring in lead optimization. Curr Top Med Chem 17:2663–2680
7. Ramesh M, Dokurugu YM, Thompson MD, Soliman ME (2017) Therapeutic, molecular and computational aspects of novel monoamine oxidase (MAO) inhibitors. Comb Chem High Throughput Screen 20:492–509
8. Abdolmaleki A, Ghasemi F, Ghasemi JB (2017) Computer-aided drug design to explore cyclodextrin therapeutics and biomedical applications. Chem Biol Drug Des 89:257–268
9. Ganesan A, Barakat K (2017) Applications of computer-aided approaches in the development of hepatitis C antiviral agents. Expert Opin Drug Discovery 12:407–425
10. Leelananda SP, Lindert S (2016) Computational methods in drug discovery. Beilstein J Org Chem 12:2694–2718
11. Hung CL, Chen CC (2014) Computational approaches for drug discovery. Drug Dev Res 75:412–418
12. Tabeshpour J, Sahebkar A, Zirak MR, Zeinali M, Hashemzaei M, Rakhshani S et al (2018) Computer-aided drug design and drug pharmacokinetic prediction: a mini-review. Curr Pharm Des 24:3014–3019
13. Zhong F, Xing J, Li X, Liu X, Fu Z, Xiong Z et al (2018) Artificial intelligence in drug design. Sci China Life Sci 61:1191–1204
14. Suryanarayanan V, Panwar U, Chandra I, Singh SK (2018) De novo design of ligands using computational methods. Methods Mol Biol 1762:71–86
15. Park H, Jung HY, Mah S, Hong S (2018) Systematic computational design and identification of low picomolar inhibitors of Aurora kinase a. J Chem Inf Model 58:700–709
16. Abdolmaleki A, Ghasemi JB, Ghasemi F (2017) Computer aided drug design for multi-target drug design: SAR/QSAR, molecular docking and pharmacophore methods. Curr Drug Targets 18:556–575
17. Zheng X, Liu Z, Li D, Wang E, Wang J (2013) Rational drug design: the search for Ras protein hydrolysis intermediate conformation inhibitors with both affinity and specificity. Curr Pharm Des 19:2246–2258
18. Jayadeepa RM, Sharma S (2011) Computational models for 5αR inhibitors for treatment of prostate cancer: review of previous works and screening of natural inhibitors of 5αR2. Curr Comput Aided Drug Des 7:231–237
19. Michel J, Essex JW (2010) Prediction of protein-ligand binding affinity by free energy simulations: assumptions, pitfalls and expectations. J Comput Aided Mol Des 24:639–658
20. Reddy MR, Erion MD (2005) Computer-aided drug design strategies used in the discovery of fructose 1, 6-bisphosphatase inhibitors. Curr Pharm Des 11:283 294
21. Irwin JJ, Shoichet BK (2005) ZINC--a free database of commercially available compounds for virtual screening. J Chem Inf Model 45:177–182
22. Irwin JJ, Sterling T, Mysinger MM, Bolstad ES, Coleman RG (2012) ZINC: a free tool to discover chemistry for biology. J Chem Inf Model 52:1757–1768
23. Wishart DS, Knox C, Guo AC, Shrivastava S, Hassanali M, Stothard P et al (2006) DrugBank: a comprehensive resource for *in silico* drug discovery and exploration. Nucleic Acids Res 34:668–672
24. Wishart DS, Knox C, Guo AC, Cheng D, Shrivastava S, Tzur D et al (2008) DrugBank: a knowledgebase for drugs, drug actions and drug targets. Nucleic Acids Res 36:901–906
25. Freitas PG, Elias TC, Pinto IA, Costa LT, de Carvalho PVSD, Omote DQ et al (2018) Computational approach to the discovery of

phytochemical molecules with therapeutic potential targets to the PKCZ protein. Lett Drug Des Discovery 15:488–499

26. Teles CB, Moreira-Dill LS, Silva Ade A, Facundo VA, de Azevedo WF Jr, da Silva LH et al (2015) A lupane-triterpene isolated from *Combretum leprosum* Mart. fruit extracts that interferes with the intracellular development of *Leishmania* (L.) *amazonensis in vitro*. BMC Complement Altern Med 15:165
27. Sá MS, de Menezes MN, Krettli AU, Ribeiro IM, Tomassini TC, Ribeiro dos Santos R et al (2011) Antimalarial activity of physalins B, D, F, and G. J Nat Prod 74:2269–2272
28. Wong CF, McCammon JA (2003) Protein flexibility and computer-aided drug design. Annu Rev Pharmacol Toxicol 43:31–45
29. Wishart DS (2008) Identifying putative drug targets and potential drug leads: starting points for virtual screening and docking. Methods Mol Biol 443:333–351
30. Śledź P, Caflisch A (2018) Protein structure-based drug design: from docking to molecular dynamics. Curr Opin Struct Biol 48:93–102
31. de Azevedo WF Jr (2011) Molecular dynamics simulations of protein targets identified in *Mycobacterium tuberculosis*. Curr Med Chem 18:1353–1366
32. Goodsell DS, Olson AJ (1990) Automated docking of substrates to proteins by simulated annealing. Proteins 8:195–202
33. Morris GM, Goodsell DS, Huey R, Olson AJ (1996) Distributed automated docking of flexible ligands to proteins: Parallel applications of AutoDock 2.4. J Comput Aided Mol Des 10:293–304
34. Morris GM, Goodsell DS, Halliday RS, Huey R, Hart WE, Belew RK et al (1998) Automated docking using a lamarckian genetic algorithm and empirical binding free energy function. J Comput Chem 19:1639–1662
35. Morris GM, Huey R, Lindstrom W, Sanner MF, Belew RK, Goodsell DS et al (2009) AutoDock4 and AutoDockTools4: Automated docking with selective receptor flexibility. J Comput Chem 30:2785–2791
36. Trott O, Olson AJ (2010) AutoDock Vina: improving the speed and accuracy of docking with a new scoring function, efficient optimization, and multithreading. J Comput Chem 31:455–461
37. Yang JM, Chen CC (2004) GEMDOCK: a generic evolutionary method for molecular docking. Proteins 55:288–304
38. Yang JM, Shen TW (2005) A pharmacophore-based evolutionary approach for screening selective estrogen receptor modulators. Proteins 59:205–220
39. Thomsen R, Christensen MH (2006) MolDock: a new technique for high-accuracy molecular docking. J Med Chem 49:3315–3321
40. Heberlé G, de Azevedo WF Jr (2011) Bio-inspired algorithms applied to molecular docking simulations. Curr Med Chem 18:1339–1352
41. De Azevedo WF Jr (2010) MolDock applied to structure-based virtual screening. Curr Drug Targets 11:327–334
42. Bitencourt-Ferreira G, de Azevedo WF Jr (2018) Development of a machine-learning model to predict Gibbs free energy of binding for protein-ligand complexes. Biophys Chem 240:63–69
43. de Ávila MB, de Azevedo WF Jr (2018) Development of machine learning models to predict inhibition of 3-dehydroquinate dehydratase. Chem Biol Drug Des 92:1468–1474
44. Russo S, de Azevedo WF (2019) Advances in the understanding of the cannabinoid receptor 1—focusing on the inverse agonists interactions. Curr Med Chem. https://doi.org/10.2174/0929867325666180417165247
45. Amaral MEA, Nery LR, Leite CE, de Azevedo Junior WF, Campos MM (2018) Pre-clinical effects of metformin and aspirin on the cell lines of different breast cancer subtypes. Invest New Drugs 36:782–796
46. Pintro VO, Azevedo WF (2017) Optimized virtual screening workflow. towards target-based polynomial scoring functions for HIV-1 protease. Comb Chem High Throughput Screen 20:820–827
47. Heck GS, Pintro VO, Pereira RR, de Ávila MB, Levin NMB, de Azevedo WF (2017) Supervised machine learning methods applied to predict ligand-binding affinity. Curr Med Chem 24:2459–2470
48. Coracini JD, de Azevedo WF Jr (2014) Shikimate kinase, a protein target for drug design. Curr Med Chem 21:592–604
49. Moraes FP, de Azevedo WF Jr (2012) Targeting imidazoline site on monoamine oxidase B through molecular docking simulations. J Mol Model 18:3877–3886
50. Soares MB, Silva CV, Bastos TM, Guimarães ET, Figueira CP, Smirlis D et al (2012) Anti-*Trypanosoma cruzi* activity of nicotinamide. Acta Trop 12:224–229
51. Vianna CP, de Azevedo WF Jr (2012) Identification of new potential *Mycobacterium tuberculosis* shikimate kinase inhibitors through

molecular docking simulations. J Mol Model 18:755–764

52. Morgan DO (1995) Principles of CDK regulation. Nature 374:131–134
53. Murray AW (1994) Cyclin-dependent kinases: regulators of the cell cycle and more. Chem Biol 1:191–195
54. Volkart PA, Bitencourt-Ferreira G, Souto AA, de Azevedo WF (2019) Cyclin-dependent kinase 2 in cellular senescence and cancer. A structural and functional review. Curr Drug Targets 20(7):716–726. https://doi.org/10.2174/1389450120666181204165344
55. Kim SH, Schulze-Gahmen U, Brandsen J, de Azevedo Júnior WF (1996) Structural basis for chemical inhibition of CDK2. Prog Cell Cycle Res 2:137–145
56. De Azevedo WF Jr, Mueller-Dieckmann HJ, Schulze-Gahmen U, Worland PJ, Sausville E, Kim SH (1996) Structural basis for specificity and potency of a flavonoid inhibitor of human CDK2, a cell cycle kinase. Proc Natl Acad Sci U S A 93:2735–2740
57. Canduri F, de Azevedo WF Jr (2005) Structural basis for interaction of inhibitors with cyclin-dependent kinase 2. Curr Comput Aided Drug Des 1:53–64
58. Krystof V, Cankar P, Frysová I, Slouka J, Kontopidis G, Dzubák P (2006) 4-arylazo-3,5-diamino-1H-pyrazole CDK inhibitors: SAR study, crystal structure in complex with CDK2, selectivity, and cellular effects. J Med Chem 49:6500–6509
59. Levin NMB, Pintro VO, Bitencourt-Ferreira G, Mattos BB, Silvério AC, de Azevedo WF Jr (2018) Development of CDK-targeted scoring functions for prediction of binding affinity. Biophys Chem 235:1–8
60. de Ávila MB, Xavier MM, Pintro VO, de Azevedo WF (2017) Supervised machine learning techniques to predict binding affinity. A study for cyclin-dependent kinase 2. Biochem Biophys Res Commun 494:305–310
61. Levin NM, Pintro VO, de Ávila MB, de Mattos BB, De Azevedo WF Jr (2017) Understanding the structural basis for inhibition of cyclin-dependent kinases. New pieces in the molecular puzzle. Curr Drug Targets 18:1104–1111
62. De Bondt HL, Rosenblatt J, Jancarik J, Jones HD, Morgan DO, Kim SH (1993) Crystal structure of cyclin-dependent kinase 2. Nature 363:595–602
63. Schulze-Gahmen U, De Bondt HL, Kim SH (1996) High-resolution crystal structures of human cyclin-dependent kinase 2 with and without ATP: bound waters and natural ligand as guides for inhibitor design. J Med Chem 39:4540–4546
64. Storn R, Price K (1997) Differential evolution—a simple and efficient heuristic for global optimization over continuous spaces. J Global Optim 11:341–359
65. Nelder JA, Mead RA (1965) Simplex method for function minimization. Comput J 7:308–313
66. Korb O, Stutzle T, Exner TE (2009) Empirical scoring functions for advanced protein-ligand docking with PLANTS. J Chem Inf Model 49:84–96
67. De Azevedo WF, Leclerc S, Meijer L, Havlicek L, Strnad M, Kim SH (1997) Inhibition of cyclin-dependent kinases by purine analogues: crystal structure of human CDK2 complexed with roscovitine. Eur J Biochem 243:518–526
68. Xavier MM, Heck GS, de Avila MB, Levin NM, Pintro VO, Carvalho NL et al (2016) SAnDReS a computational tool for statistical analysis of docking results and development of scoring functions. Comb Chem High Throughput Screen 19:801–812

Chapter 11

Docking with GemDock

Gabriela Bitencourt-Ferreira and Walter Filgueira de Azevedo Jr.

Abstract

GEMDOCK is a protein-ligand docking software that makes use of an elegant biologically inspired computational methodology based on the differential evolution algorithm. As any docking program, GEMDOCK has two major features to predict the binding of a small-molecule ligand to the binding site of a protein target: the search algorithm and the scoring function to evaluate the generated poses. The GEMDOCK scoring function uses a piecewise potential energy function integrated into the differential evolutionary algorithm. GEMDOCK has been applied to a wide range of protein systems with docking accuracy similar to other docking programs such as Molegro Virtual Docker, AutoDock4, and AutoDock Vina. In this chapter, we explain how to carry out protein-ligand docking simulations with GEMDOCK. We focus this tutorial on the protein target cyclin-dependent kinase 2.

Key words GEMDOCK, Molecular docking, Cyclin-dependent kinase 2, Drug design, Protein-ligand interactions

1 Introduction

The goal in any protein-ligand docking simulation is to move a small organic molecule to the minimum energy position into the binding site of a protein target [1–11]. From the computational point of view, this is a typical optimization problem, which depends on the number of degrees of freedom in the ligand and the protein target. In most of the computational tools developed to carry protein-ligand docking simulations, the flexibility of the ligand is mandatory, and the protein flexibility is optional. Addition of flexibility to the protein target occurs on the rotatable angles of the side chains of the amino acids found in the binding site of the biomolecule. Such care with adding rotatable angles to the protein-ligand system is due to the computational cost of the increase in the

Walter Filgueira de Azevedo Jr. (ed.), *Docking Screens for Drug Discovery*, Methods in Molecular Biology, vol. 2053,
https://doi.org/10.1007/978-1-4939-9752-7_11,

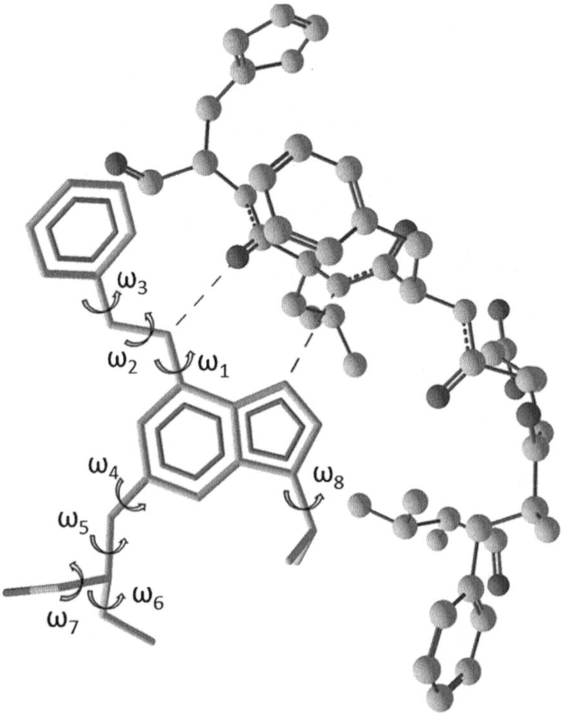

Fig. 1 ATP-binding pocket of CDK2 where we show the main residues which participate in intermolecular interactions. Rotatable angles of the ligands are indicated as ω. We do not show the rotatable angles of the amino acids. We show the intermolecular hydrogen bonds as dashed lines. We generated this figure with Molegro Virtual Docker [17]

number of degrees of freedom [12–15]. To illustrate our point, let us consider the structure of human cyclin-dependent kinase 2 (CDK2) in complex with the inhibitor roscovitine (PDB access code: 2A4L) [16]. Figure 1 shows the binding pocket of CDK2 and the structure of the inhibitor.

If we consider the rotatable angles of the inhibitor, we have a total of eight angles, indicated as ωs in Fig. 1. Taking only the rotatable angles of the side chains of the amino acids that form the molecular fork of the CDK2, we have an additional 12 degrees of freedom to be added to the system. In summary, if we consider the flexibility of the side chains of the amino acids participating in intermolecular interactions with the ligand, we have a computational model closer to the biological reality. On the other hand, we elevate the complexity of the system [12–15], which increases the computational cost of the protein-ligand docking simulation.

Among the most successful search algorithms used for protein-ligand docking simulations, the biologically inspired algorithms have been particularly successful [12]. For instance, we have genetic algorithm implemented in the program AutoDock [18–21], differential evolution in GEMDOCK [22–24] and, ant colony optimization in the Molecular Virtual Docker [17], to mention a few. Our focus here is on the use of GEMDOCK for protein-ligand docking simulations. The program GEMDOCK has been successfully employed in molecular docking simulations for a wide range of protein systems. It has been cited in more than 170 scientific publications (search carried out on January 12, 2019).

Furthermore, evaluation of its docking performance indicated redocking root mean square deviation (RMSD) < 2.0 Å for 79% of crystallographic structures used as a benchmark [22–24]. GEMDOCK is an acronym for Generic Evolutionary Method for molecular DOCKing, and its first version was released in 2004 [22]. The details about the implementation of the differential evolution [25] and a piecewise empirical scoring function are described elsewhere [22–24]. We describe here how to carry out docking simulations employing GEMDOCK. To illustrate its use, we consider the crystallographic structure of CDK2 in complex with roscovitine.

2 Biological System

In this tutorial, we show how to perform protein-ligand docking simulations of cyclin-dependent kinase 2 (CDK2) (EC 2.7.11.22) with GEMDOCK 2 [22–24]. This drug target has been intensively studied for the development of anticancer treatment [26–35]. The first crystallographic structure of CDK2 was determined in 1993 at the University of California, Berkeley [36]. Structural analysis of the CDK2 showed a typical bilobal architecture of serine/threonine protein kinases (EC 2.7.11.1). CDK2 structure has an N-terminal domain that is mainly composed of a distorted beta sheet and a short alpha helix. A helix bundle forms the C-terminal in the CDK2 structure. The two lobes of the CDK2 structure allow the binding of the ATP molecule [37]. In this tutorial, we carry out redock of the roscovitine against the structure of CDK2. This inhibitor is bound to the ATP-binding pocket of CDK2, which characterized it as a competitive inhibitor [16].

3 Graphical Tutorial

Our goal here is to carry out redocking simulation against the crystallographic structure 2A4L. Redocking simulations are used to validate a docking protocol. It is generally accepted that docking protocols that generate docking RMSD below 2.0 Å are acceptable [8].

In the flowchart below (Fig. 2), we see the main steps to redock a ligand in the structure of a protein using GEMDOCK 2.1 [22–24] and SAnDReS [38]. In the first step, we download the atomic coordinates of the complex we are going to use to test a docking protocol (redocking simulation). Following, we set up the directory where all files will be stored. Next, we prepare the binding. To do so, we need the PDB file for the protein structure. Then, we prepare the ligands. We may carry out docking simulations with more than one ligand. To do the docking simulation, we need the set up the docking parameters, and then we start the docking. After finishing the docking simulations, we may carry out the statistical analysis of the docking results with the program SAnDReS [38].

To run this tutorial, we consider that you have GEMDOCK installed on your computer, and it is open, as shown in Fig. 3. We used version 2.1 of GEMDOCK, but this tutorial should work for

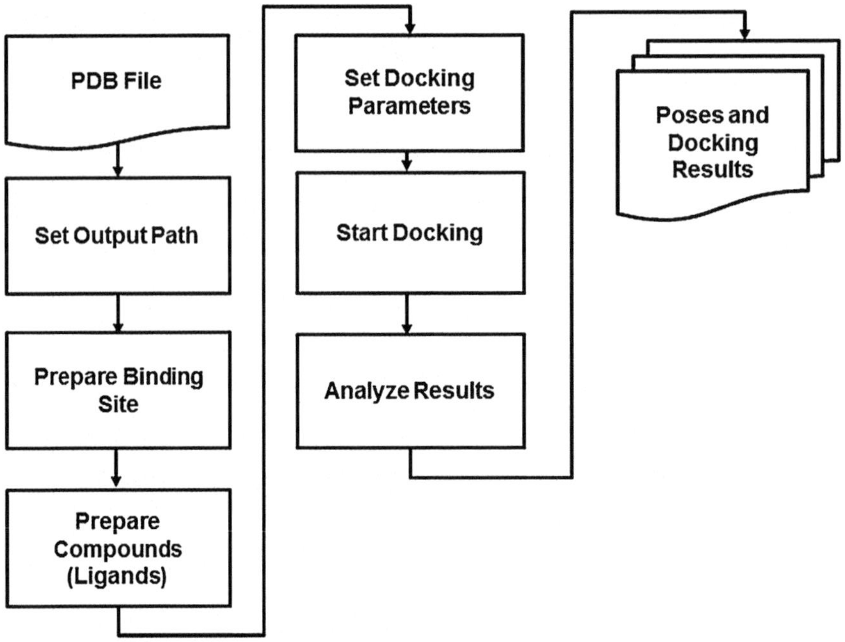

Fig. 2 Flowchart showing all the steps of this tutorial

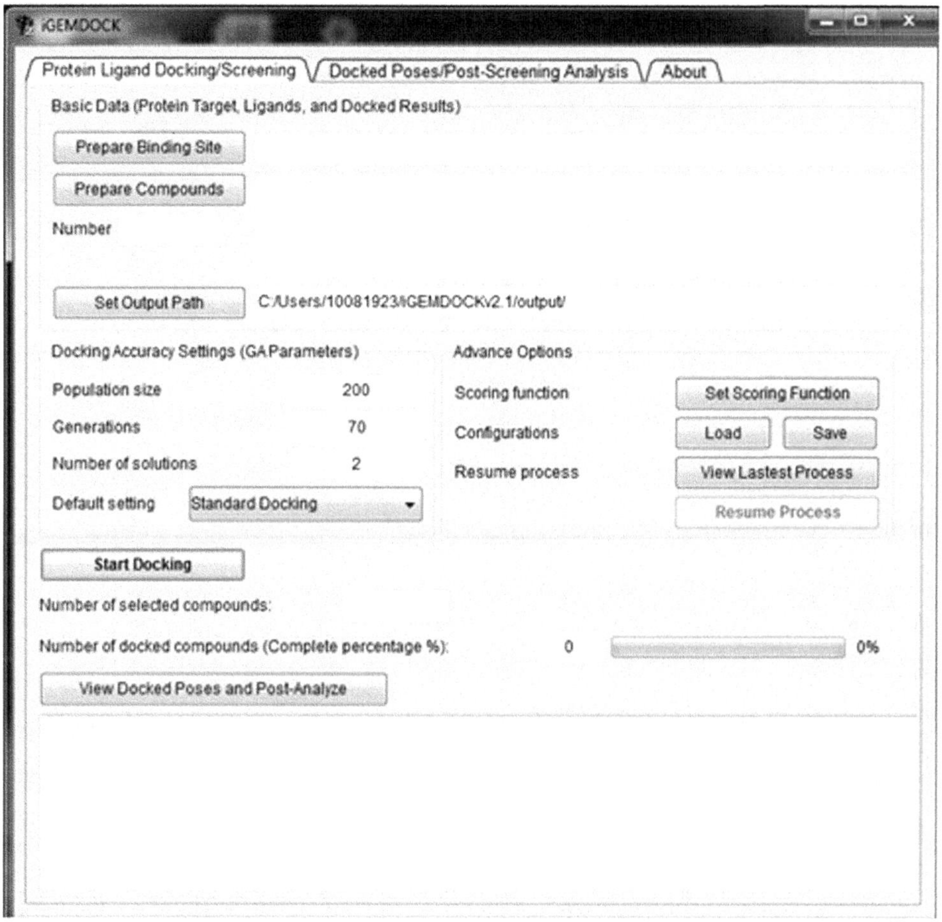

Fig. 3 GEMDOCK main window

earlier versions. We used GEMDOCK for Windows; it is mostly the same for the Linux version. Figure 3 shows the setup window of GEMDOCK. Then, we access the Protein Data Bank (PDB) [39–41] (www.rcsb.org/pdb) and download the atomic coordinates for CDK2 in complex with roscovitine (PDB access code: 2A4L). Next, we split the original PDB file into two files, one for roscovitine (*lig.pdb*) and another for the CDK2 (*prot.pdb*).

We initially set the output path clicking on the "Set Output Path" button indicated in Fig. 3. We browse and choose the folder where the PDB files (*prot.pdb* and *lig.pdb*) named here as 2A4L. GEMDOCK has a new output path. In this directory, we have the PDB files for the binding site (*prot.pdb*) and the ligand (*lig.pdb*). Now we upload the coordinates for the binding site, we click on the "Prepare Binding Site" button. On the new window, we click on

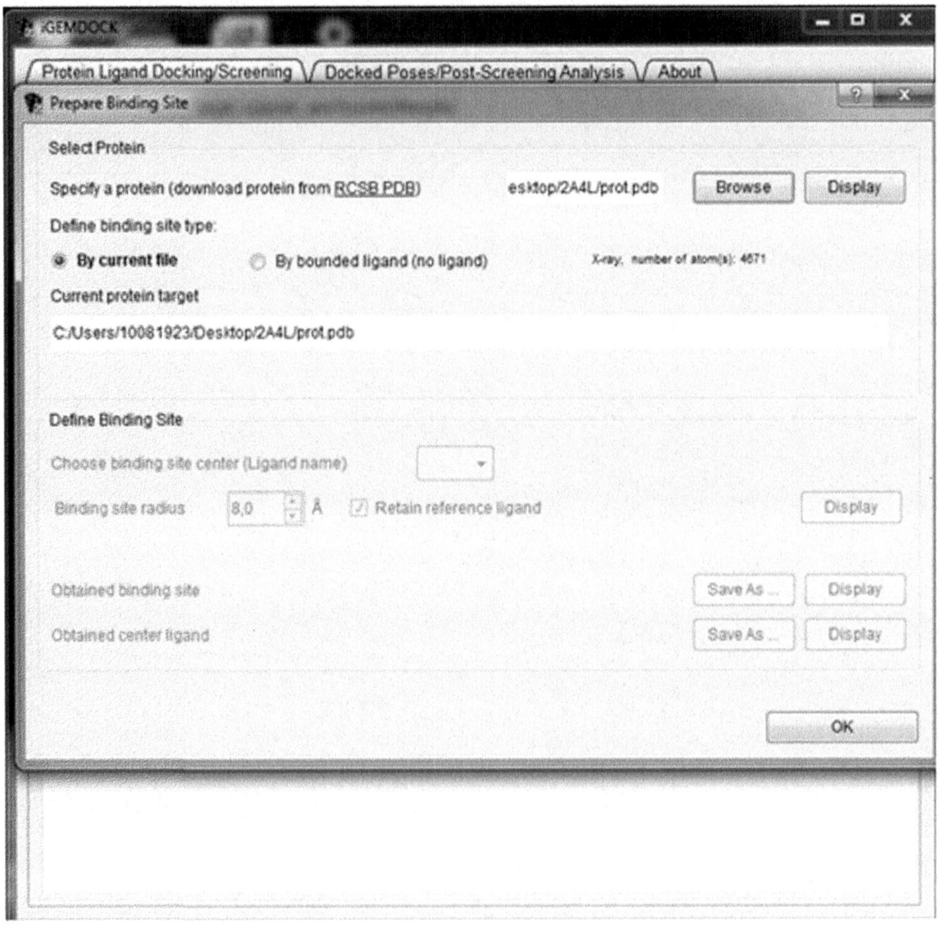

Fig. 4 "Prepare Binding Site" menu

the Browse button. We find the same folder indicated for the output path and click on the *prot.pdb* file and on the Open button. The protein target is shown in Fig. 4. We can click on the OK button.

Now, we can prepare the compounds (ligands). Click on the "Prepare Compounds" button (*see* Fig. 3). On this new window (Fig. 5), we can select either ligand files or a Folder with ligand files. Here we are interested in redock a ligand; we click on the Ligands button to select one ligand PDB file. We go to the same folder and select the lig.pdb file, and click on the Open button. We return to the previous window. We can see the chosen ligand file; then we click on the OK button. It is noteworthy that we may use this step to load multiple ligands, for instance, to carry out virtual screening

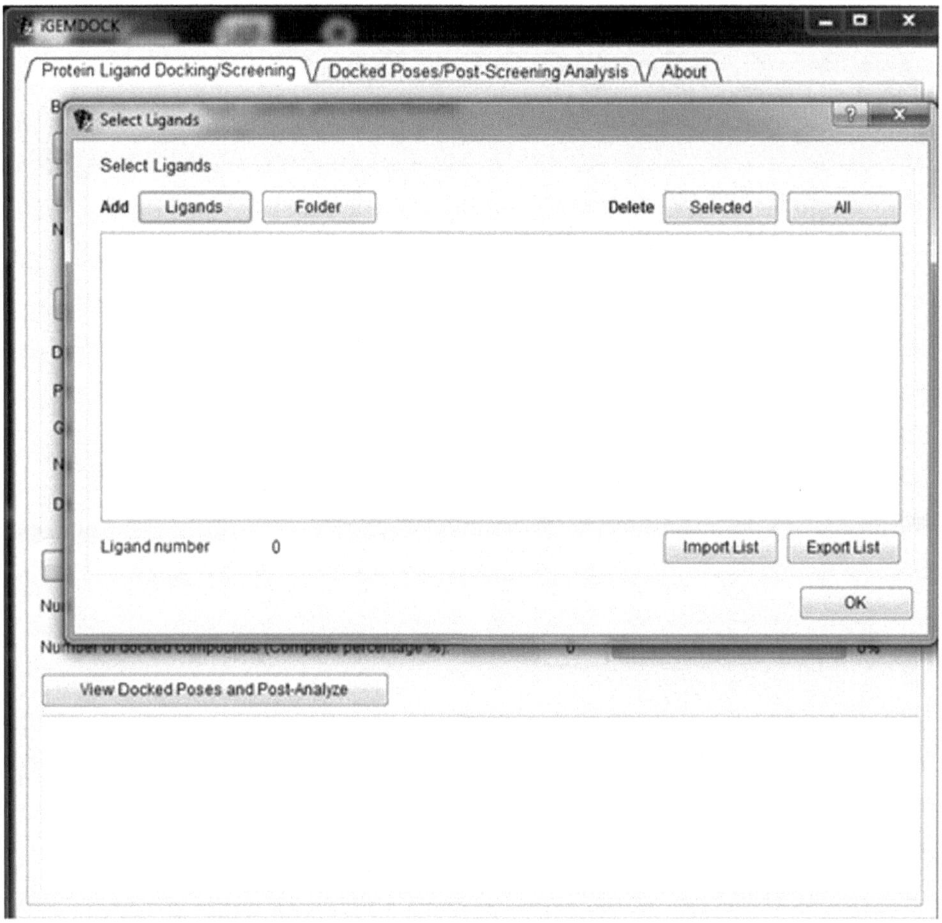

Fig. 5 "Select Ligands" menu

simulations, where we try to fit several ligands to the binding site and select the one with the lowest scoring function value. We are back to the main window, where we can see the selected files (Fig. 6). Now we can choose the docking parameters.

Besides the Population Size, Generation, and Number of solutions, we can define the docking settings, as indicated below (Fig. 7). We leave the "Standard Docking" option. We changed the number of solutions to 10, so we will have 10 poses at the end of our simulation. Since we changed the number of solutions, GEMDOCK updated the default setting to "Custom." We are ready to go. Click on the "Start Docking" button. Then, we click on the OK button. If everything is fine, you can follow the docking progress on the window below (Fig. 8).

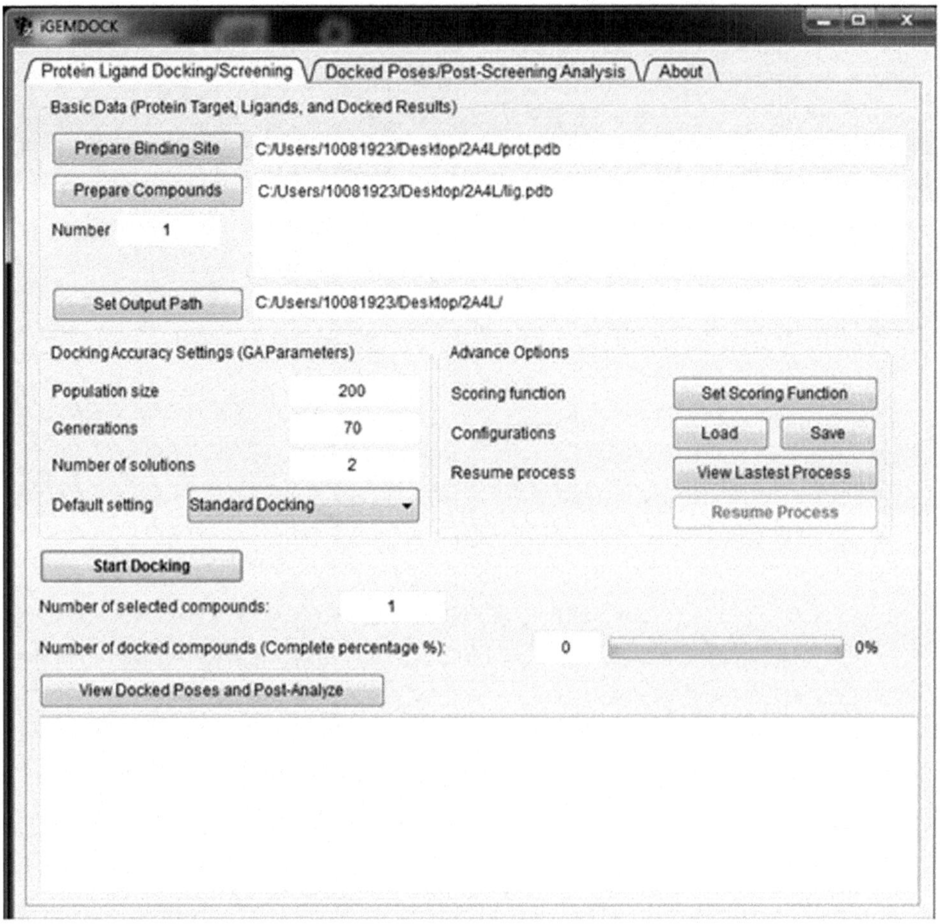

Fig. 6 The main window with the prot.pdb and lig.pdb files

During the docking simulation, GEMDOCK shows the fitness function value for the best pose in each generation. GEMDOCK also indicates where to find each pose. GEMDOCK creates a folder named "docked_Pose" where all poses will be stored. Once finished, GEMDOCK shows the "Docking process finish" Message (Fig. 9). Click on the OK button. We have all the poses in the docked Pose folder. To analyze the docking results, we click on "Docked Poses/Post-Screening Analysis" button (*see* Fig. 9). Now we have to define the binding site. In the new window, click on the "Binding Site" button (Fig. 10). In the new pop-up window, we have to click on the Browse button. Following, we select the *prot.pdb* file and click on the Open button. Then we click on the OK button.

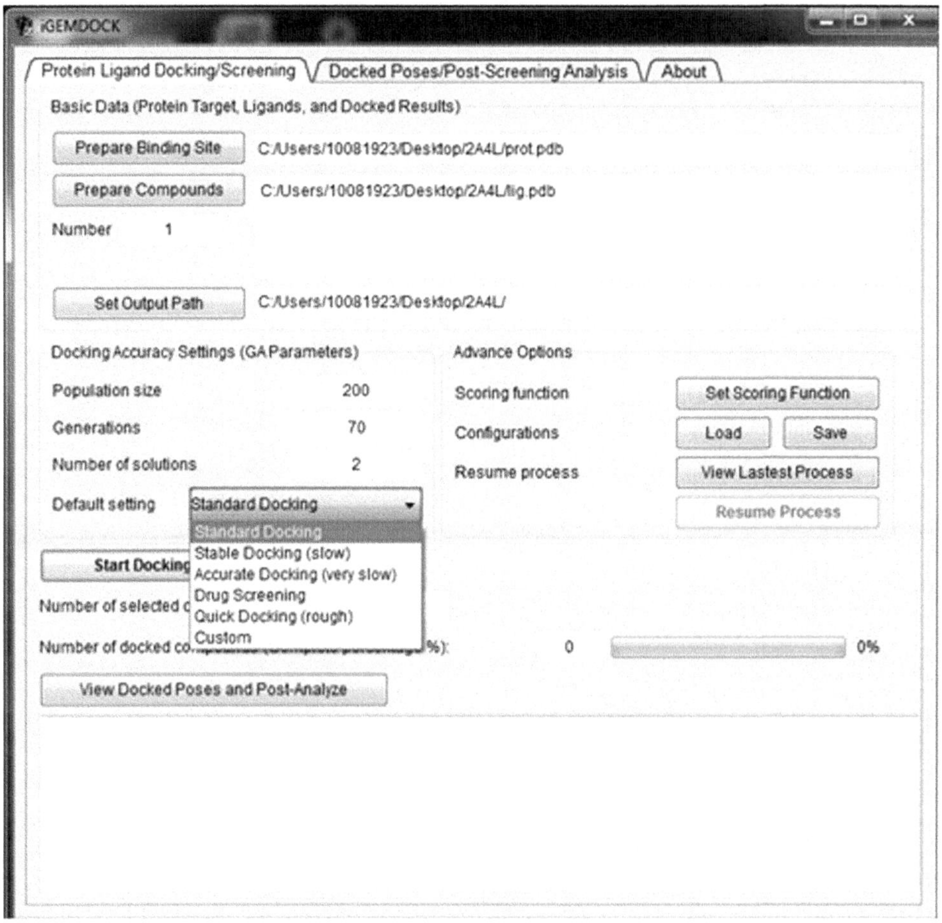

Fig. 7 Setting up of docking parameters

Now we have to upload the pose. We click on the "Docked Poses" button. We may now select the folder where the poses are and click on the Folder button. Following, we choose the *docked_Pose* folder and click on the OK button. In Fig. 11, we can see all 10 poses. We mark all 10 poses and click on the OK button. Following, we have to click on the "Set Output Path" button. Then, we select the 2A4L folder and click on the OK button. Now we click on the "Interaction Profile" button (Fig. 12). We finally have our docking results on the screen. We may save the results on a *.xls* file clicking on the Excel button (Fig. 13). We save the docking results as *gemdock.xls*.

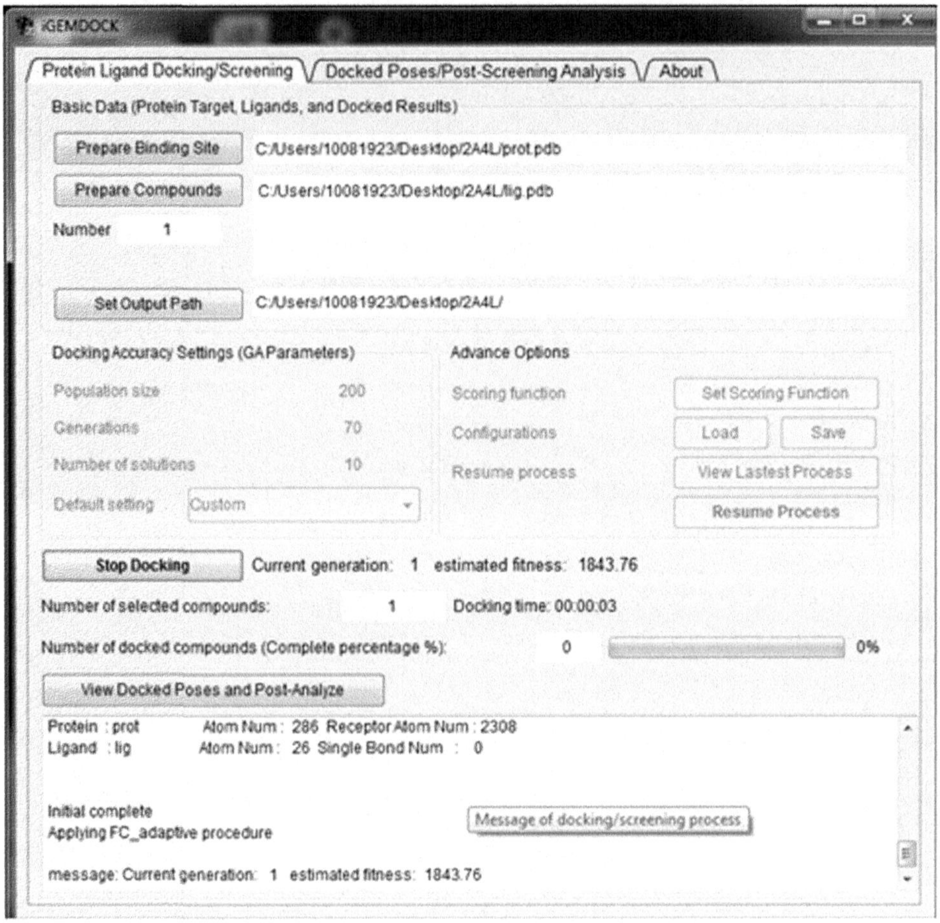

Fig. 8 Evolution of the docking simulation

To analyze docking results generated using GEMDOCK 2.1, we may use free software SAnDReS [38]. SAnDReS is an integrated computational environment for statistical analysis of docking simulations and application of machine-learning techniques to predict ligand binding affinity. In Fig. 14, we have the main GUI window of SAnDReS 1.0.2. To use SAnDReS to analyze docking results generated with GEMDOCK, we need to have the docking results file in the CSV format. Once in this format, for instance, gemdock.csv, we may type the filename in the "Redocking CSV File" field. To analyze the docking results, click on *Docking Hub->Statistical Analysis of Scoring Functions* vs. *RMSD* (Fig. 15). Then click on the Yes button.

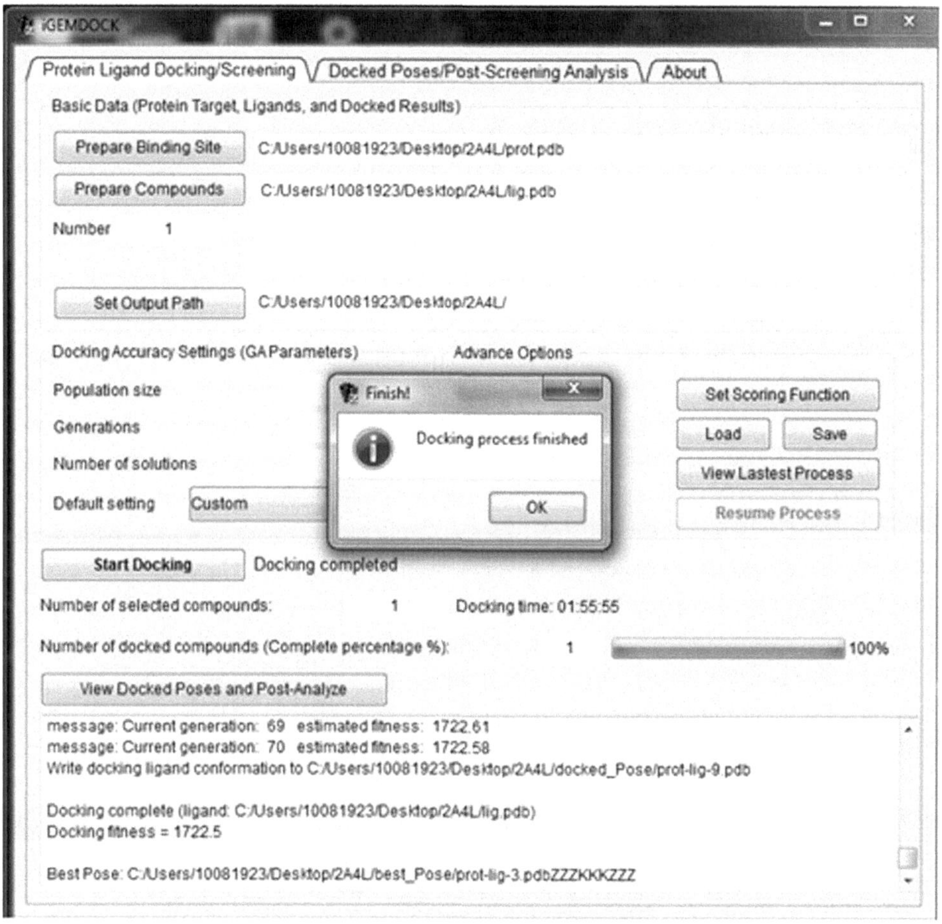

Fig. 9 The main window when GEMDOCK finishes the docking simulation

SAnDReS generates a CSV file with the statistical analysis (strmsd.csv) and shows the partial results on the main GUI window. SAnDReS also creates individual CSV files for each scoring function, as shown in the column in the black rectangle (Fig. 16) To generate plots, click on *Docking Hub->Prepare Files to Plot Redock Results*. On the new pop-up window, select the plot parameters, click on the Generate Files button, and then click on the Close button. Then, click on the "Plot Redock Results (Scatter Plot)" button. In the new pop-up window, click on the "Plot pltcsv File" button. SAnDReS shows the generated plot file on the screen (Fig. 17). All generated data are on the Project Directory. Click the Exit button to finish SAnDReS. As we can see in Fig. 17, we have a successful docking simulation, with docking RMSD of 0.559 Å. We may apply the same procedure to find potential new inhibitors for CDK2 using a dataset of small organic molecules available in the ZINC database [42, 43].

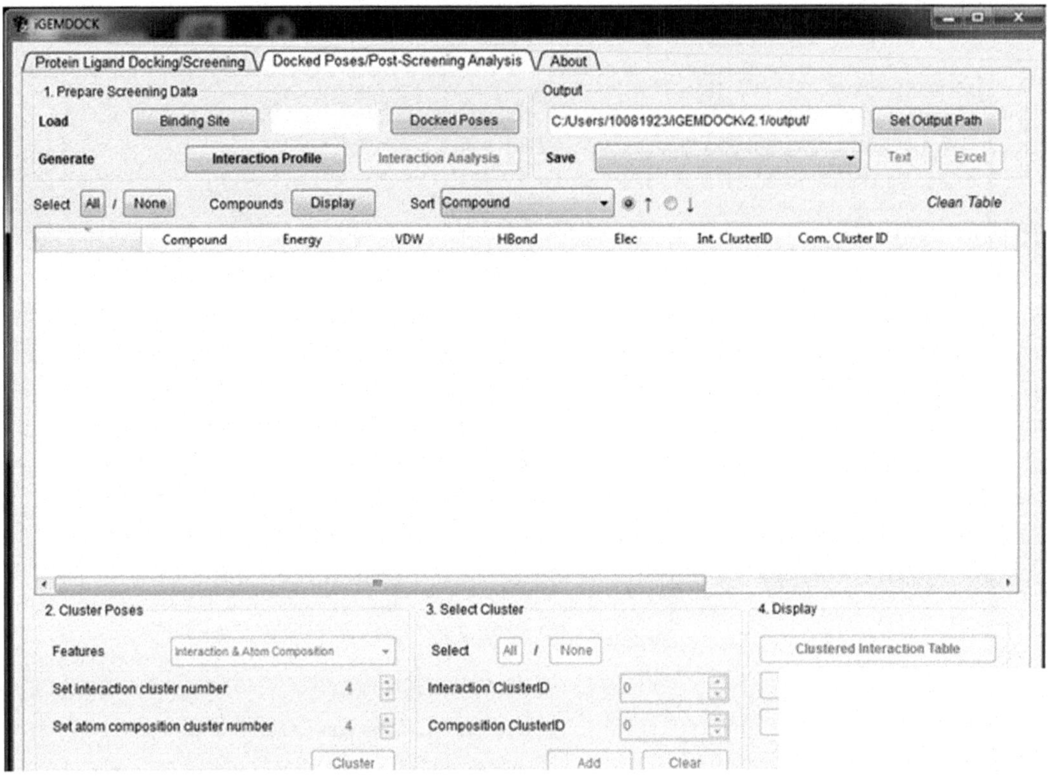

Fig. 10 "Docked Poses/Post-Screening Analysis" window

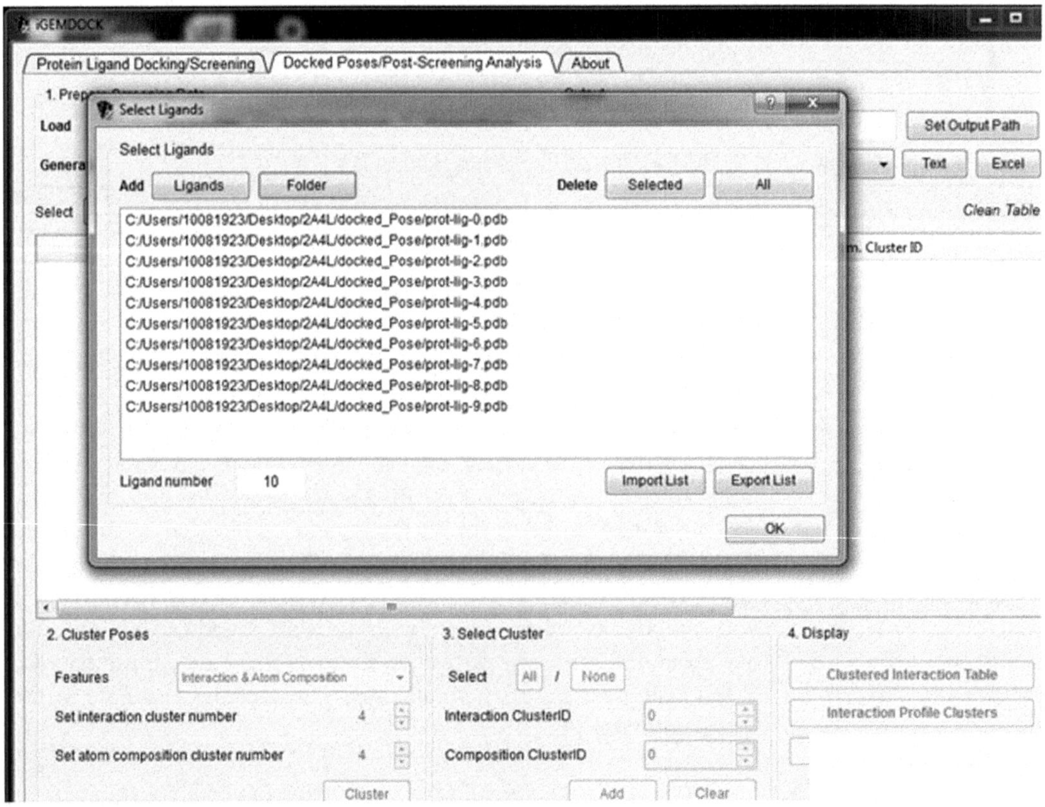

Fig. 11 All poses generated for this docking simulation

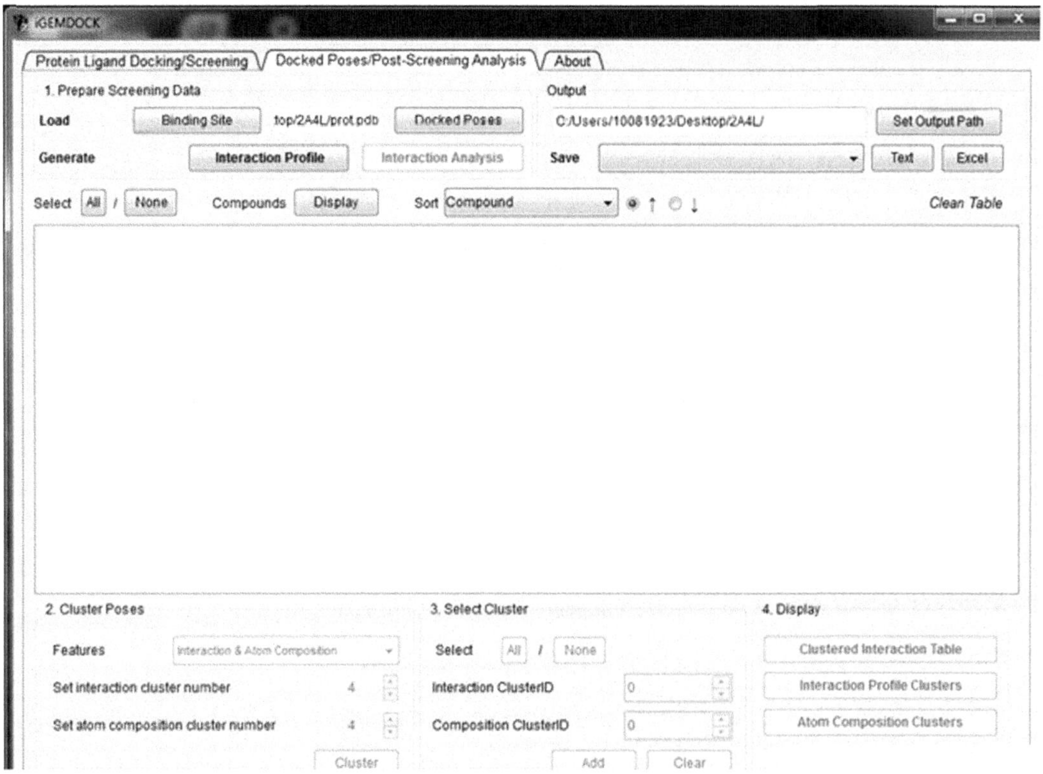

Fig. 12 "Docked Poses/Post-Screening Analysis" window

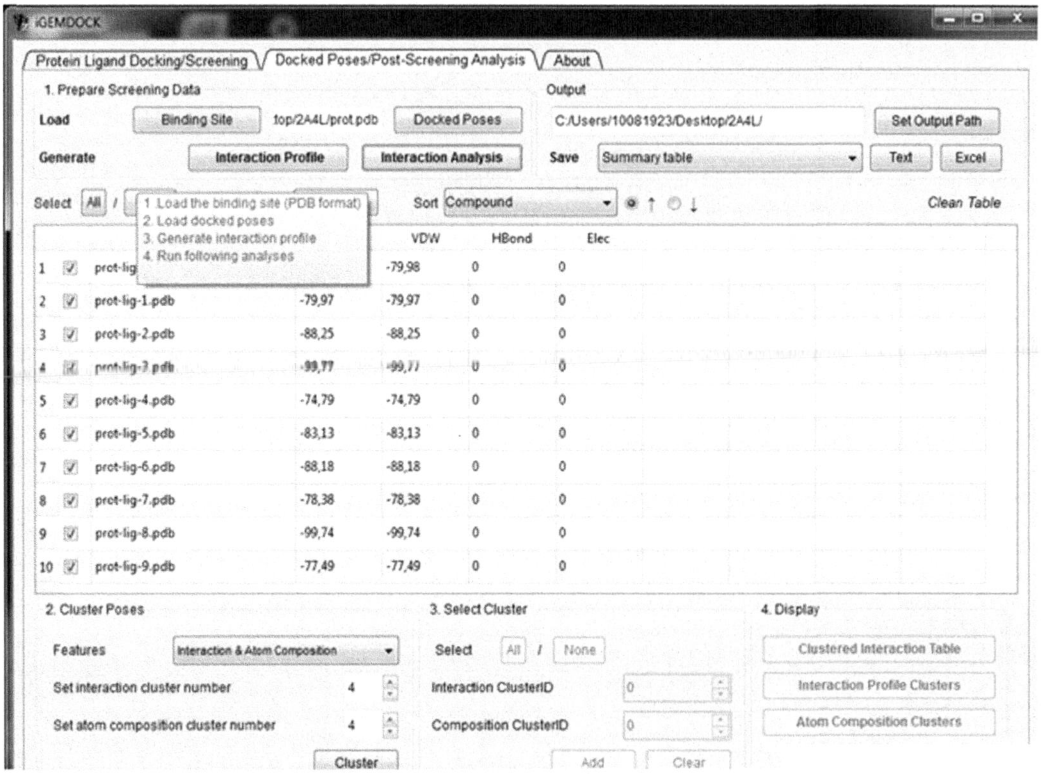

Fig. 13 Docking results and scoring function values

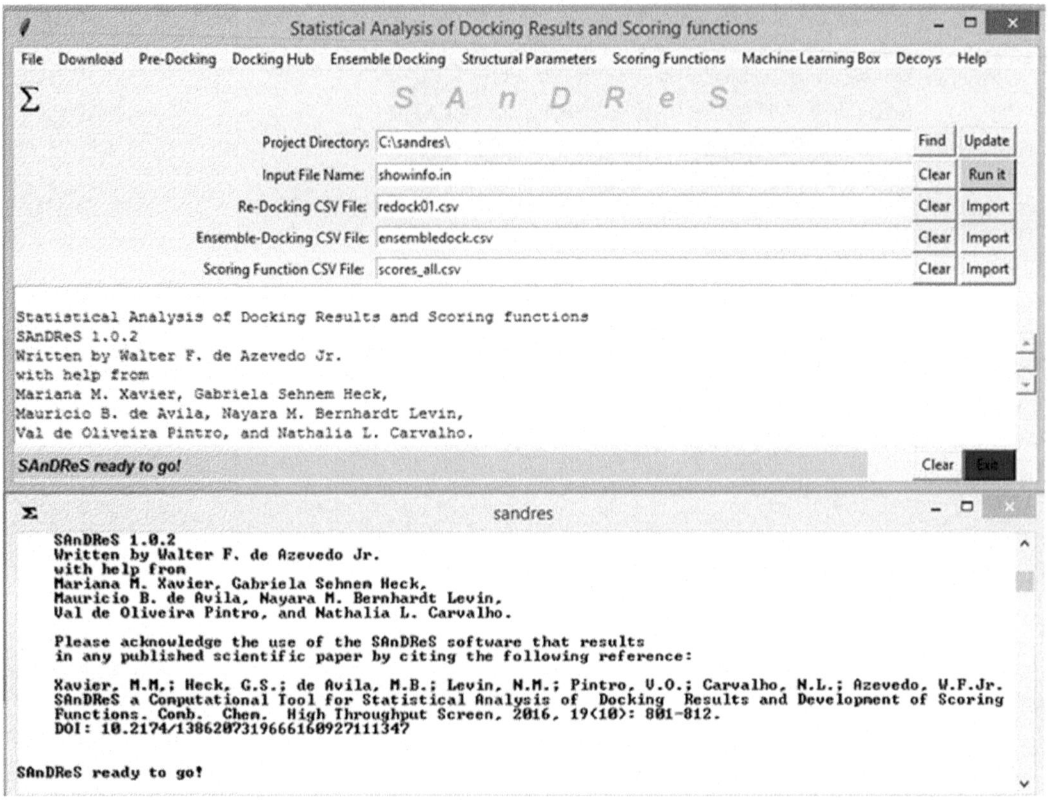

Fig. 14 SAnDReS main GUI

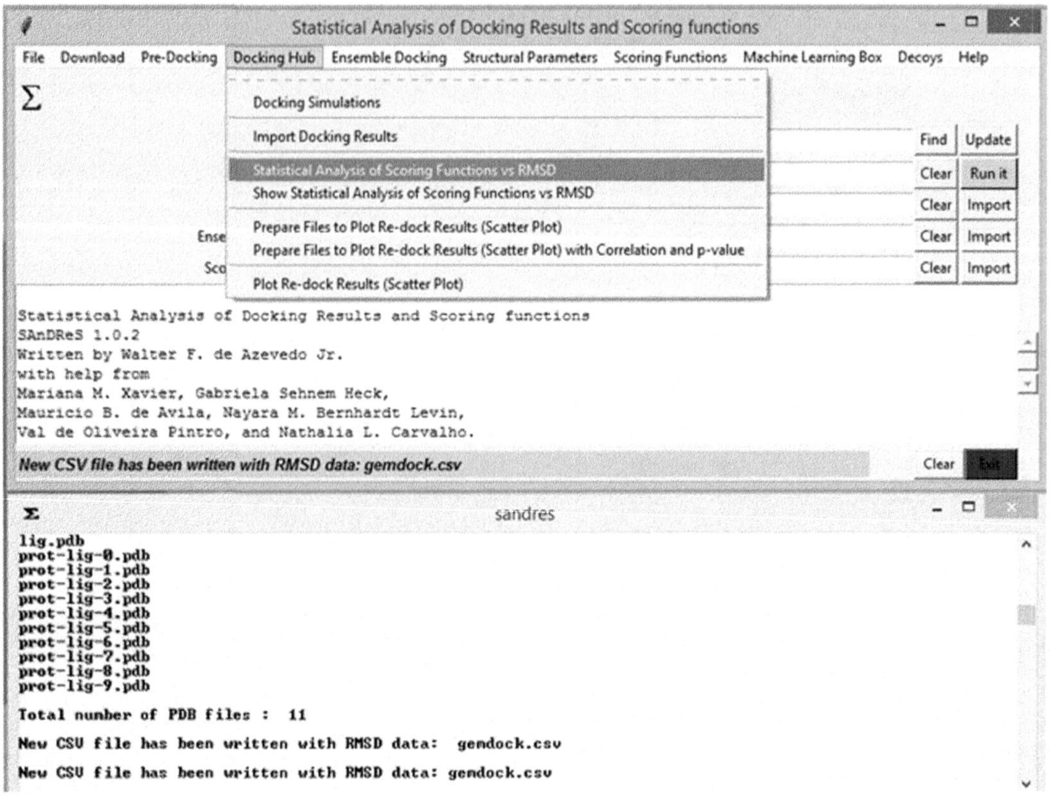

Fig. 15 Procedure of starting the statistical analysis of docking results

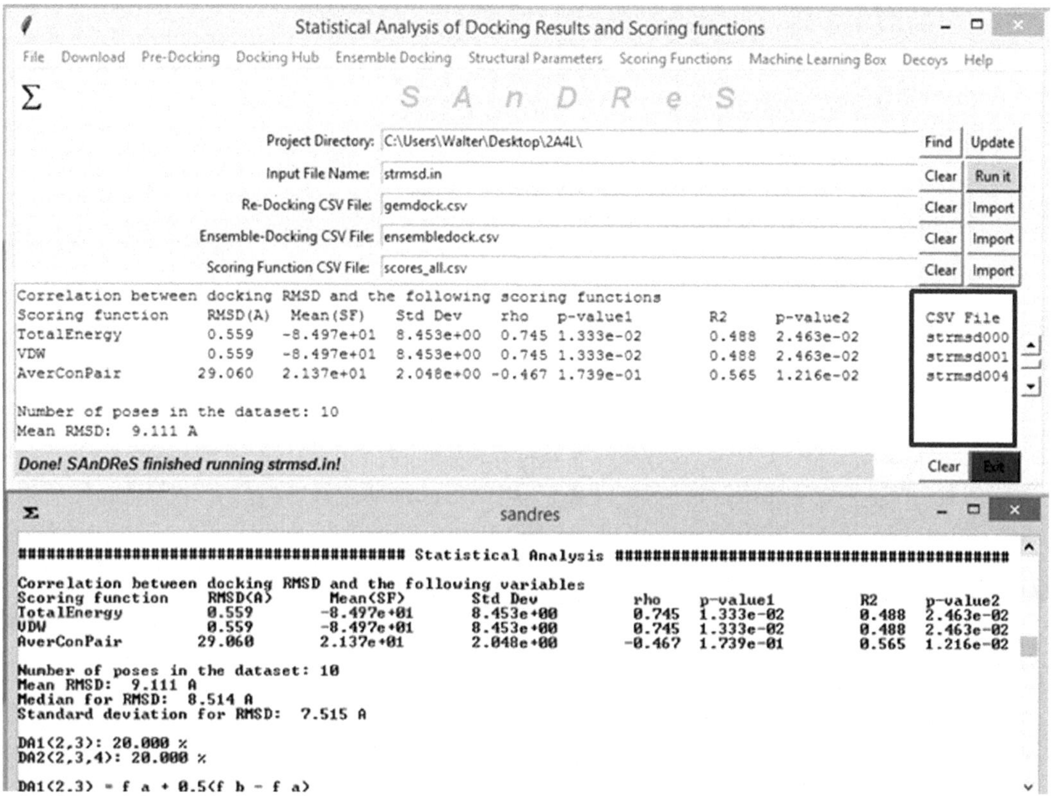

Fig. 16 Statistical analysis of the docking results generated with GEMDOCK

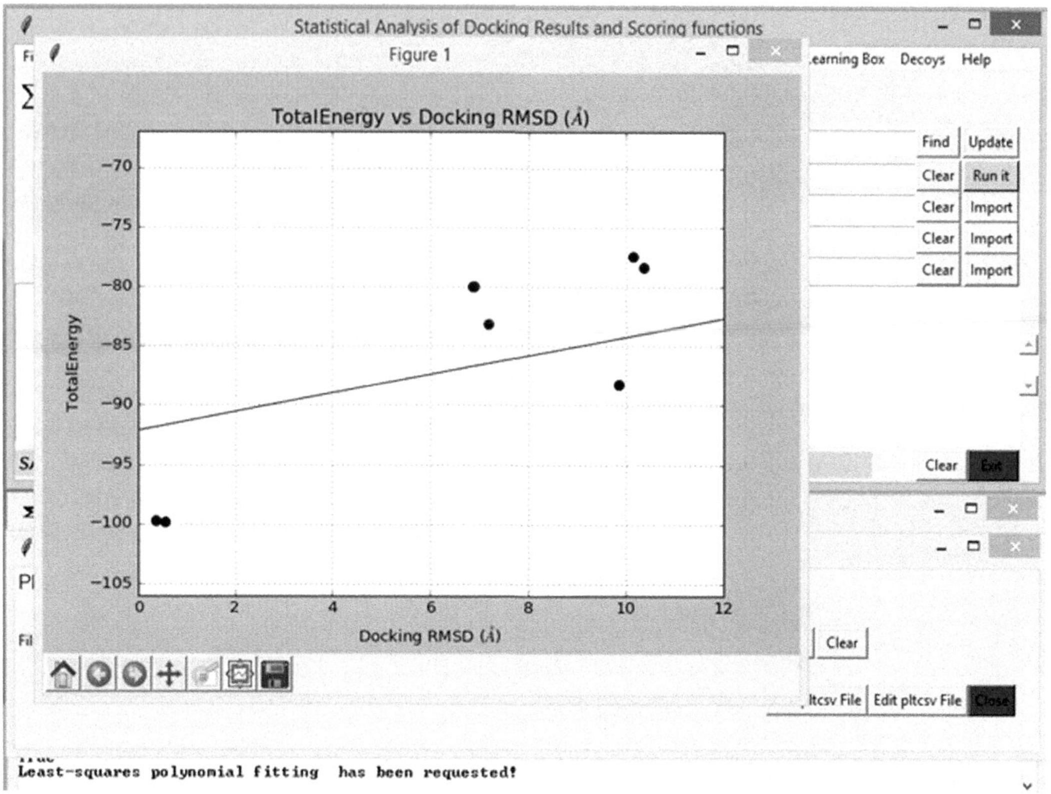

Fig. 17 Scatter plot between docking RMSD and total energy scoring function

4 Availability

All files necessary to run this tutorial are available at https://azevedolab.net/resources/2A4L.zip. The program SAnDReS is available to download at https://github.com/azevedolab/sandres.

5 Colophon

We used the program Molegro Virtual Docker [17] to generate Fig. 1. We employed the program GemDock to create Figs. 3–13. We created Fig. 2 using Microsoft PowerPoint 2016. We used the program SAnDReS [38] to generate Figs. 14–17. We performed molecular docking simulations described in this chapter using a Desktop PC with 4 GB memory, a 1 TB hard disk, and an Intel® Core® i3-2120 at 3.30 GHz processor running Windows 8.1.

6 Final Remarks

Analysis of protein-ligand interactions is a fundamental problem in computer-aided drug design. Assessment of structural and binding data related to protein-ligand complexes helps in the establishment of the structural basis for the binding affinity of the ligand for a broad spectrum of proteins [44–87]. The primary computational approach to address structures of protein-ligand complexes is molecular docking simulation. In this chapter, we discussed the use of differential evolution implemented in the GEMDOCK program to address protein-ligand docking simulations. GEMDOCK is an integrated computational tool to carry out protein-ligand docking simulations. It combines a differential evolution algorithm with an elegant piecewise scoring function that allows the user to carry out all step necessary for docking simulation with the GEMDOCK. We described in details how to carry out docking simulations with GEMDOCK.

Furthermore, we explained how to use the program SAnDReS to evaluate the docking results generated with GEMDOCK. The integration of GEMDOCK and SAnDReS allows a fast and reliable docking simulation. The robust statistical analysis interface of SAnDReS facilitates the analysis of the docking results, allowing the user to test different docking protocols and compare their performance.

Acknowledgments

This work was supported by grants from CNPq (Brazil) (308883/2014-4). This study was financed in part by the Coordenação de Aperfeiçoamento de Pessoal de Nivel Superior—Brasil (CAPES)—Finance Code 001. GB-F acknowledges support from PUCRS/BPA fellowship. WFA is a researcher for CNPq (Brazil) (Process Numbers: 308883/2014-4 and 309029/2018-0).

References

1. Saikia S, Bordoloi M (2019) Molecular docking: challenges, advances and its use in drug discovery perspective. Curr Drug Targets 20:501. https://doi.org/10.2174/1389450119666181022153016
2. Krüger J, Thiel P, Merelli I, Grunzke R, Gesing S (2016) Portals and web-based resources for virtual screening. Curr Drug Targets 17:1649–1660
3. Abdolmaleki A, Ghasemi JB, Ghasemi F (2017) Computer aided drug design for multi-target drug design: SAR/QSAR, molecular docking and pharmacophore methods. Curr Drug Targets 18:556–575
4. Scotti L, Mendonca Junior FJ, Ishiki HM, Ribeiro FF, Singla RK, Barbosa Filho JM et al (2017) Docking studies for multi-target drugs. Curr Drug Targets 18:592–604
5. Sulimov VB, Kutov DC, Sulimov AV (2019) Advances in docking. Curr Med Chem. https://doi.org/10.2174/0929867325666180904115000
6. Freitas PG, Elias TC, Pinto IA, Costa LT, de Carvalho PVSD, Omote DQ et al (2018) Computational approach to the discovery of phytochemical molecules with therapeutic potential targets to the PKCZ protein. Lett Drug Des Discov 15:488–499
7. de Avila MB, de Azevedo WF (2014) Data mining of docking results. Application to 3-dehydroquinate dehydratase. Curr Bioinforma 9:361–379
8. Azevedo LS, Moraes FP, Xavier MM, Pantoja EO, Villavicencio B, Finck JA et al (2012) Recent progress of molecular docking simulations applied to development of drugs. Curr Bioinforma 7:352–365
9. De Azevedo WF Jr (2010) Structure-based virtual screening. Curr Drug Targets 11:261–263
10. De Azevedo WF Jr (2010) MolDock applied to structure-based virtual screening. Curr Drug Targets 11:327–334
11. Dias R, de Azevedo WF Jr (2008) Molecular docking algorithms. Curr Drug Targets 9:1040–1047
12. Heberlé G, de Azevedo WF Jr (2011) Bio-inspired algorithms applied to molecular docking simulations. Curr Med Chem 18:1339–1352
13. Mirzaei H, Zarbafian S, Villar E, Mottarella S, Beglov D, Vajda S et al (2015) Energy minimization on manifolds for docking flexible molecules. J Chem Theory Comput 11:1063–1076
14. Higo J, Dasgupta B, Mashimo T, Kasahara K, Fukunishi Y, Nakamura H (2015) Virtual-system-coupled adaptive umbrella sampling to compute free-energy landscape for flexible molecular docking. J Comput Chem 36:1489–1501
15. Hoffer L, Chira C, Marcou G, Varnek A, Horvath D (2015) S4MPLE—sampler for multiple protein-ligand entities: methodology and rigid-site docking benchmarking. Molecules 20:8997–9028
16. De Azevedo WF, Leclerc S, Meijer L, Havlicek L, Strnad M, Kim SH (1997) Inhibition of cyclin-dependent kinases by purine analogues: crystal structure of human cdk2 complexed with roscovitine. Eur J Biochem 243:518–526
17. Thomsen R, Christensen MH (2006) MolDock: a new technique for high-accuracy molecular docking. J Med Chem 49:3315–3321
18. Goodsell DS, Olson AJ (1990) Automated docking of substrates to proteins by simulated annealing. Proteins 8:195–202
19. Morris GM, Goodsell DS, Huey R, Olson AJ (1996) Distributed automated docking of flexible ligands to proteins: parallel applications of AutoDock 2.4. J Comput Aided Mol Des 10:293–304
20. Morris GM, Goodsell DS, Halliday RS, Huey R, Hart WE, Belew RK et al (1998)

Automated docking using a Lamarckian genetic algorithm and empirical binding free energy function. J Comput Chem 19:1639–1662

21. Morris GM, Huey R, Lindstrom W, Sanner MF, Belew RK, Goodsell DS et al (2009) AutoDock4 and AutoDockTools4: automated docking with selective receptor flexibility. J Comput Chem 30:2785–2791
22. Yang JM (2004) Development and evaluation of a generic evolutionary method for protein-ligand docking. J Comput Chem 25:843–857
23. Yang JM, Chen CC (2004) GEMDOCK: a generic evolutionary method for molecular docking. Proteins 55:288–304
24. Hsu KC, Chen YF, Lin SR, Yang JM (2011) iGEMDOCK: a graphical environment of enhancing GEMDOCK using pharmacological interactions and post-screening analysis. BMC Bioinformatics 12(Suppl 1):33
25. Storn R, Price KV (1997) Differential evolution: a simple and efficient heuristic for global optimization over continuous spaces. J Glob Optim 11:341–369
26. Morgan DO (1995) Principles of CDK regulation. Nature 374:131–134
27. Murray AW (1994) Cyclin-dependent kinases: regulators of the cell cycle and more. Chem Biol 1:191–195
28. Volkart PA, Bitencourt-Ferreira G, Souto AA, de Azevedo WF (2019) Cyclin-dependent kinase 2 in cellular senescence and cancer. A structural and functional review. Curr Drug Targets 20:716–726. https://doi.org/10.2174/1389450120666181204165344
29. Kim SH, Schulze-Gahmen U, Brandsen J, de Azevedo Júnior WF (2005) Structural basis for chemical inhibition of CDK2. Prog Cell Cycle Res 2:137–145
30. De Azevedo WF Jr, Mueller-Dieckmann HJ, Schulze-Gahmen U, Worland PJ, Sausville E, Kim SH (1996) Structural basis for specificity and potency of a flavonoid inhibitor of human CDK2, a cell cycle kinase. Proc Natl Acad Sci U S A 93:2735–2740
31. Canduri F, de Azevedo WF Jr (2005) Structural basis for interaction of inhibitors with cyclin-dependent kinase 2. Curr Comput Aided Drug Des 1:53–64
32. Krystof V, Cankar P, Frysová I, Slouka J, Kontopidis G, Dzubák P et al (2006) 4-arylazo-3,5-diamino-1H-pyrazole CDK inhibitors: SAR study, crystal structure in complex with CDK2, selectivity, and cellular effects. J Med Chem 49:6500–6509
33. Levin NMB, Pintro VO, Bitencourt-Ferreira G, Mattos BB, Silvério AC, de Azevedo WF Jr (2018) Development of CDK-targeted scoring functions for prediction of binding affinity. Biophys Chem 235:1–8
34. de Ávila MB, Xavier MM, Pintro VO, de Azevedo WF (2017) Supervised machine learning techniques to predict binding affinity. A study for cyclin-dependent kinase 2. Biochem Biophys Res Commun 494:305–310
35. Levin NM, Pintro VO, de Ávila MB, de Mattos BB, De Azevedo WF Jr (2017) Understanding the structural basis for inhibition of cyclin-dependent kinases. New pieces in the molecular puzzle. Curr Drug Targets 18:1104–1111
36. De Bondt HL, Rosenblatt J, Jancarik J, Jones HD, Morgan DO, Kim SH (1993) Crystal structure of cyclin-dependent kinase 2. Nature 363:595–602
37. Schulze-Gahmen U, De Bondt HL, Kim SH (1996) High-resolution crystal structures of human cyclin-dependent kinase 2 with and without ATP: bound waters and natural ligand as guides for inhibitor design. J Med Chem 39:4540–4546
38. Xavier MM, Heck GS, de Avila MB, Levin NM, Pintro VO, Carvalho NL et al (2016) SAnDReS a computational tool for statistical analysis of docking results and development of scoring functions. Comb Chem High Throughput Screen 19:801–812
39. Berman HM, Westbrook J, Feng Z, Gilliland G, Bhat TN, Weissig H et al (2000) The Protein Data Bank. Nucleic Acids Res 28:235–242
40. Berman HM, Battistuz T, Bhat TN, Bluhm WF, Bourne PE, Burkhardt K et al (2002) The Protein Data Bank. Acta Crystallogr D Biol Crystallogr 58:899–907
41. Westbrook J, Feng Z, Chen L, Yang H, Berman HM (2003) The Protein Data Bank and structural genomics. Nucleic Acids Res 31:489–491
42. Irwin JJ, Shoichet BK (2005) ZINC—a free database of commercially available compounds for virtual screening. J Chem Inf Model 45:177–182
43. Irwin JJ, Sterling T, Mysinger MM, Bolstad ES, Coleman RG (2012) ZINC: a free tool to discover chemistry for biology. J Chem Inf Model 52:1757
44. Canduri F, Fadel V, Basso LA, Palma MS, Santos DS, de Azevedo WF Jr (2005) New catalytic mechanism for human purine nucleoside phosphorylase. Biochem Biophys Res Commun 327(3):646–649
45. Filgueira de Azevedo W Jr, Canduri F, Simões de Oliveira J, Basso LA, Palma MS, Pereira JH et al (2002) Molecular model of shikimate

kinase from *Mycobacterium tuberculosis*. Biochem Biophys Res Commun 295:142–148

46. Canduri F, Teodoro LG, Fadel V, Lorenzi CC, Hial V, Gomes RA et al (2001) Structure of human uropepsin at 2.45 A resolution. Acta Crystallogr D Biol Crystallogr 57:1560–1570
47. Pereira JH, Canduri F, de Oliveira JS, da Silveira NJ, Basso LA, Palma MS et al (2003) Structural bioinformatics study of EPSP synthase from *Mycobacterium tuberculosis*. Biochem Biophys Res Commun 312:608–614
48. de Azevedo WF Jr, Dias R (2008) Experimental approaches to evaluate the thermodynamics of protein-drug interactions. Curr Drug Targets 9:1071–1076
49. Delatorre P, Rocha BA, Souza EP, Oliveira TM, Bezerra GA, Moreno FB et al (2007) Structure of a lectin from *Canavalia gladiata* seeds: new structural insights for old molecules. BMC Struct Biol 7:52
50. de Azevedo WF Jr, Canduri F, dos Santos DM, Pereira JH, Bertacine Dias MV, Silva RG et al (2003) Crystal structure of human PNP complexed with guanine. Biochem Biophys Res Commun 312:767–772
51. Canduri F, de Azevedo WF (2008) Protein crystallography in drug discovery. Curr Drug Targets 9:1048–1053
52. Marques MR, Pereira JH, Oliveira JS, Basso LA, de Azevedo WF Jr, Santos DS et al (2007) The inhibition of 5-enolpyruvylshikimate-3-phosphate synthase as a model for development of novel antimicrobials. Curr Drug Targets 8:445–457
53. Filgueira de Azevedo W Jr, dos Santos GC, dos Santos DM, Olivieri JR, Canduri F, Silva RG et al (2003) Docking and small angle X-ray scattering studies of purine nucleoside phosphorylase. Biochem Biophys Res Commun 309:923–928
54. Canduri F, Perez PC, Caceres RA, de Azevedo WF Jr (2007) Protein kinases as targets for antiparasitic chemotherapy drugs. Curr Drug Targets 8:389–398
55. Dias MV, Borges JC, Ely F, Pereira JH, Canduri F, Ramos CH et al (2006) Structure of chorismate synthase from *Mycobacterium tuberculosis*. J Struct Biol 154:130–143
56. Dias MV, Ely F, Palma MS, de Azevedo WF Jr, Basso LA, Santos DS (2007) Chorismate synthase: an attractive target for drug development against orphan diseases. Curr Drug Targets 8:437–444
57. Silva RG, Pereira JH, Canduri F, de Azevedo WF Jr, Basso LA, Santos DS (2005) Kinetics and crystal structure of human purine nucleoside phosphorylase in complex with 7-methyl-6-thio-guanosine. Arch Biochem Biophys 442:49–58
58. Timmers LF, Caceres RA, Vivan AL, Gava LM, Dias R, Ducati RG et al (2008) Structural studies of human purine nucleoside phosphorylase: towards a new specific empirical scoring function. Arch Biochem Biophys 479:28–38
59. de Azevedo WF Jr (2011) Molecular dynamics simulations of protein targets identified in *Mycobacterium tuberculosis*. Curr Med Chem 18:1353–1366
60. de Azevedo WF Jr (2011) Protein targets for development of drugs against *Mycobacterium tuberculosis*. Curr Med Chem 18:1255–1257
61. Caceres RA, Saraiva Timmers LF, Dias R, Basso LA, Santos DS, de Azevedo WF Jr (2008) Molecular modeling and dynamics simulations of PNP from *Streptococcus agalactiae*. Bioorg Med Chem 16:4984–4993
62. Dias MV, Faím LM, Vasconcelos IB, de Oliveira JS, Basso LA, Santos DS et al (2007) Effects of the magnesium and chloride ions and shikimate on the structure of shikimate kinase from *Mycobacterium tuberculosis*. Acta Crystallogr Sect F Struct Biol Cryst Commun 63:1–6
63. de Azevedo WF Jr, Ward RJ, Canduri F, Soares A, Giglio JR, Arni RK (1998) Crystal structure of piratoxin-I: a calcium-independent, myotoxic phospholipase A2-homologue from *Bothrops pirajai* venom. Toxicon 36:1395–1406
64. Dias R, Timmers LF, Caceres RA, de Azevedo WF Jr (2008) Evaluation of molecular docking using polynomial empirical scoring functions. Curr Drug Targets 9:1062–1070
65. da Silveira NJ, Uchôa HB, Canduri F, Pereira JH, Camera JC Jr, Basso LA et al (2004) Structural bioinformatics study of PNP from *Schistosoma mansoni*. Biochem Biophys Res Commun 322:100–104
66. de Azevedo WF Jr, Dias R (2008) Evaluation of ligand-binding affinity using polynomial empirical scoring functions. Bioorg Med Chem 16:9378–9382
67. Bezerra GA, Oliveira TM, Moreno FB, de Souza EP, da Rocha BA, Benevides RG et al (2007) Structural analysis of *Canavalia maritima* and *Canavalia gladiata* lectins complexed with different dimannosides: new insights into the understanding of the structure-biological activity relationship in legume lectins. J Struct Biol 160:168–176
68. Canduri F, Fadel V, Dias MV, Basso LA, Palma MS, Santos DS et al (2005) Crystal structure of human PNP complexed with hypoxanthine

and sulfate ion. Biochem Biophys Res Commun 326:335–338

69. Timmers LF, Pauli I, Caceres RA, de Azevedo WF Jr (2008) Drug-binding databases. Curr Drug Targets 9:1092–1099
70. Delatorre P, Rocha BA, Gadelha CA, Santi-Gadelha T, Cajazeiras JB, Souza EP et al (2006) Crystal structure of a lectin from *Canavalia maritima* (ConM) in complex with trehalose and maltose reveals relevant mutation in ConA-like lectins. J Struct Biol 154:280–286
71. Rádis-Baptista G, Moreno FB, de Lima Nogueira L, Martins AM, de Oliveira Toyama D, Toyama MH et al (2006) Crotacetin, a novel snake venom C-type lectin homolog of convulxin, exhibits an unpredictable antimicrobial activity. Cell Biochem Biophys 44:412–423
72. Breda A, Basso LA, Santos DS, de Azevedo WF Jr (2008) Virtual screening of drugs: score functions, docking, and drug design. Curr Comput Aided Drug Des 4:265–272
73. Nolasco DO, Canduri F, Pereira JH, Cortinóz JR, Palma MS, Oliveira JS et al (2004) Crystallographic structure of PNP from *Mycobacterium tuberculosis* at 1.9A resolution. Biochem Biophys Res Commun 324:789–794
74. Arcuri HA, Canduri F, Pereira JH, da Silveira NJ, Camera Júnior JC, de Oliveira JS et al (2004) Molecular models for shikimate pathway enzymes of *Xylella fastidiosa*. Biochem Biophys Res Commun 320:979–991
75. Soares MB, Silva CV, Bastos TM, Guimarães ET, Figueira CP, Smirlis D et al (2012) Anti-*Trypanosoma cruzi* activity of nicotinamide. Acta Trop 12:224–229
76. Rocha BA, Delatorre P, Oliveira TM, Benevides RG, Pires AF, Sousa AA et al (2011) Structural basis for both pro- and anti-inflammatory response induced by mannose-specific legume lectin from *Cymbosema roseum*. Biochimie 93:806–816
77. Ducati RG, Basso LA, Santos DS, de Azevedo WF Jr (2010) Crystallographic and docking studies of purine nucleoside phosphorylase from *Mycobacterium tuberculosis*. Bioorg Med Chem 18:4769–4774
78. Manhani KK, Arcuri HA, da Silveira NJ, Uchôa HB, de Azevedo WF Jr, Canduri F (2005) Molecular models of protein kinase 6 from *Plasmodium falciparum*. J Mol Model 12:42–48
79. Arcuri HA, Borges JC, Fonseca IO, Pereira JH, Neto JR, Basso LA et al (2008) Structural studies of shikimate 5-dehydrogenase from *Mycobacterium tuberculosis*. Proteins 72:720–730
80. Marques MR, Vaso A, Neto JR, Fossey MA, Oliveira JS, Basso LA et al (2008) Dynamics of glyphosate-induced conformational changes of *Mycobacterium tuberculosis* 5-enolpyruvylshikimate-3-phosphate synthase (EC 2.5.1.19) determined by hydrogen-deuterium exchange and electrospray mass spectrometry. Biochemistry 47:7509–7522
81. Cavada BS, Moreno FB, da Rocha BA, de Azevedo WF Jr, Castellón RE, Goersch GV et al (2006) cDNA cloning and 1.75 A crystal structure determination of PPL2, an endochitinase and N-acetylglucosamine-binding hemagglutinin from *Parkia platycephala* seeds. FEBS J 273:3962–3974
82. Arcuri HA, Zafalon GF, Marucci EA, Bonalumi CE, da Silveira NJ, Machado JM (2010) SKPDB: a structural database of shikimate pathway enzymes. BMC Bioinformatics 11:12
83. Moreno FB, de Oliveira TM, Martil DE, Viçoti MM, Bezerra GA, Abrego JR et al (2008) Identification of a new quaternary association for legume lectins. J Struct Biol 161:133–143
84. Russo S, de Azevedo WF (2019) Advances in the understanding of the cannabinoid receptor 1 - focusing on the inverse agonists interactions. Curr Med Chem. https://doi.org/10.2174/0929867325666180417165247
85. Amaral MEA, Nery LR, Leite CE, de Azevedo Junior WF, Campos MM (2018) Pre-clinical effects of metformin and aspirin on the cell lines of different breast cancer subtypes. Investig New Drugs 36:782–796
86. Bitencourt-Ferreira G, de Azevedo WF Jr (2018) Development of a machine-learning model to predict Gibbs free energy of binding for protein-ligand complexes. Biophys Chem 240:63–69
87. de Ávila MB, Bitencourt-Ferreira G, de Azevedo WF Jr (2019) Structural basis for inhibition of enoyl-[acyl carrier protein] reductase (InhA) from *Mycobacterium tuberculosis*. Curr Med Chem. https://doi.org/10.2174/0929867326666181203125229

Chapter 12

Docking with SwissDock

Gabriela Bitencourt-Ferreira and Walter Filgueira de Azevedo Jr.

Abstract

Protein-ligand docking simulation is central in drug design and development. Therefore, the development of web servers intended to docking simulations is of pivotal importance. SwissDock is a web server dedicated to carrying out protein-ligand docking simulation intuitively and elegantly. SwissDock is based on the protein-ligand docking program EADock DSS and has a simple and integrated interface. The SwissDock allows the user to upload structure files for a protein and a ligand, and returns the results by e-mail. To facilitate the upload of the protein and ligand files, we can prepare these input files using the program UCSF Chimera. In this chapter, we describe how to use UCSF Chimera and SwissDock to perform protein-ligand docking simulations. To illustrate the process, we describe the molecular docking of the competitive inhibitor roscovitine against the structure of human cyclin-dependent kinase 2.

Key words SwissDock, Molecular docking, Cyclin-dependent kinase 2, Drug design, Protein-ligand interactions

1 Introduction

Protein-ligand docking simulations are one of the most used computational approaches in the computer-aided drug design [1–10]. Applications of protein-ligand docking simulations have the potential of identifying ligands for a specific protein target. Such results may speedup drug design and development since it is possible to carry out docking simulations of thousands of potential ligands against a protein target; this procedure is named virtual screening [11–20]. The success of the identification of inhibitors of HIV-1 protease illustrates the potential of such in silico approaches [21–30].

In parallel with the development of new computational tools to perform docking simulations, we witnessed an explosion in the number of experimental structures of protein targets. Most of these structures present ligands complexed with the protein. Such richness of information has the potential to be applied to validate protein-ligand docking programs and also to develop empirical scoring functions targeted at specific protein systems. These

Walter Filgueira de Azevedo Jr. (ed.), *Docking Screens for Drug Discovery*, Methods in Molecular Biology, vol. 2053,
https://doi.org/10.1007/978-1-4939-9752-7_12,

computational methodologies improve docking accuracy and can generate scoring functions calibrated to biological systems of interest [31–40].

The development of web servers dedicated to molecular docking simulations opens the possibility to carry out analysis of intermolecular interactions using your browser. Such facilities are convenient for many research groups interested in some aspects of the docking simulation but not necessarily willing to dedicate time and resources to install specific molecular docking tools. Although many docking programs are freeware, there are docking packages that can cost thousands of dollars for a single machine license. Among the most used web servers dedicated to protein-ligand docking simulations, we have the following: DockingServer (http://www.dockingserver.com/web), Blaster [41], DockingAtUTMB (http://docking.utmb.edu/), Pardock (http://www.scfbio-iitd.res.in/dock/pardock.jsp), PatchDock (http://bioinfo3d.cs.tau.ac.il/PatchDock/), MetaDock (http://dock.bioinfo.pl/), PPDock (http://140.112.135.49/ppdock/index.html) and MEDock (http://medock.ee.ncku.edu.tw/), and SwissDock (http://www.swissdock.ch/docking) [42].

Among these webservers that are freely available to perform protein-ligand docking simulations, the SwissDock is the most used for molecular docking with over 380 citations in the web of science (search carried out on January 12, 2019). SwissDock has overall performance similar to other docking programs such as AutoDock [43–46], Molegro Virtual Docker [47–49], and AutoDock Vina [50]. The web server SwissDock uses the protein-ligand docking program EADock DSS [51], whose algorithm contains the following steps:

1. Generation of several binding modes centered in a virtual box (local docking) or close to docking cavities (blind docking).
2. Evaluation of the protein-ligand binding energies using a CHARMM-based scoring function.
3. Selection and clustering of the lowest energy poses.
4. Download the most favorable clusters.

In this chapter, we describe in detail how to carry out protein-ligand docking simulations using SwissDock. To prepare all files necessary to perform docking with SwissDock, we use the program UCSF Chimera [52]. To illustrate the application of UCSF Chimera and SwissDock, we describe the redocking simulation of an inhibitor against the structure of human cyclin-dependent kinase 2 (CDK2).

2 Biological System

In this tutorial, we show how to perform protein-ligand docking simulations of cyclin-dependent kinase 2 (CDK2) (EC 2.7.11.22) with SwissDock [42]. CDK2 is involved in the control of cell cycle progression, and its inhibition has been shown to stop cell cycle, thereby leading to cell apoptosis. Such a mechanism has a high potential of being used in the treatment of cancer [53–60]. Due to its importance, CDK2 has been submitted to intensive structural and functional studies. There are over 400 crystallographic structures of CDK2 at the Protein Data Bank (PDB) (search carried out on January 12, 2019). Here, we perform our docking simulations with the structure 2A4l [61].

3 Graphical Tutorial

In the flowchart shown in Fig. 1, we see the main steps to redock a ligand in the structure of a protein using UCSF Chimera and SwissDock. For redocking purposes, the first step is the downloading of a protein structure in complex with a small molecule not covalently bound to the protein. Following this, we prepare the coordinate files with the program UCSF Chimera. Then, we are ready to carry out docking simulations with SwissDock. We upload the protein and ligand files, and then, we perform the docking

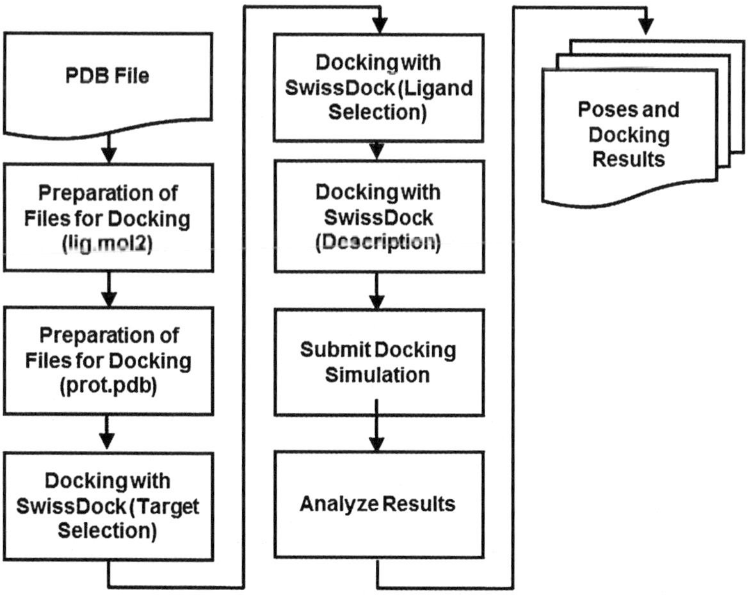

Fig. 1 Flowchart describing all steps to carry out protein-ligand docking simulations using UCSF Chimera and SwissDock

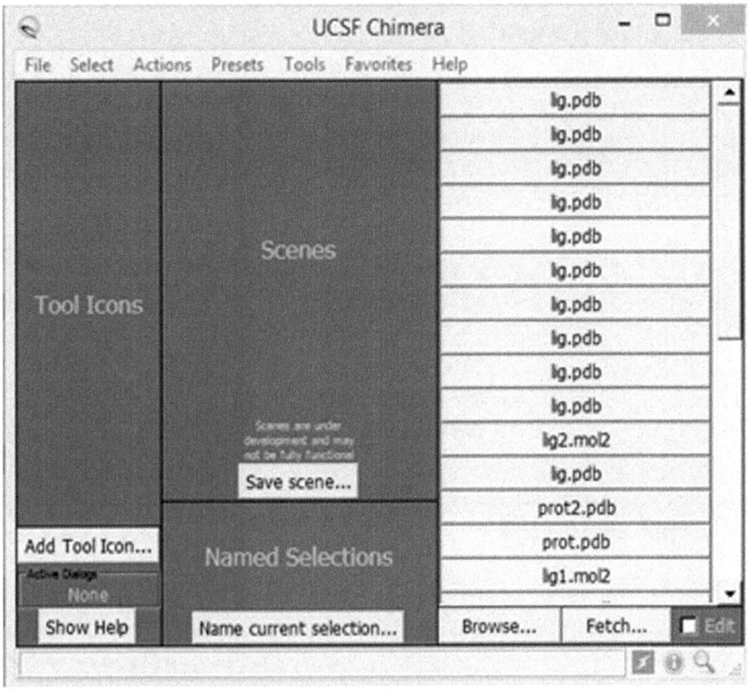

Fig. 2 Main window of UCSF Chimera

simulation. The final steps involve analysis of the docking results. In the following text, we describe all the steps in detail.

We consider that you have UCSF Chimera installed on your computer, and it is open, as shown in Fig. 2. We used the version 1.11.2, but this tutorial should work for earlier versions. We used UCSF Chimera for Windows; it is mostly the same for the Linux and Mac OS X versions. To load a new structure file, click on *File-> Open...* Then browse the folder where the lig.pdb and prot.pdb files are. You can download a zipped folder with these files by clicking here: https://azevedolab.net/resources/SwiisDock_2A4L_files.zip. Now we choose lig.pdb file and click on the Open button (Fig. 3). There we go, a beautiful-looking view of the roscovitine molecule (Fig. 4).

To add hydrogen atoms, click on *Tools->Structure Editing-> AddH*. On the new pop-up window (Fig. 5), we select the hydrogen option and click on the OK button. The hydrogen atoms have been added to the roscovitine structure. The hydrogen atoms are indicated in white in the molecular structure. Now we are ready to save this structure as a mol2 file. Click on *File->Save as Mol2...* We keep the same root name for the ligand (lig). Then we click on the Save button. Now we close this session to start taking care of the *prot.pdb* file, then we click on *File->Close Session*.

We reopen the UCSF Chimera and click on *File->Open...* As previously seen in this tutorial, browse to the folder where lig.pdb

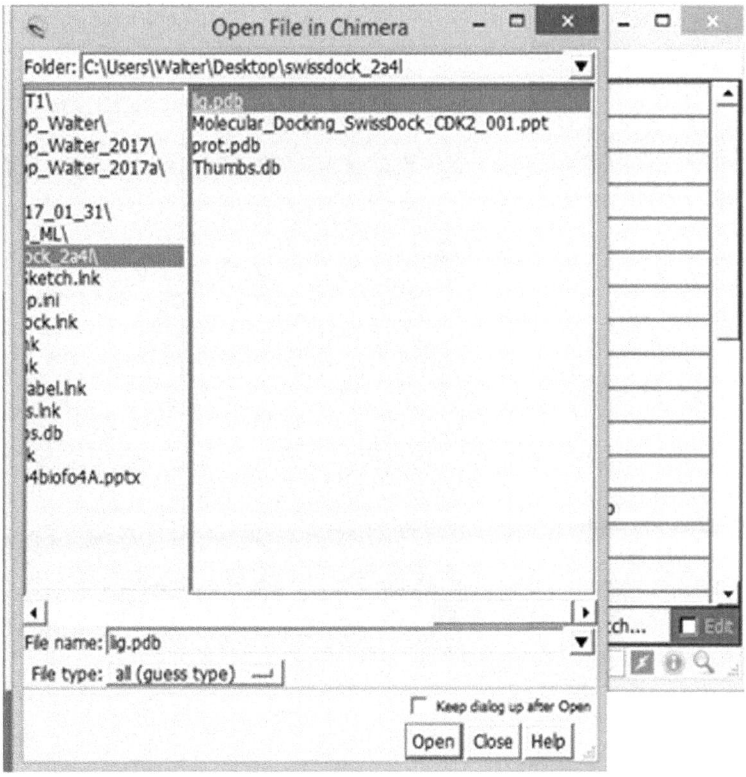

Fig. 3 "Open File in Chimera" window

and prot.pdb files are. Then click on prot.pdb and the Open button. We have the ribbon representation of the CDK2 structure (Fig. 6). To prepare the protein file for docking, click on *Tools-> Structure Editing->Dock Prep.* On the new pop-up window, unmark "Write Mol2 file" option and click OK (Fig. 7).

On the "Add Hydrogen for Dock Prep" window, we leave the default parameters and click on the OK button. On the "Assign Charges for Dock" window, we leave the default parameters and click OK. Once finished, click on *File-Save PDB...* We are going to keep the same filename and overwrite the original file. Click on the Save button. Then, click Yes. Now we close the program. Click on *File->Quit.* To carry out docking simulation, we go to http://www.swissdock.ch/docking. We have the entry page of SwissDock (Fig. 8).

To perform docking with SwissDock, firstly we select the target, click on "upload file" option. Then, we click on the "Choose File" button. We go to the folder where the structures are and upload *prot.pdb* file. Then, SwissDock will carry out a preliminary analysis of the structure (Fig. 9). It may take a few seconds... If everything goes fine, you will get the "Successful setup" message. To upload lig.mol2, click on "upload file" option. Click on the

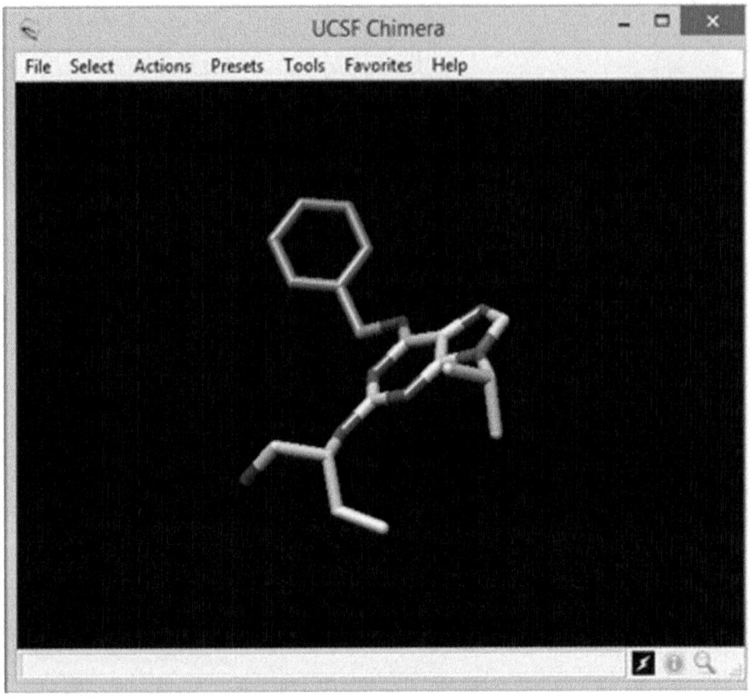

Fig. 4 Structure of the inhibitor roscovitine on the UCSF Chimera

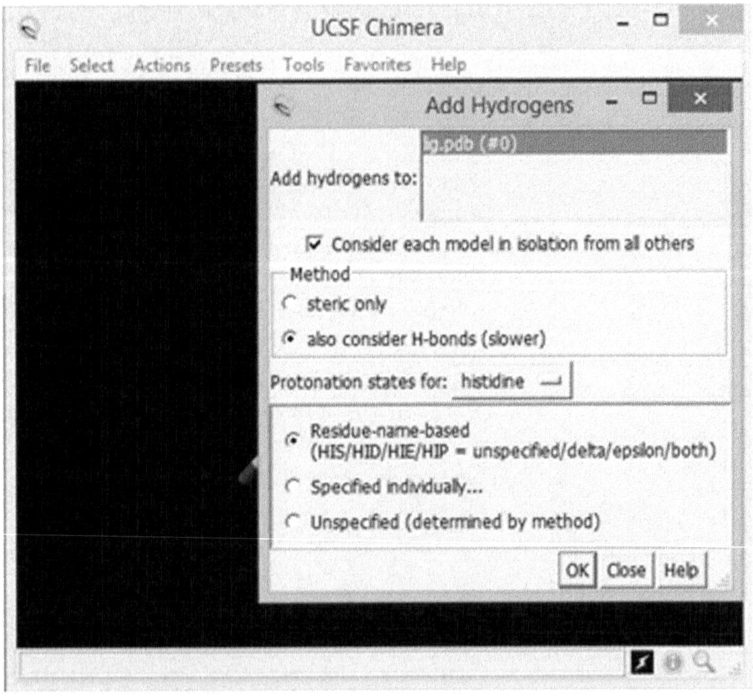

Fig. 5 "Add Hydrogens" window

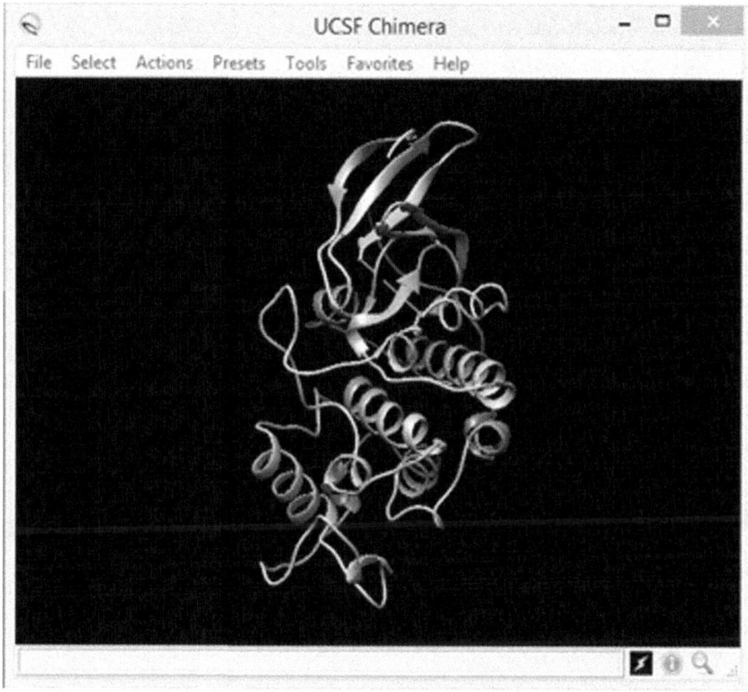

Fig. 6 Ribbon structure of CDK2

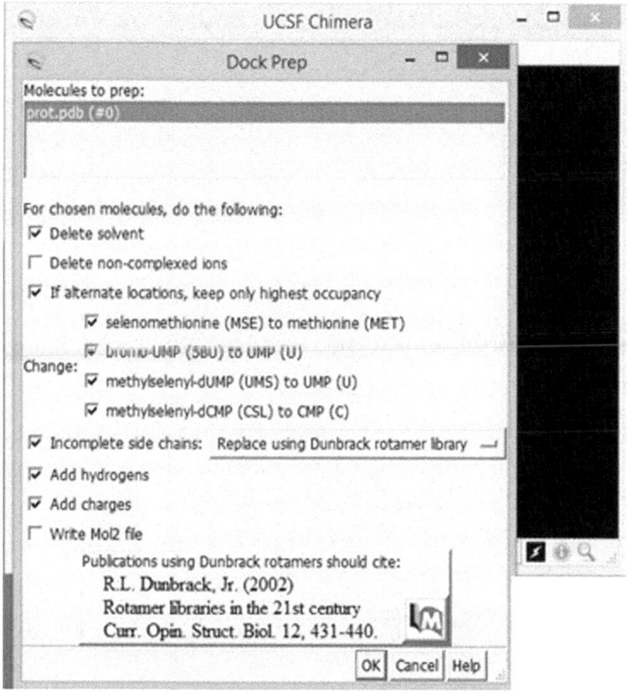

Fig. 7 "Dock prep" window

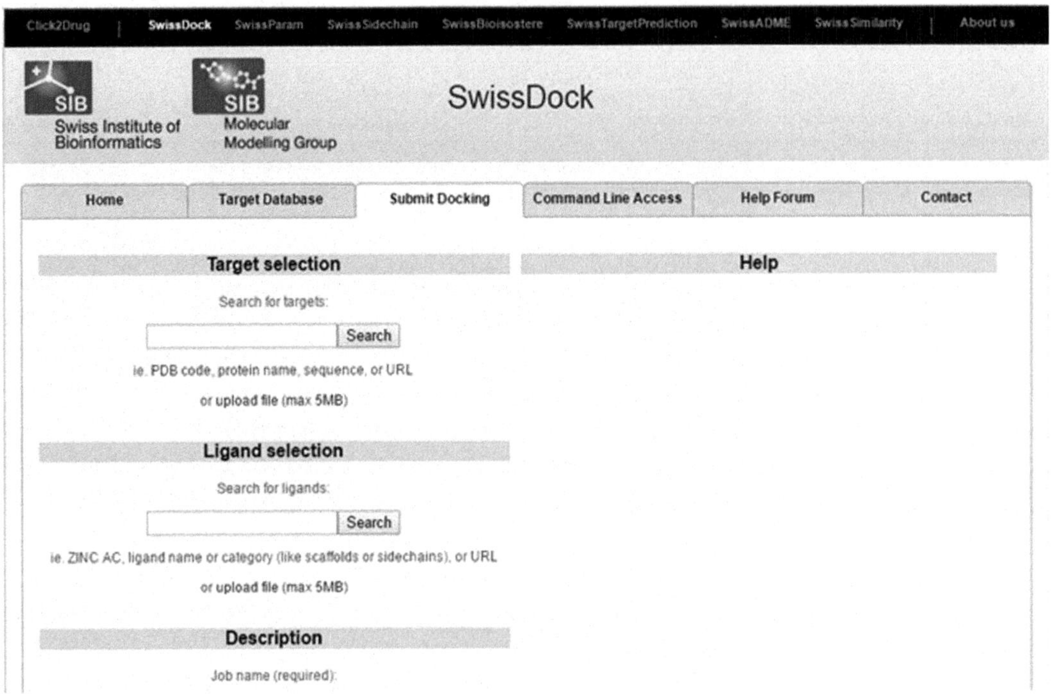

Fig. 8 Entry page of SwissDock

Fig. 9 SwissDock checks the target structure

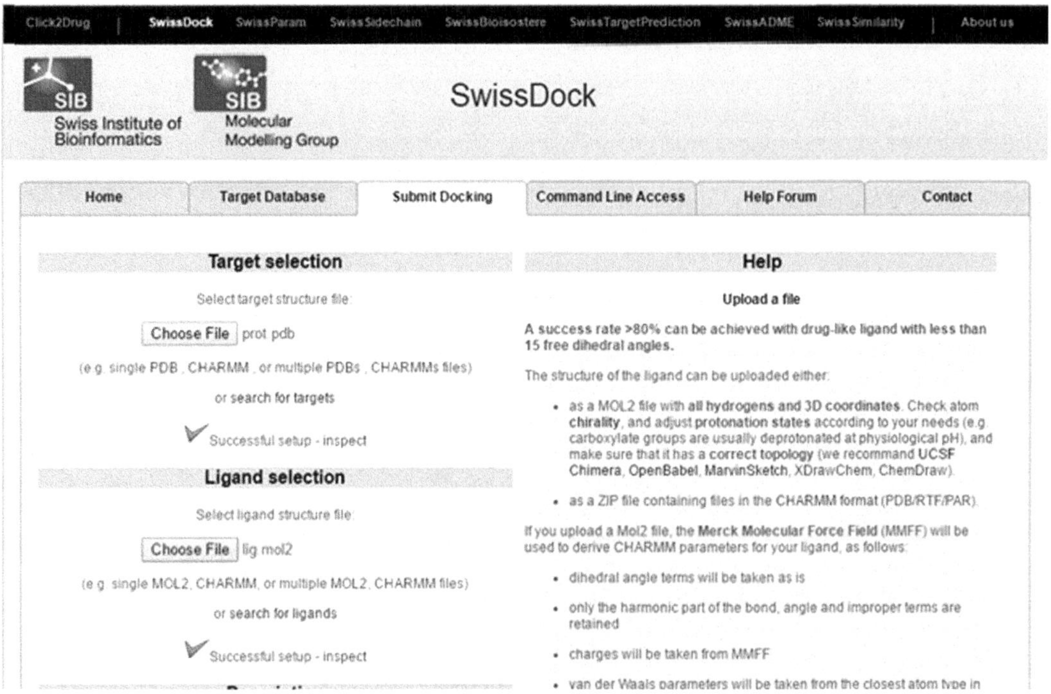

Fig. 10 The web server SwissDock analyses all input files before docking simulations

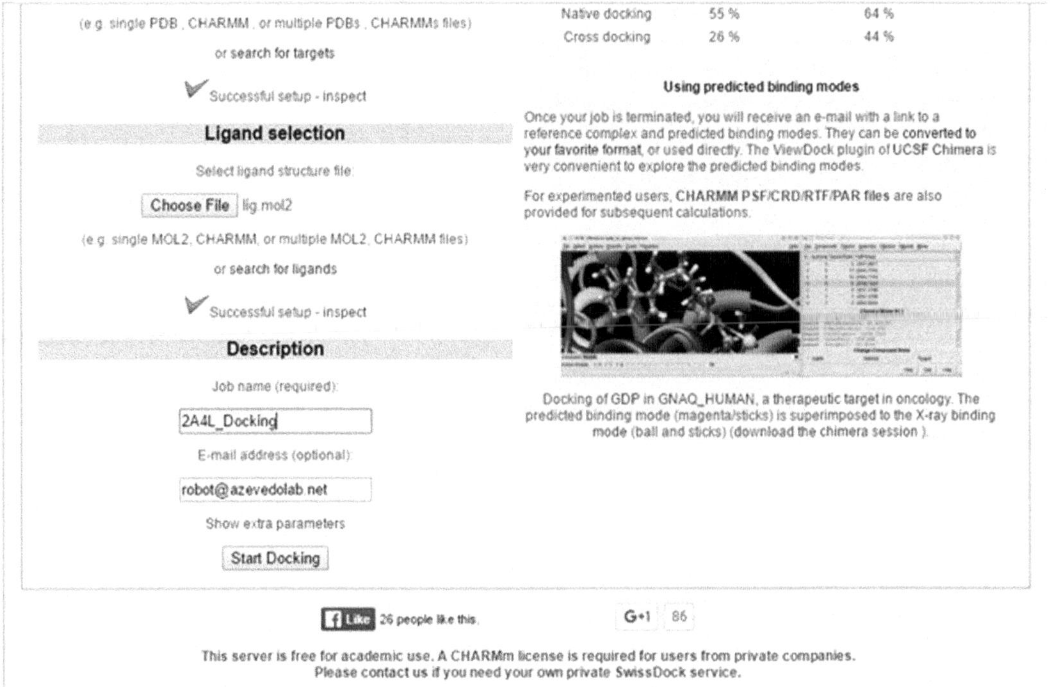

Fig. 11 Description part of SwissDock

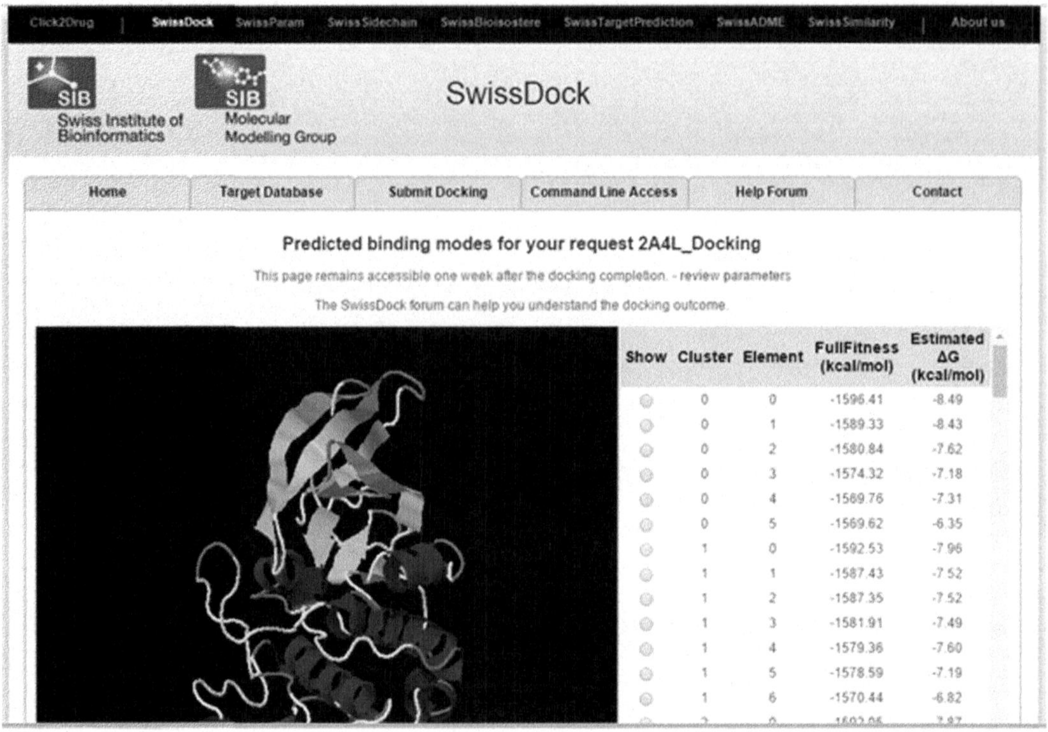

Fig. 12 Docking results

"Choose File" button. Go to the folder where the structures are and upload *lig.mol2* file. SwissDock also carries out a preliminary analysis of the lig.mol2 file. If everything goes fine, you will get the "Successful setup" messages, as shown in Fig. 10. We should keep in mind that we need to get the "Successful setup" message for both molecules.

The SwissDock server sends an email with a link to download the results. On the captured screen (Fig. 12), we have the results for this tutorial. SwissDock shows an interactive table with the calculated binding affinity for each pose. We may download CSV file (clusters.dock4.csv) and zipped file (predictions file) with the docking results. We have to unzip the zipped folder and copy *cluster.dock4.pdb* file to the same folder where *lig.mol2* file is. To analyze docking results generated using SwissDock, we may use the free software SAnDReS [40]. The program SAnDReS is an integrated computational environment for statistical analysis of docking simulations and application of machine-learning techniques to predict ligand-binding affinity.

4 Availability

All files necessary to run this tutorial are available at https://azevedolab.net/resources/swissdock_2a4l.zip.

5 Colophon

We created Fig. 1 using Microsoft PowerPoint 2016. We used the program UCSF Chimera [52] to generate Figs. 2–7. We captured screen from SwissDock site (http://www.swissdock.ch/docking) [42] to make Figs. 8, 9, 10, 11, and 12. We performed molecular docking simulations described in this chapter using a Desktop PC with 4GB of memory, a 1 TB hard disk, and an Intel® Core® i3-2120 at 3.30 GHz processor running Windows 8.1.

6 Final Remarks

SwissDock is a fully integrated computational tool dedicated to carrying out docking simulation through a web interface. Here we perform docking simulations using the complex CDK2-roscovitine. We present all docking processes in detail, which allows even inexperienced users to obtain their results. Since the SwissDock evaluates protein-ligand binding energy using a scoring function based on the CHARMM22 force field [51], several energy terms are determined in each docking simulation. These energy terms may be used to develop a targeted-scoring function, which calibrates the energy terms specific for the biological systems of interest.

Acknowledgments

This work was supported by grants from CNPq (Brazil) (308883/2014-4). This study was financed in part by the Coordenação de Aperfeiçoamento de Pessoal de Nivel Superior—Brasil (CAPES)—Finance Code 001. GB-F acknowledges support from PUCRS/BPA fellowship. WFA is a senior researcher for CNPq (Brazil) (Process Numbers: 308883/2014-4 and 309029/2018-0).

References

1. Aarthy M, Singh SK (2018) Discovery of potent inhibitors for the inhibition of dengue envelope protein: an *in silico* approach. Curr Top Med Chem 18:1585–1602
2. Saikia S, Bordoloi M (2018) Molecular docking: challenges, advances and its use in drug discovery perspective. Curr Drug Targets 20:501–521. https://doi.org/10.2174/1389450119666181022153016

3. Pereira F, Aires-de-Sousa J (2018) Computational methodologies in the exploration of marine natural product leads. Mar Drugs 16:236
4. Sehgal SA, Hammad MA, Tahir RA, Akram HN, Ahmad F (2018) Current therapeutic molecules and targets in neurodegenerative diseases based on *in silico* drug design. Curr Neuropharmacol 16:649–663
5. Zloh M, Kirton SB (2018) The benefits of in silico modeling to identify possible small-molecule drugs and their off-target interactions. Future Med Chem 10:423–432
6. Ishiki HM, Filho JMB, da Silva MS, Scotti MT, Scotti L (2018) Computer-aided drug design applied to Parkinson targets. Curr Neuropharmacol 16:865–880
7. Śledź P, Caflisch A (2018) Protein structure-based drug design: from docking to molecular dynamics. Curr Opin Struct Biol 48:93–102
8. Baig MH, Ahmad K, Rabbani G, Danishuddin M, Choi I (2018) Computer aided drug design and its application to the development of potential drugs for neurodegenerative disorders. Curr Neuropharmacol 16:740–748
9. Sahlgren C, Meinander A, Zhang H, Cheng F, Preis M, Xu C et al (2017) Tailored approaches in drug development and diagnostics: from molecular design to biological model systems. Adv Healthc Mater 6(21). https://doi.org/10.1002/adhm.201700258
10. Ramesh M, Dokurugu YM, Thompson MD, Soliman ME (2017) Therapeutic, molecular and computational aspects of novel monoamine oxidase (MAO) inhibitors. Comb Chem High Throughput Screen 20:492–509
11. Kim J, Yang G, Ha J (2017) Targeting of AMP-activated protein kinase: prospects for computer-aided drug design. Expert Opin Drug Discov 12:47–59
12. Guedes RA, Serra P, Salvador JA, Guedes RC (2016) Computational approaches for the discovery of human proteasome inhibitors: an overview. Molecules 21:927
13. Fukunishi Y, Mashimo T, Misoo K, Wakabayashi Y, Miyaki T, Ohta S et al (2016) Miscellaneous topics in computer-aided drug design: synthetic accessibility and GPU computing, and other topics. Curr Pharm Des 22:3555–3568
14. Baig MH, Ahmad K, Roy S, Ashraf JM, Adil M, Siddiqui MH et al (2016) Computer aided drug design: success and limitations. Curr Pharm Des 22:572–581
15. Cardamone F, Pizzi S, Iacovelli F, Falconi M, Desideri A (2017) Virtual screening for the development of dual-inhibitors targeting topoisomerase IB and tyrosyl-DNA phosphodiesterase 1. Curr Drug Targets 18:544–555
16. Macalino SJ, Gosu V, Hong S, Choi S (2015) Role of computer-aided drug design in modern drug discovery. Arch Pharm Res 38:1686–1701
17. Scotti L, Scotti MT (2015) Computer aided drug design studies in the discovery of secondary metabolites targeted against age-related neurodegenerative diseases. Curr Top Med Chem 15:2239–2252
18. Tian S, Wang J, Li Y, Li D, Xu L, Hou T (2015) The application of in silico drug-likeness predictions in pharmaceutical research. Adv Drug Deliv Rev 86:2–10
19. Mallipeddi PL, Kumar G, White SW, Webb TR (2014) Recent advances in computer-aided drug design as applied to anti-influenza drug discovery. Curr Top Med Chem 14:1875–1889
20. Azevedo LS, Moraes FP, Xavier MM, Pantoja EO, Villavicencio B, Finck JA et al (2012) Recent progress of molecular docking simulations applied to development of drugs. Curr Bioinforma 7:352–365
21. Srivastava HK, Bohari MH, Sastry GN (2012) Modeling anti-HIV compounds: the role of analogue-based approaches. Curr Comput Aided Drug Des 8:224–248
22. Ghosh AK, Osswald HL, Prato G (2016) Recent progress in the development of HIV-1 protease inhibitors for the treatment of HIV/AIDS. J Med Chem 59:5172–5208
23. Zhan P, Pannecouque C, De Clercq E, Liu X (2016) Anti-HIV drug discovery and development: current innovations and future trends. J Med Chem 59:2849–2878
24. Forli S, Olson AJ (2015) Computational challenges of structure-based approaches applied to HIV. Curr Top Microbiol Immunol 389:31–51
25. Ghosh AK, Brindisi M (2015) Organic carbamates in drug design and medicinal chemistry. J Med Chem 58:2895–2940
26. Patel RV, Park SW (2014) Journey describing the discoveries of anti-HIV triterpene acid families targeting HIV-entry/fusion, protease functioning and maturation stages. Curr Top Med Chem 14:1940–1966
27. Fang Z, Song Y, Zhan P, Zhang Q, Liu X (2014) Conformational restriction: an effective tactic in 'follow-on'-based drug discovery. Future Med Chem 6:885–901

28. Schimer J, Konvalinka J (2014) Unorthodox inhibitors of HIV protease: looking beyond active-site-directed peptidomimetics. Curr Pharm Des 20:3389–3397
29. Pang X, Liu Z, Zhai G (2014) Advances in non-peptidomimetic HIV protease inhibitors. Curr Med Chem 21:1997–2011
30. Thomas SE, Mendes V, Kim SY, Malhotra S, Ochoa-Montaño B, Blaszczyk M et al (2017) Structural biology and the design of new therapeutics: from HIV and cancer to mycobacterial infections: a paper dedicated to John Kendrew. J Mol Biol 429:2677–2693
31. Fradera X, Mestres J (2004) Guided docking approaches to structure-based design and screening. Curr Top Med Chem 4:687–700
32. Bitencourt-Ferreira G, de Azevedo WF Jr (2018) Development of a machine-learning model to predict Gibbs free energy of binding for protein-ligand complexes. Biophys Chem 240:63–69
33. de Ávila MB, de Azevedo WF Jr (2018) Development of machine learning models to predict inhibition of 3-dehydroquinate dehydratase. Chem Biol Drug Des 92:1468–1474
34. Amaral MEA, Nery LR, Leite CE, de Azevedo Junior WF, Campos MM (2018) Pre-clinical effects of metformin and aspirin on the cell lines of different breast cancer subtypes. Investig New Drugs 36:782–796
35. Levin NMB, Pintro VO, Bitencourt-Ferreira G, Mattos BB, Silvério AC, de Azevedo WF Jr (2018) Development of CDK-targeted scoring functions for prediction of binding affinity. Biophys Chem 235:1–8
36. Freitas PG, Elias TC, Pinto IA, Costa LT, de Carvalho PVSD, Omote DQ et al (2018) Computational approach to the discovery of phytochemical molecules with therapeutic potential targets to the PKCZ protein. Lett Drug Des Discov 15:488–499
37. Pintro VO, Azevedo WF (2017) Optimized virtual screening workflow. Towards target-based polynomial scoring functions for HIV-1 protease. Comb Chem High Throughput Screen 20:820–827
38. de Ávila MB, Xavier MM, Pintro VO, de Azevedo WF (2017) Supervised machine learning techniques to predict binding affinity. A study for cyclin-dependent kinase 2. Biochem Biophys Res Commun 494:305–310
39. Heck GS, Pintro VO, Pereira RR, de Ávila MB, Levin NMB, de Azevedo WF (2017) Supervised machine learning methods applied to predict ligand-binding affinity. Curr Med Chem 24:2459–2470
40. Xavier MM, Heck GS, de Avila MB, Levin NM, Pintro VO, Carvalho NL et al (2016) SAnDReS a computational tool for statistical analysis of docking results and development of scoring functions. Comb Chem High Throughput Screen 19:801–812
41. Irwin JJ, Shoichet BK, Mysinger† MM, Huang N, Colizzi F, Wassam P et al (2011) Automated docking screens: a feasibility study. J Med Chem 52:5712–5720
42. Grosdidier A, Zoete V, Michielin O (2011) SwissDock, a protein-small molecule docking web service based on EADock DSS. Nucleic Acids Res 39:270–277
43. Goodsell DS, Olson AJ (1990) Automated docking of substrates to proteins by simulated annealing. Proteins 8:195–202
44. Morris GM, Goodsell DS, Huey R, Olson AJ (1996) Distributed automated docking of flexible ligands to proteins: parallel applications of AutoDock 2.4. J Comput Aided Mol Des 10:293–304
45. Morris GM, Goodsell DS, Halliday RS, Huey R, Hart WE, Belew RK et al (1998) Automated docking using a Lamarckian genetic algorithm and empirical binding free energy function. J Comput Chem 19:1639–1662
46. Morris GM, Huey R, Lindstrom W, Sanner MF, Belew RK, Goodsell DS et al (2009) AutoDock4 and AutoDockTools4: automated docking with selective receptor flexibility. J Comput Chem 30:2785–2791
47. Thomsen R, Christensen MH (2006) MolDock: a new technique for high-accuracy molecular docking. J Med Chem 49:3315–3321
48. Heberlé G, de Azevedo WF Jr (2011) Bio-inspired algorithms applied to molecular docking simulations. Curr Med Chem 18:1339–1352
49. De Azevedo WF Jr (2010) MolDock applied to structure-based virtual screening. Curr Drug Targets 11:327–334
50. Trott O, Olson AJ (2010) AutoDock Vina: improving the speed and accuracy of docking with a new scoring function, efficient optimization, and multithreading. J Comput Chem 31:455–461
51. Grosdidier A, Zoete V, Michielin O (2011) Fast docking using the CHARMM force field with EADock DSS. J Comput Chem 32:2149–2159
52. Pettersen EF, Goddard TD, Huang CC, Couch GS, Greenblatt DM, Meng EC et al (2004) UCSF Chimera—a visualization system

for exploratory research and analysis. J Comput Chem 25:1605–1612
53. Morgan DO (1995) Principles of CDK regulation. Nature 374:131–134
54. Murray AW (1994) Cyclin-dependent kinases: regulators of the cell cycle and more. Chem Biol 1:191–195
55. Volkart PA, Bitencourt-Ferreira G, Souto AA, de Azevedo WF (2018) Cyclin-dependent kinase 2 in cellular senescence and cancer. A structural and functional review. Curr Drug Targets 20:716–726. https://doi.org/10.2174/1389450120666181204165344
56. Kim SH, Schulze-Gahmen U, Brandsen J, de Azevedo Júnior WF (1996) Structural basis for chemical inhibition of CDK2. Prog Cell Cycle Res 2:137–145
57. De Azevedo WF Jr, Mueller-Dieckmann HJ, Schulze-Gahmen U, Worland PJ, Sausville E, Kim SH (1996) Structural basis for specificity and potency of a flavonoid inhibitor of human CDK2, a cell cycle kinase. Proc Natl Acad Sci U S A 93:2735–2740
58. Canduri F, de Azevedo WF Jr (2005) Structural basis for interaction of inhibitors with cyclin-dependent kinase 2. Curr Comput Aided Drug Des 1:53–64
59. Krystof V, Cankar P, Frysová I, Slouka J, Kontopidis G, Dzubák P et al (2006) 4-arylazo-3,5-diamino-1H-pyrazole CDK inhibitors: SAR study, crystal structure in complex with CDK2, selectivity, and cellular effects. J Med Chem 49:6500–6509
60. Levin NM, Pintro VO, de Ávila MB, de Mattos BB, De Azevedo WF Jr (2017) Understanding the structural basis for inhibition of cyclin-dependent kinases. New pieces in the molecular puzzle. Curr Drug Targets 18:1104–1111
61. De Azevedo WF, Leclerc S, Meijer L, Havlicek L, Strnad M, Kim SH (1997) Inhibition of cyclin-dependent kinases by purine analogues: crystal structure of human cdk2 complexed with roscovitine. Eur J Biochem 243:518–526

Chapter 13

Molecular Docking Simulations with ArgusLab

Gabriela Bitencourt-Ferreira and Walter Filgueira de Azevedo Jr.

Abstract

Molecular docking is the major computational technique employed in the early stages of computer-aided drug discovery. The availability of free software to carry out docking simulations of protein-ligand systems has allowed for an increasing number of studies using this technique. Among the available free docking programs, we discuss the use of ArgusLab (http://www.arguslab.com/arguslab.com/ArgusLab.html) for protein-ligand docking simulation. This easy-to-use computational tool makes use of a genetic algorithm as a search algorithm and a fast scoring function that allows users with minimal experience in the simulations of protein-ligand simulations to carry out docking simulations. In this chapter, we present a detailed tutorial to perform docking simulations using ArgusLab.

Key words ArgusLab, Molecular docking, Protein-ligand interactions, Cyclin-dependent kinase 2, Drug design, Molecular recognition

1 Introduction

Molecular docking simulation of biomolecular systems is a dynamic topic of research in the computational simulation of protein targets for drug development. This type of simulation has a pivotal role in the discovery of potential new drugs through computational studies [1–21]. The basic idea in the development of modern protein-ligand docking programs is to have an integrated environment with at least one search algorithm and a computational method to estimate the binding energy of the ligand in the complex with a protein structure. This computational technique to determine the binding affinity is named scoring function and can be calibrated to calculate the free energy of binding, the log of the inhibition constant, or log of the dissociation constant [7], to mention the most commonly applied binding affinities. As the input of any docking program, we have the atomic coordinates of the target, our protein structure, and the ligand coordinates. The docking program generates a complex comprising the protein and the

Walter Filgueira de Azevedo Jr. (ed.), *Docking Screens for Drug Discovery*, Methods in Molecular Biology, vol. 2053, https://doi.org/10.1007/978-1-4939-9752-7_13,

ligand. In addition, it estimates the binding affinity of the protein-ligand complex [17].

It is customary, in any docking study, to start the procedure as a validation step. We use a crystallographic structure of a protein-ligand complex and recover the crystallographic position of the ligand through docking simulation. The position obtained from the docking simulation is named pose. The primary parameter applied to evaluate the docking quality is the root mean square deviation (RMSD) determined by the following equation,

$$\mathrm{RMSD} = \sqrt{\frac{\sum_{i=1}^{N} \left[(x_{x,i} - x_{p,i})^2 + (y_{x,i} - y_{p,i})^2 + (z_{x,i} - z_{p,i})^2 \right]}{N}} \quad (1)$$

where x_x, y_x, and z_x are the crystallographic atomic coordinates for the ligand and x_p, y_p, and z_p are the atomic coordinates for the pose. When we calculate the summation, we consider the N non-hydrogen atoms in the ligand structure. So, it is clear that the ideal would be an RMSD = 0.0 Å. Most of the researchers involved in the development of docking programs consider that RMSDs < 2.0 Å are acceptable [22].

Since the majority of the docking programs generate more than one pose, it is customary to evaluate the docking accuracy of all poses created for a docking simulation. The following equation defines docking accuracy (DA) as follows:

$$\mathrm{DA} = f_l + 0.5(f_l - f_h) \quad (2)$$

where f_l is the fraction poses for which the docking RMSD is less than l, and f_h is the fraction poses for which the docking RMSD is less than h, where $l < h$ [23, 24].

In this chapter, we describe a detailed tutorial explaining the use of molecular docking simulation of a protein-ligand system. Due to the user-friendly interface and free availability of the program, we chose the ArgusLab software [25] to carry our molecular docking simulations. So far, we have only windows version of the ArgusLab, but the developer has announced the creation of an iPad version, intended to be an educational platform for teaching protein-ligand docking simulations (http://www.arguslab.com/arguslab.com/ArgusLab_for_iPad.html).

ArgusLab has been applied to a broad spectrum of protein-ligand systems [25–50], ranging from enzymes (acetylcholine esterase [AChE]) [50] to copper chaperone protein [41], and metabotropic glutamate receptors (mGluRs) [27]. It has been reported that ArgusLab can carry out protein-ligand docking simulations with similar docking performance when compared with others protein-ligand docking programs such as AutoDock [27, 43, 48], Autodock Vina, ArgusLab, Molegro Virtual Docker,

Hex-Cuda [50], and GOLD [25]. Nevertheless, application of ArgusLab scoring function showed poor predictive performance for analysis of binding affinity of estrogen receptor β when compared with molecular mechanics-generalized born surface area (MM-GBSA) re-scoring available in the program Glide [33].

2 ArgusLab

In this tutorial, you will learn how to carry out docking simulation using the ArgusLab [25] docking program. This docking software is freely available at www.arguslab.com. We used the atomic coordinates of cyclin-dependent kinase 2 (CDK2) in complex with 3-amino-6-(4-{[2-(dimethylamino)ethyl]sulfamoyl}phenyl)-n-pyridin-3-ylpyrazine-2-carboxamide (PDB access code: 4ACM) [51].

3 Biological System

In this chapter, we show how to carry out molecular docking simulation of CDK2 (EC 2.7.11.22) with ArgusLab [25]. Figure 1 shows the electrostatic molecular surface of the ATP-binding pocket with the structure of the inhibitor 3-amino-6-(4-{[2-(dimethylamino)ethyl]sulfamoyl}phenyl)-n-pyridin-3-ylpyrazine-2-carboxamide (PDB access code: 4ACM) bound to CDK2 crystallographic structure [51]. CDK2 has been intensively studied as a target for the development of anticancer drugs

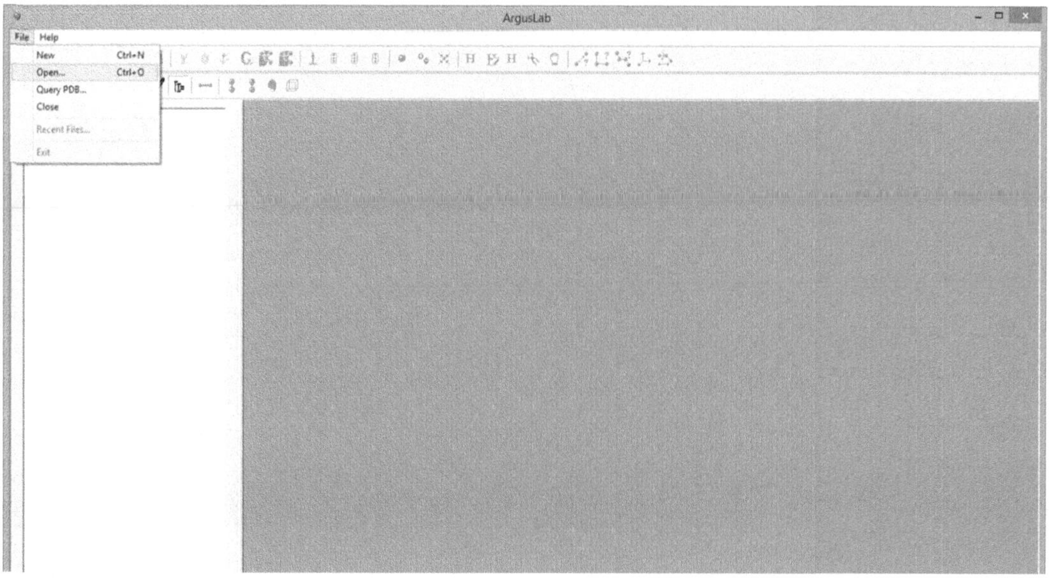

Fig. 1 Main menu of ArgusLab

[52–61]. The first structure of human CDK2 was obtained in 1993 [62]. Analysis of the CDK2 structure indicated a typical bilobal architecture of serine/threonine protein kinases.

4 Graphical Tutorial

To run this tutorial, you need to have ArgusLab [25] installed on your computer. To obtain the coordinates to be used in this tutorial, you may go to the Protein Data Bank (PDB) [63–65] (www.rcsb.org/pdb) and download the atomic coordinates for CDK2 in complex with an inhibitor (PDB access code: 4ACM) [51].

Considering that you installed ArgusLab, and it is running on your desktop, to open a PDB file, click *File>Open...*, as shown in Fig. 1. Then browse the folder where you have the PDB file. You will have the structure in the graphical screen. On the left, you have the Tree View tool. Click on the "+" to expand the tree (Fig. 2).

Expand the Tree View of 4ACM and open up the Residues/Misc. folder to show the ligands (Fig. 3). You should be able to see the directory tree of ArgusLab, where the ligands of the structure 4ACM are evident(Fig. 4). The active ligand in "1302 7YG" will be used in the docking simulations. Left click on "1302 7YG" in the Tree View to select the active ligand. It should appear in yellow (Fig. 5).

Now click on *Edit>Hide Unselected*, as shown in Fig. 6. You will have only the active ligand on the screen. To center the ligand, click on the button of the main menu indicated in Fig. 7. To add

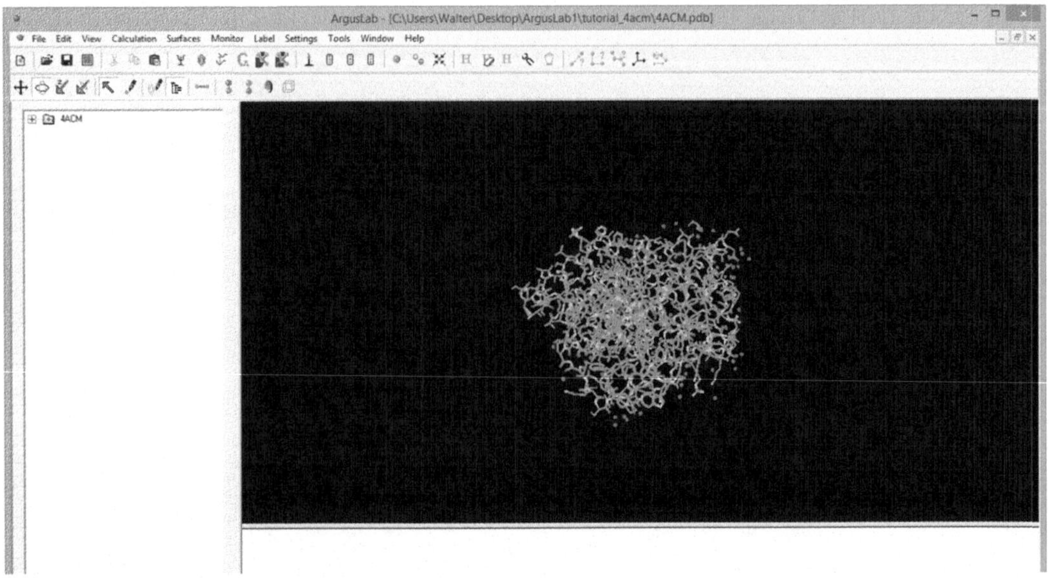

Fig. 2 Graphical window of ArgusLab with CDK2 structure

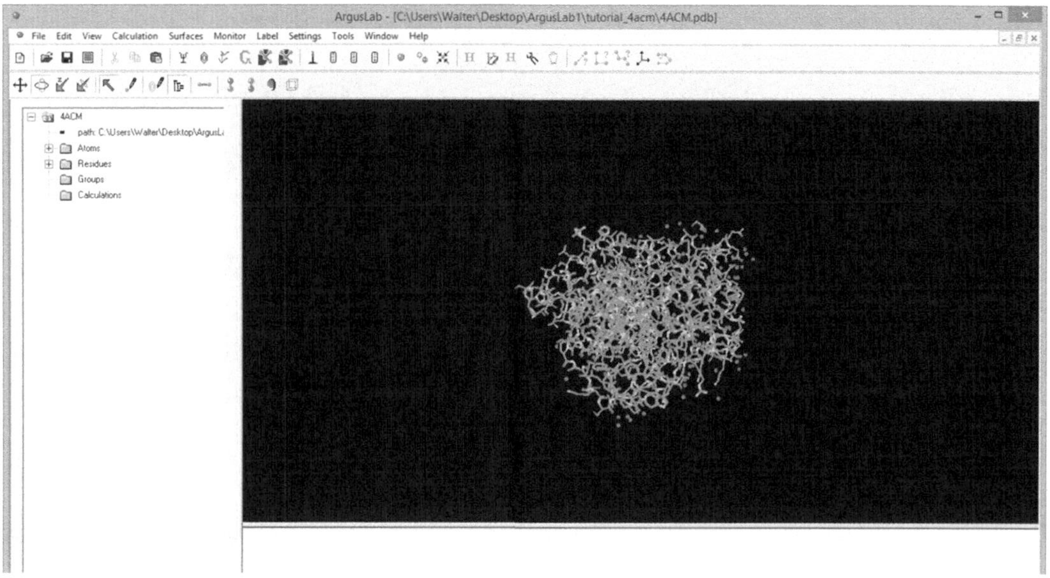

Fig. 3 Directory tree of ArgusLab

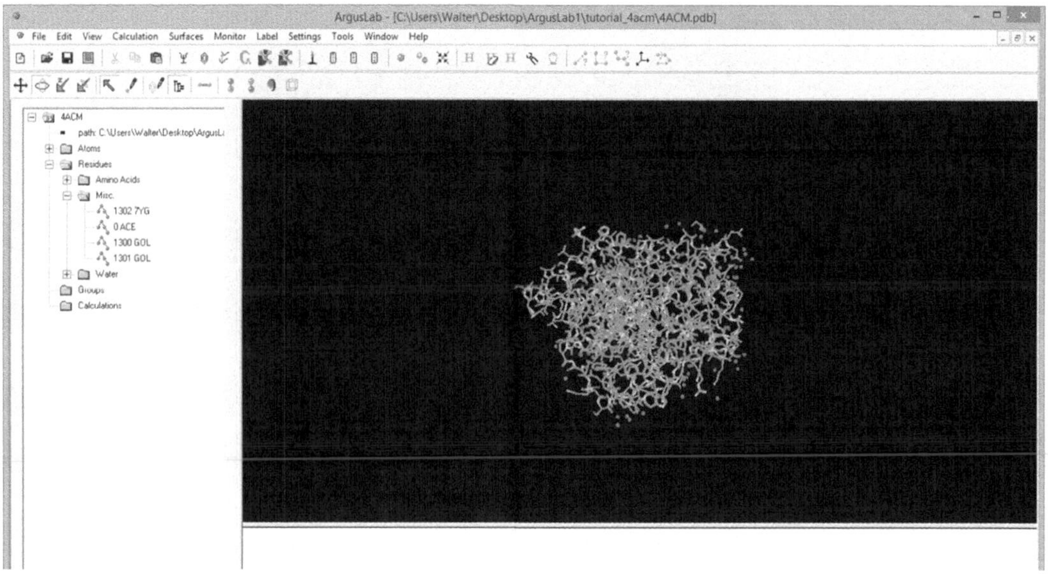

Fig. 4 Directory tree of ArgusLab, where we can see the ligands of the structure 4ACM

hydrogens to the ligand, click <Crtl>H keys. In Fig. 8, we have the ligand with hydrogens added to the structure. Here we have the ligand with hydrogens attached to the structure (Fig. 8). Right click on "1302 7YG" on the Tree View and select "Make a Ligand Group from This Residues" option (Fig. 9). Then, expand the Groups folder in the Tree View. Now we have access to the ligand

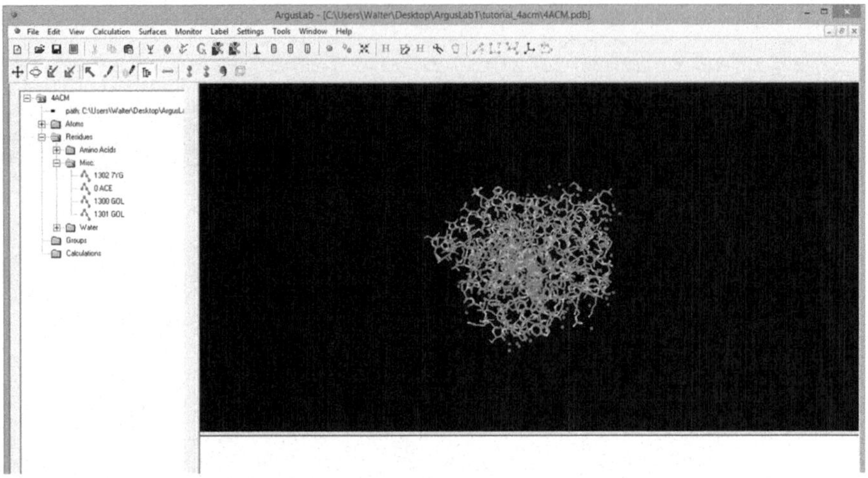

Fig. 5 Graphical window of ArgusLab with the ligand structure

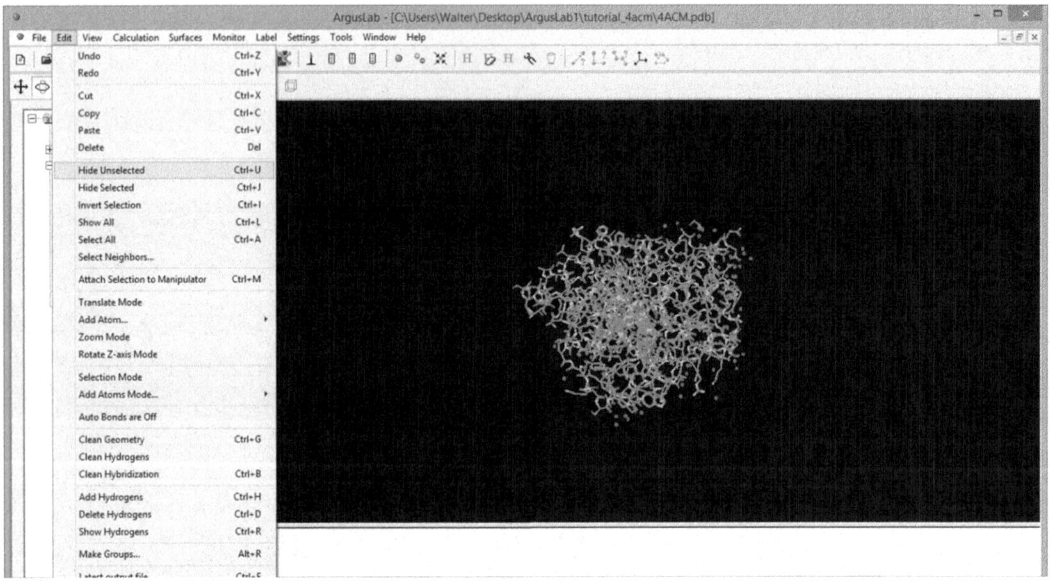

Fig. 6 Edit menu of ArgusLab

in the Groups folder (7YG). Left-click on "17YG" in the Groups folder to select the atoms of the ligand on the screen. Copy (Ctrl +C) and paste (Ctrl+V) the selected ligand. Expand the Misc. folder, and you will see the copy of the ligand named "2184 7YG" (Fig. 10). Right click on "2184 7YG" on the Tree View and select "Make a Ligand Group from This Residues" option, as shown in Fig. 11. Now we have two ligands in the Groups folder named "1 7YG" and "2 7YG." Now we have to rename these

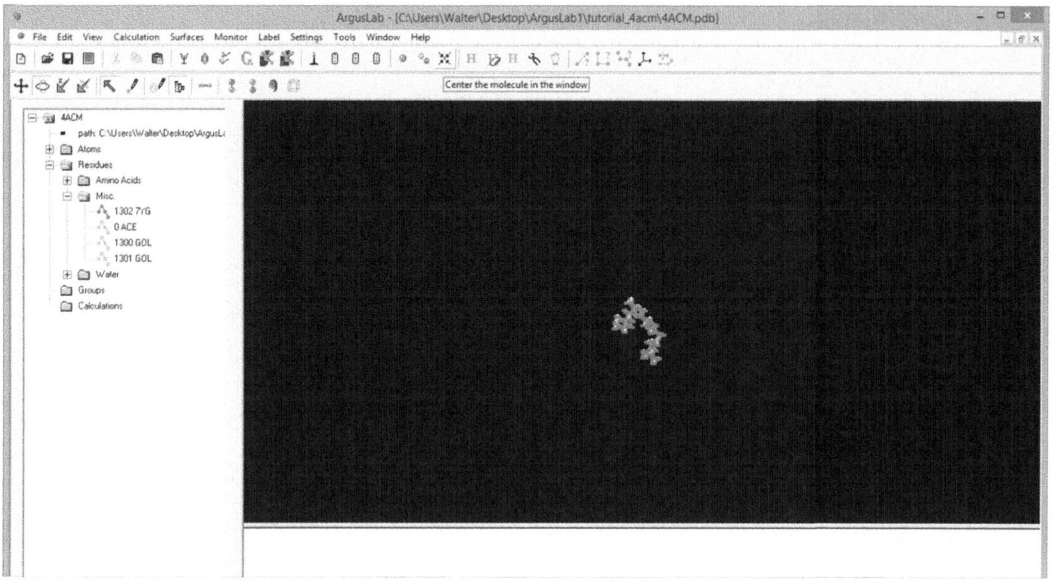

Fig. 7 Main menu of ArguLab, where we highlight the "Center the molecule in the window" button

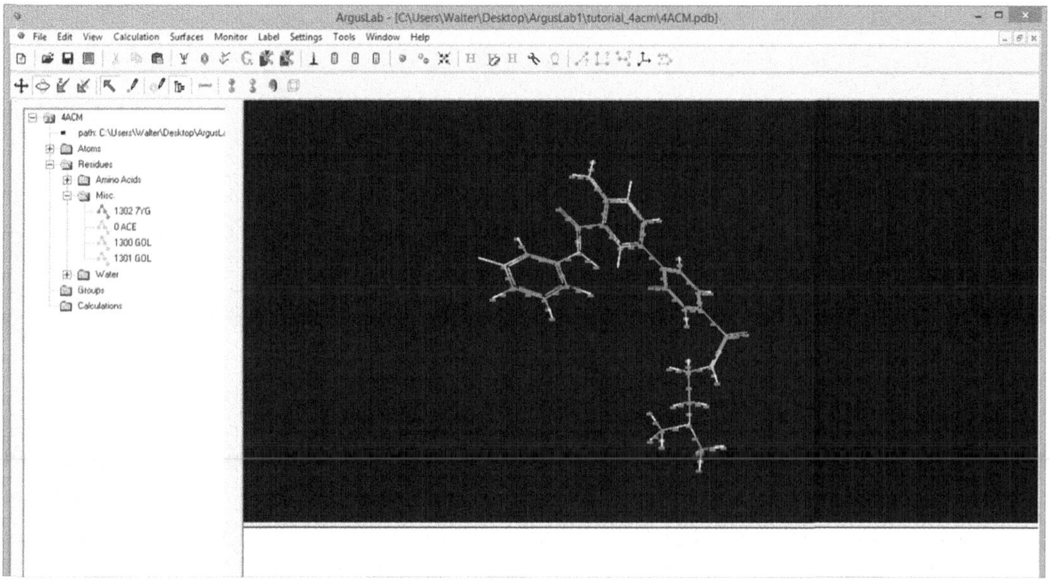

Fig. 8 Graphical window of ArgusLab, where we can see the ligand structure

ligands to "ligand-xray" and "ligand," respectively. Right-click on "1 7YG" in the Groups folder and select "Modify group..." option, as shown in Fig. 12.

In the "Modify group..." dialog box, type in the "ligand-xray" (Fig. 13). Don't change the Group type. Do the same to the "2 7YG" and rename to "ligand." Right-click on the ligand and select

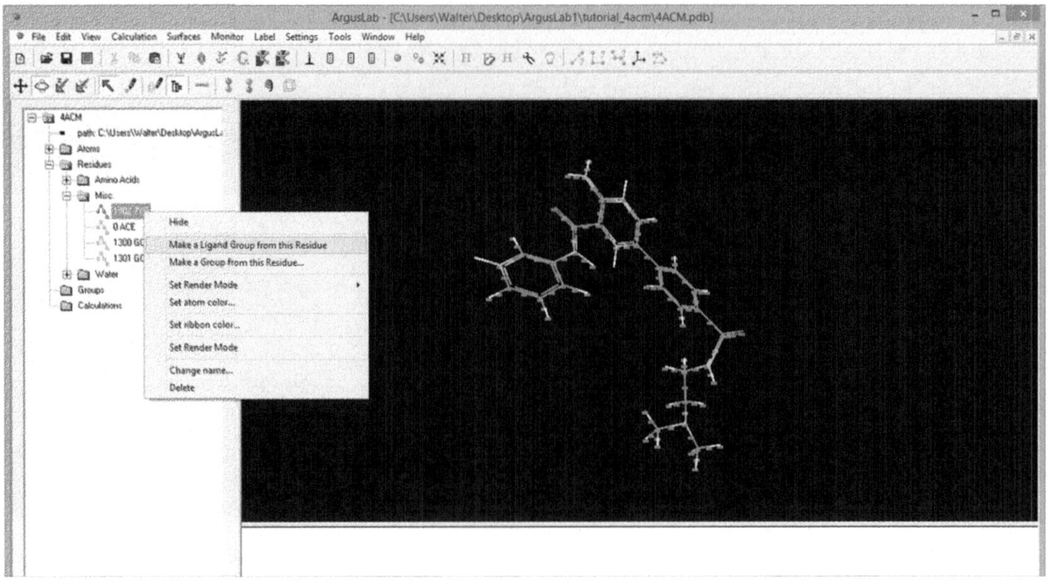

Fig. 9 Tree view of ArgusLab, where we select the option "Make a Ligand Group from this Residue"

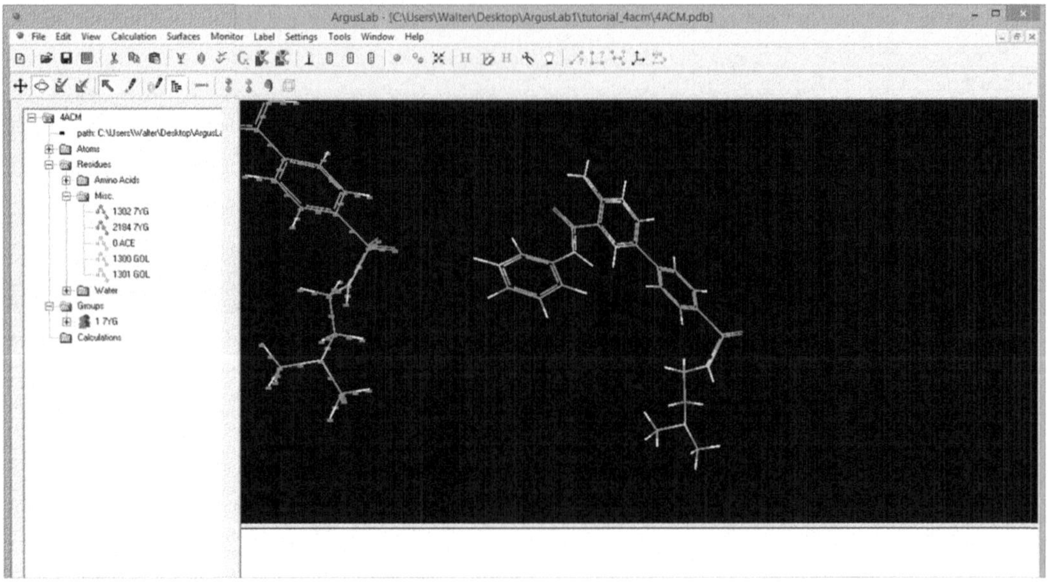

Fig. 10 Graphical window of ArgusLab, where we can see the copy of the ligand structure

"Set Render Mode" and choose "Cylinder med" option, as shown in Fig. 14. You will have the view of the window of ArgusLab, where you can see the copy of the ligand structure (Fig. 15). Then, right-click on the ligand-xray in the Groups folder and choose "Make a BindingSite Group for this Group," as shown in Fig. 16.

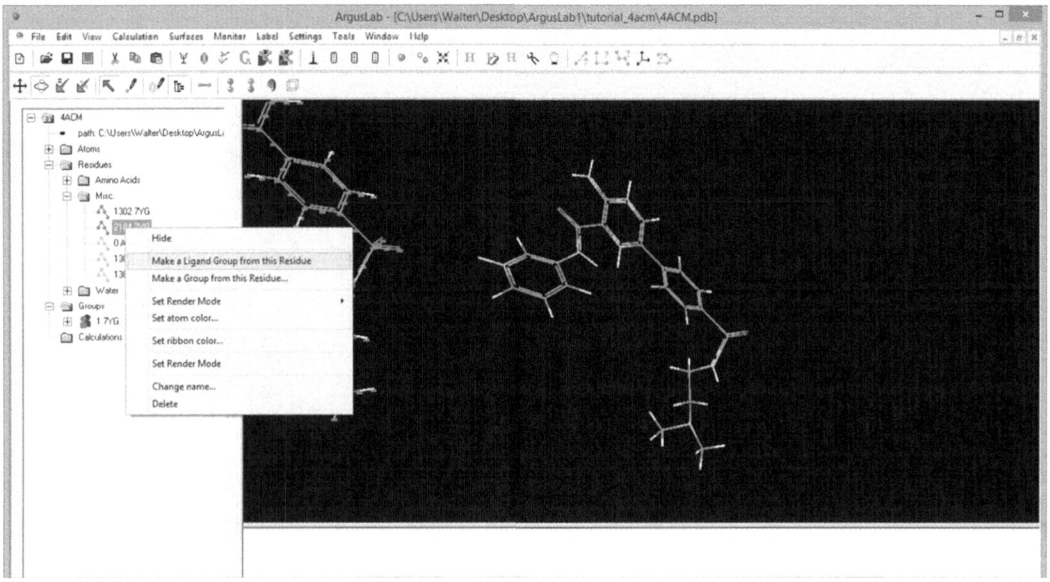

Fig. 11 Tree view of ArgusLab, where we select the option "Make a Ligand Group from this Residue"

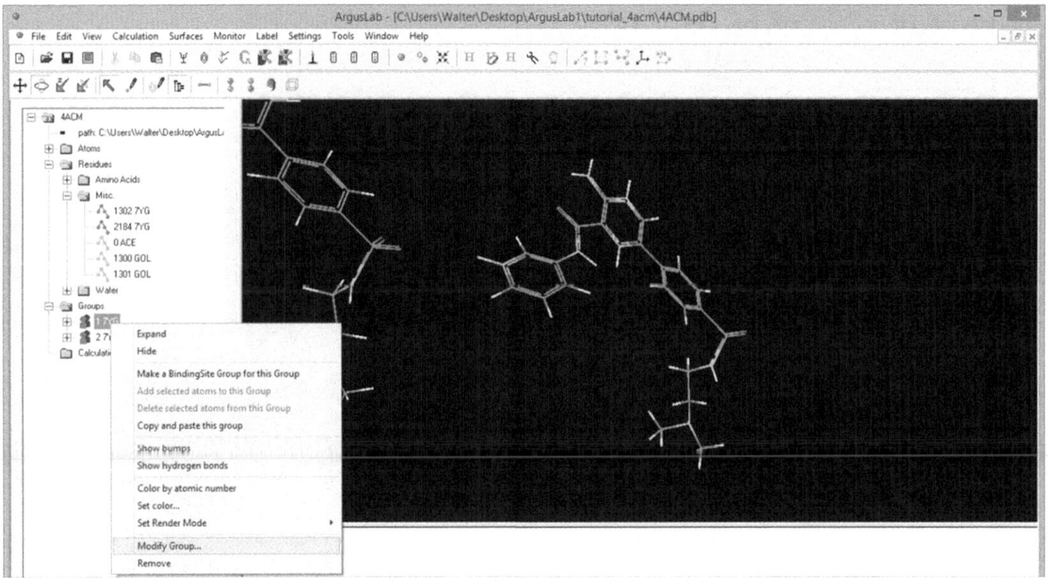

Fig. 12 Tree view of ArgusLab, where we select the option "Modify Group..."

Now we have the binding site as shown in Fig. 17. Center the molecules as explained before. In the main menu, click on *Calculation>Dock a Ligand...* (Fig. 18).

Then, we have the dialog box to enter docking parameters (Fig. 19). Select "4ACM: ligand" on Ligand drop box. Then press "Calculate Size" button. Next, we press the "Advanced..."

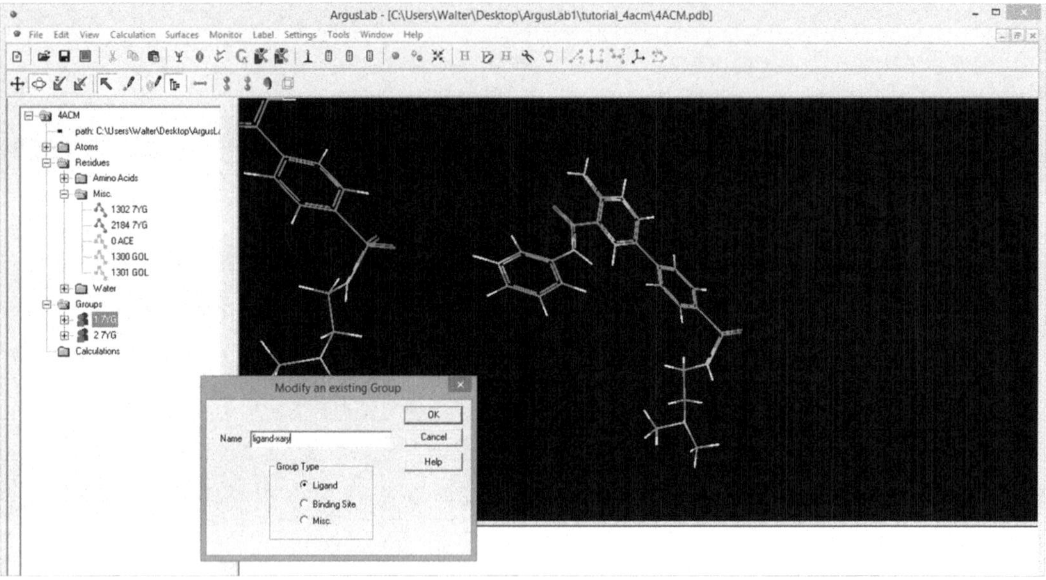

Fig. 13 A pop-up window of the ArgusLab, where we can modify an existing group in the structure

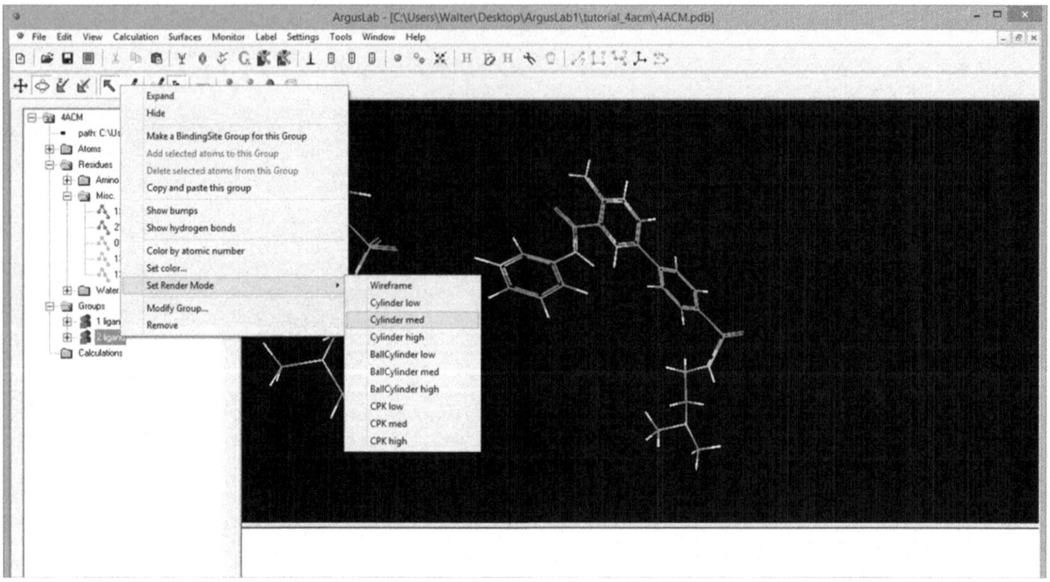

Fig. 14 Main menu of the ArgusLab

button and change "Max. number of poses" to 500. We then press "OK" button. To start docking simulation, we press "Start" button. After a few seconds, we will see the message "Docking run: elapsed time...," as shown in Fig. 20. In the Tree View tool, select ligand and ligand-xray by holding down the "Ctrl" key and left-clicking on both groups. You will have the screen shown in Fig. 21.

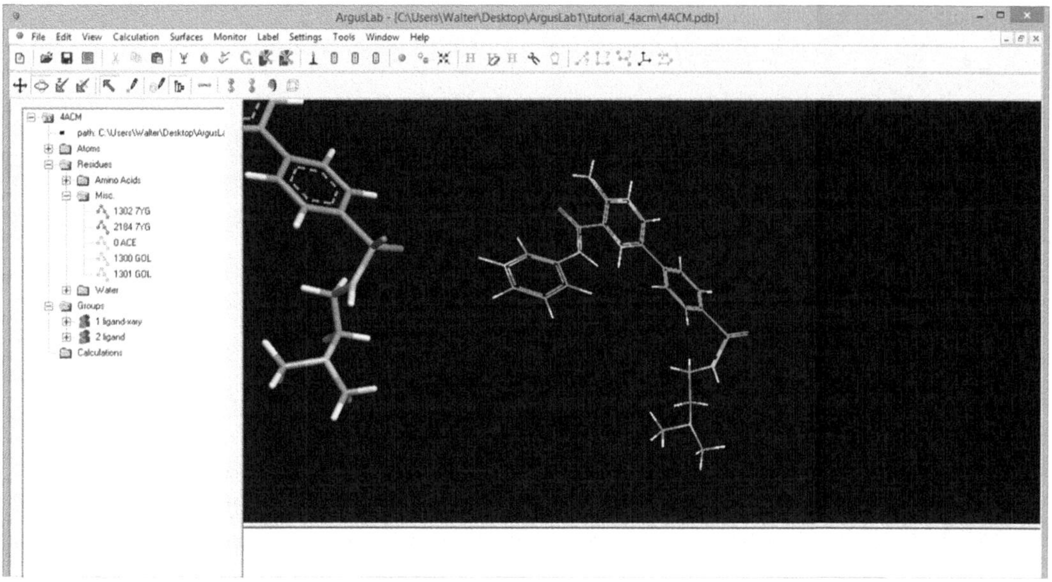

Fig. 15 Graphical window of ArgusLab, where we can see the copy of the ligand structure

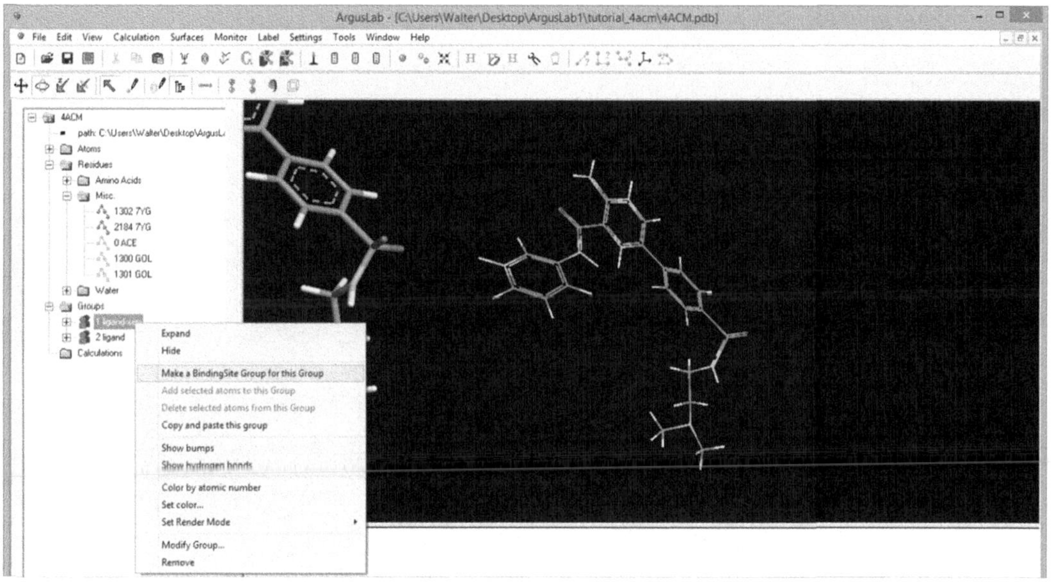

Fig. 16 Tree view of ArgusLab, where we select the option "Make a BindingSite Group for this Group"

Right-click on the "Groups" folder tab in the Tree View and select "Calc RMSD position between two similar Groups," as shown in Fig. 22. Then, we have a pop-up window with the docking RMSD (2.360842 Å). In the main menu, click on *File>Save as...* . Then choose ArgusLab Files (∗.agl). Next, repeat the procedure and save the file in the PDB format. In the Tree View, expand Calculations

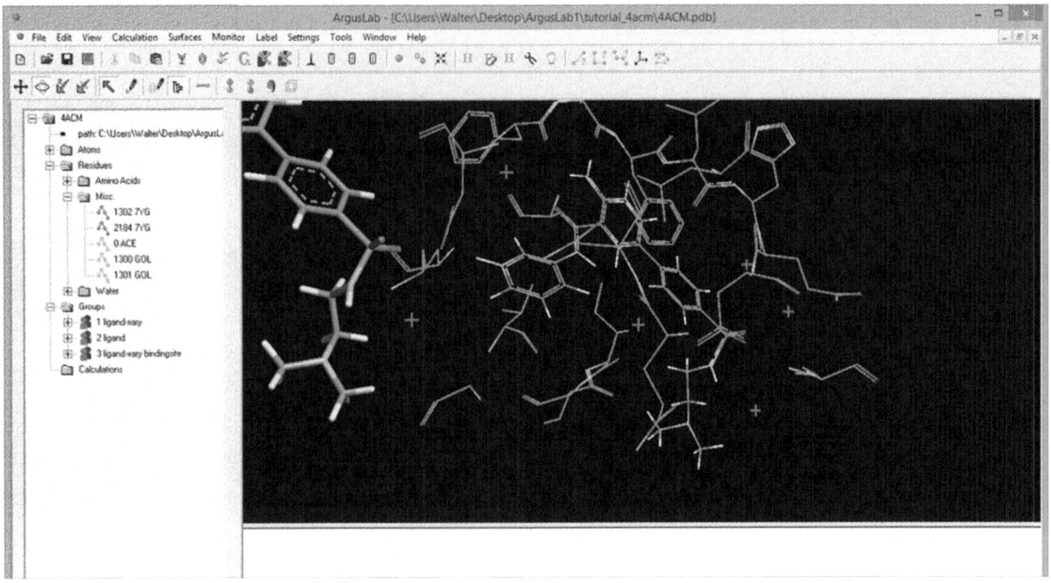

Fig. 17 Graphical window of ArgusLab, where we can see the binding site

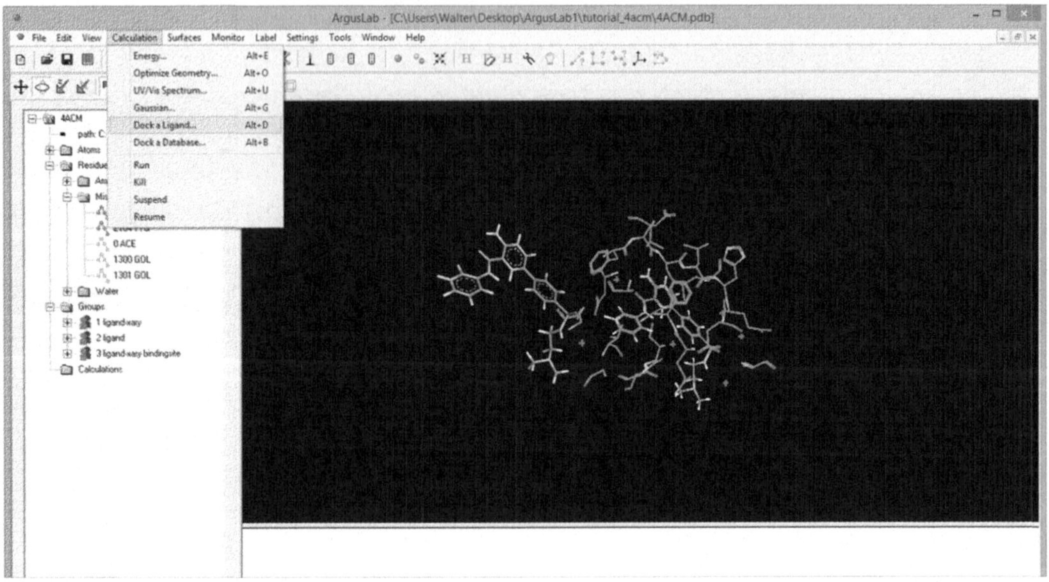

Fig. 18 Main menu of ArgusLab

folder. Then, right-click on "ArgusDock..." and select "Save to file... ." Alternative docking protocol using a Lamarckian genetic algorithm is available in the ArgusLab.

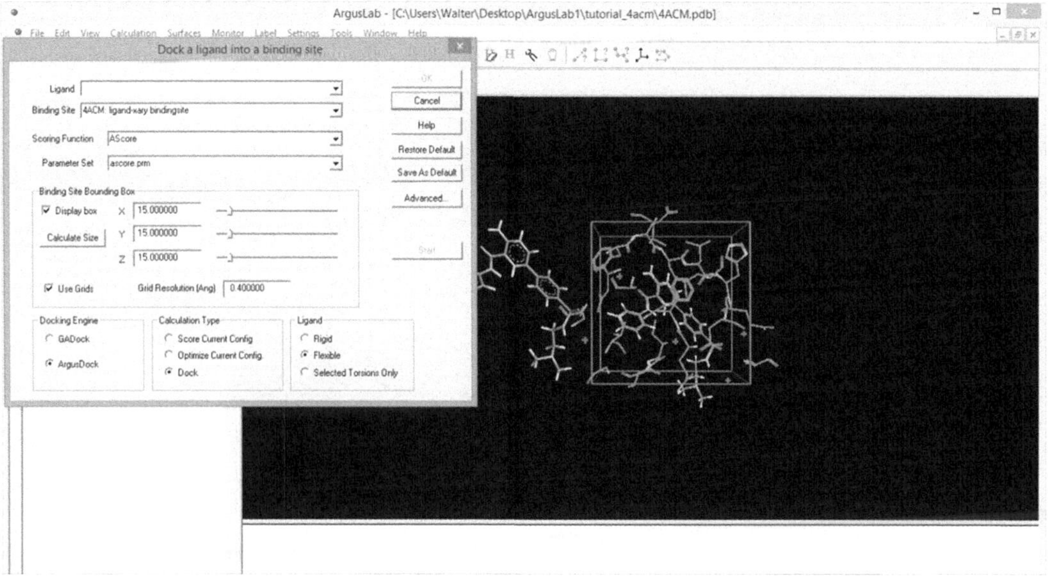

Fig. 19 Pop-up window of the ArgusLab for definition of the docking parameters

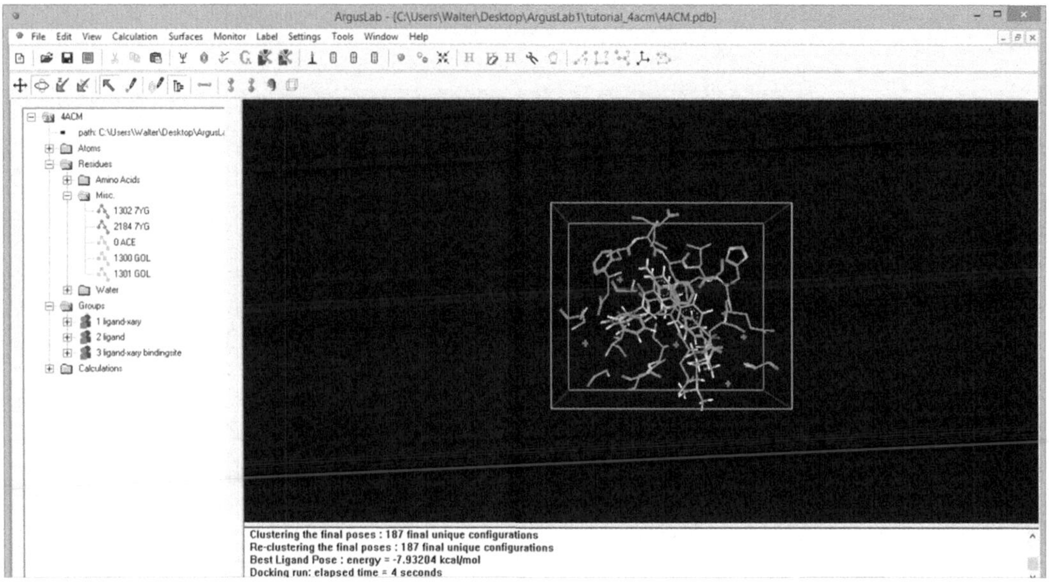

Fig. 20 Main menu of ArgusLab, where we see that the program finished the docking simulation

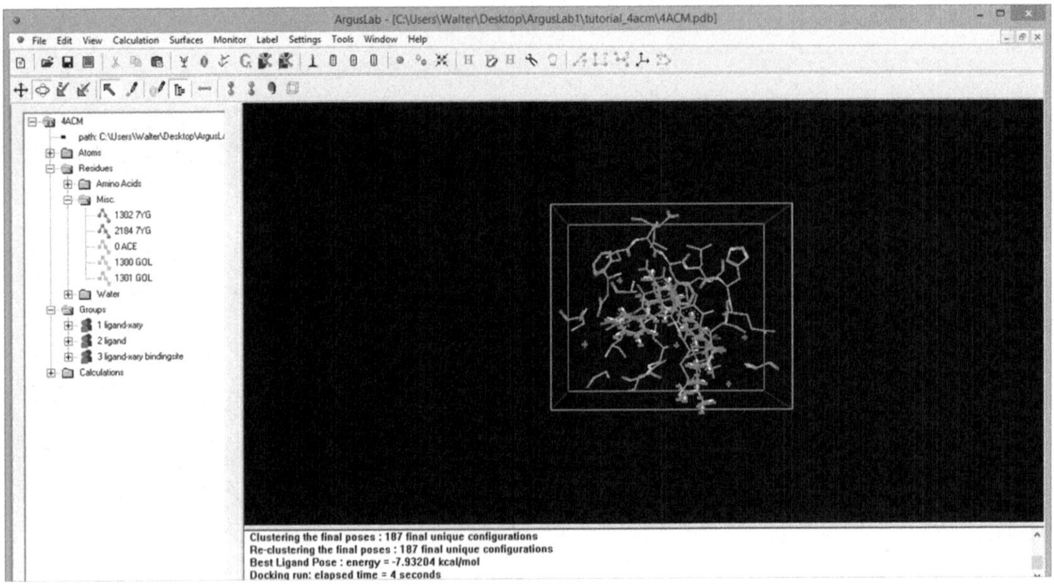

Fig. 21 Graphical window of ArgusLab, where we see the docking results

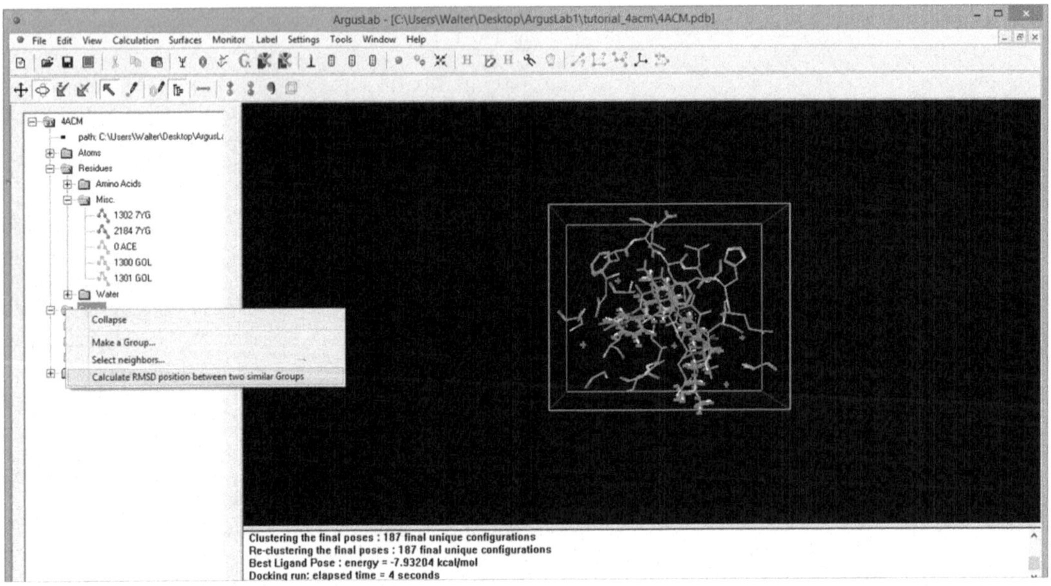

Fig. 22 Tree view of ArgusLab, where we select the option "Calculate RMSD position between two similar Groups"

5 Availability

The ArgusLab is available for downloading at http://www.arguslab.com/arguslab.com/ArgusLab_files/arguslab.zip.

6 Colophon

We used the program ArgusLab [25] to generate Figs. 1–22. We performed molecular docking simulations described in this chapter using a Desktop PC with 4GB of memory, a 1 TB hard disk, and an Intel® Core® i3-2120 at 3.30 GHz processor running Windows 8.1.

7 Final Remarks

Molecular docking simulations of biological systems open the possibility to generate the protein-ligand complex structure. Such simulations can identify potential new drugs. The use of the program ArgusLab to create protein-ligand complexes has been successfully applied to a wide range of biological systems [25–50], which further validate the importance of this program in the simulation of such complex systems.

Acknowledgments

This work was supported by grants from CNPq (Brazil) (308883/2014-4). This study was financed in part by the Coordenação de Aperfeiçoamento de Pessoal de Nivel Superior—Brasil (CAPES)—Finance Code 001. GB-F acknowledges support from PUCRS/BPA fellowship. WFA is a senior researcher for CNPq (Brazil) (Process Numbers: 308883/2014-4 and 309029/2018-0).

References

1. Filgueira de Azevedo W Jr, dos Santos GC, dos Santos DM, Olivieri JR, Canduri F, Silva RG et al (2003) Docking and small angle X-ray scattering studies of purine nucleoside phosphorylase. Biochem Biophys Res Commun 309:923–928
2. da Silveira NJ, Arcuri HA, Bonalumi CE, de Souza FP, Mello IM, Rahal P et al (2005) Molecular models of NS3 protease variants of the hepatitis C virus. BMC Struct Biol 5:1
3. Silveira NJ, Uchôa HB, Pereira JH, Canduri F, Basso LA, Palma MS et al (2005) Molecular models of protein targets from *Mycobacterium tuberculosis*. J Mol Model 11:160–166
4. da Silveira NJ, Bonalumi CE, Uchõa HB, Pereira JH, Canduri F, de Azevedo WF (2006) DBMODELING: a database applied to the study of protein targets from genome projects. Cell Biochem Biophys 44:366–374
5. da Silveira NJF, Bonalumi CE, Arcuri HA, de Azevedo WF Jr (2007) Molecular modeling databases: a new way in the search of proteins targets for drug development. Curr Bioinforma 2:1–10
6. Marques MR, Pereira JH, Oliveira JS, Basso LA, de Azevedo WF Jr, Santos DS et al (2007) The inhibition of 5-enolpyruvylshikimate-3-phosphate synthase as a model for development of novel antimicrobials. Curr Drug Targets 8:445–457
7. Breda A, Basso LA, Santos DS, de Azevedo WF Jr (2008) Virtual screening of drugs: score functions, docking, and drug design. Curr Comput Aided Drug Des 4:265–272
8. de Azevedo WF Jr, Dias R (2008) Computational methods for calculation of ligand-binding affinity. Curr Drug Targets 9:1031–1039

9. Dias R, de Azevedo WF Jr (2008) Molecular docking algorithms. Curr Drug Targets 9:1040–1047
10. Arcuri HA, Zafalon GF, Marucci EA, Bonalumi CE, da Silveira NJ, Machado JM et al (2010) SKPDB: a structural database of shikimate pathway enzymes. BMC Bioinformatics 11:12
11. De Azevedo WF Jr (2010) MolDock applied to structure-based virtual screening. Curr Drug Targets 11:327–334
12. Ducati RG, Basso LA, Santos DS, de Azevedo WF Jr (2010) Crystallographic and docking studies of purine nucleoside phosphorylase from ***Mycobacterium*** *tuberculosis*. Bioorg Med Chem 18:4769–4774
13. Heberlé G, de Azevedo WF Jr (2011) Bio-inspired algorithms applied to molecular docking simulations. Curr Med Chem 18:1339–1352
14. Rocha BA, Delatorre P, Oliveira TM, Benevides RG, Pires AF, Sousa AA et al (2011) Structural basis for both pro- and anti-inflammatory response induced by mannose-specific legume lectin from *Cymbosema roseum*. Biochimie 93:806–816
15. Vianna CP, de Azevedo WF Jr (2012) Identification of new potential *Mycobacterium tuberculosis* shikimate kinase inhibitors through molecular docking simulations. J Mol Model 18:755–764
16. Moraes FP, de Azevedo WF Jr (2012) Targeting imidazoline site on monoamine oxidase B through molecular docking simulations. J Mol Model 18:3877–3886
17. Azevedo LS, Moraes FP, Xavier MM, Pantoja EO, Villavicencio B, Finck JA et al (2012) Recent progress of molecular docking simulations applied to development of drugs. Curr Bioinforma 7:352–365
18. Coracini JD, de Azevedo WF Jr (2014) Shikimate kinase, a protein target for drug design. Curr Med Chem 21:592–604
19. de Avila MB, de Azevedo WF (2014) Data mining of docking results. Application to 3-dehydroquinate dehydratase. Curr Bioinforma 9:361–379
20. Teles CB, Moreira-Dill LS, Silva Ade A, Facundo VA, de Azevedo WF Jr, da Silva LH et al (2015) A lupane-triterpene isolated from ***Combretum leprosum*** Mart. fruit extracts that interferes with the intracellular development of ***Leishmania*** (L.) ***amazonensis in vitro***. BMC Complement Altern Med 15:165
21. de Azevedo WF Jr (2016) Opinion paper: targeting multiple cyclin-dependent kinases (CDKs): a new strategy for molecular docking studies. Curr Drug Targets 17:2
22. Xavier MM, Heck GS, de Avila MB, Levin NM, Pintro VO, Carvalho NL et al (2016) SAnDReS a computational tool for statistical analysis of docking results and development of scoring functions. Comb Chem High Throughput Screen 19:801–812
23. Ballante F, Marshall GR (2016) An automated strategy for binding-pose selection and docking assessment in structure-based drug design. J Chem Inf Model 56:54–72
24. Vieth M, Hirst JD, Kolinski A, Brooks CL III (1998) Assessing energy functions for flexible docking. J Comput Chem 19:1612–1622
25. Joy S, Nair PS, Hariharan R, Pillai MR (2006) Detailed comparison of the protein-ligand docking efficiencies of GOLD, a commercial package and ArgusLab, a licensable freeware. In Silico Biol 6:601–605
26. Sami AJ, Haider MK (2007) Identification of novel catalytic features of endo-beta-1,4-glucanase produced by mulberry longicorn beetle *Apriona germari*. J Zhejiang Univ Sci B 8:765–770
27. Yanamala N, Tirupula KC, Klein-Seetharaman J (2008) Preferential binding of allosteric modulators to active and inactive conformational states of metabotropic glutamate receptors. BMC Bioinformatics 1:16
28. Naz A, Bano K, Bano F, Ghafoor NA, Akhtar N (2009) Conformational analysis (geometry optimization) of nucleosidic antitumor antibiotic showdomycin by Arguslab 4 software. Pak J Pharm Sci 22:78–82
29. Singh KD, Muthusamy K (2009) In silico genome analysis and drug efficacy test of influenza A virus (H1N1) 2009. Indian J Microbiol 49:358–364
30. Duverna R, Ablordeppey SY, Lamango NS (2010) Biochemical and docking analysis of substrate interactions with polyisoprenylated methylated protein methyl esterase. Curr Cancer Drug Targets 10:634–648
31. Sridhar GR, Rao AA, Srinivas K, Nirmala G, Lakshmi G, Suryanarayna D et al (2010) Butyrylcholinesterase in metabolic syndrome. Med Hypotheses 75:648–651
32. Parasuraman S, Raveendran R (2011) Effect of cleistanthin A and B on adrenergic and cholinergic receptors. Pharmacogn Mag 7:243–247
33. Balaji B, Ramanathan M (2012) Prediction of estrogen receptor β ligands potency and selectivity by docking and MM-GBSA scoring methods using three different scaffolds. J Enzyme Inhib Med Chem 27:832–844

34. Hussain Basha S, Prasad RN (2012) In-silico screening of pleconaril and its novel substituted derivatives with neuraminidase of H1N1 influenza strain. BMC Res Notes 5:105
35. Elavarasan S, Bhakiaraj D, Chellakili B, Elavarasan T, Gopalakrishnan M (2012) One pot synthesis, structural and spectral analysis of some symmetrical curcumin analogues catalyzed by calcium oxide under microwave irradiation. Spectrochim Acta A Mol Biomol Spectrosc 97:717–721
36. Sridhar GR, Nageswara Rao PV, Kaladhar DS, Devi TU, Kumar SV (2012) In silico docking of HNF-1a receptor ligands. Adv Bioinforma 2012:705435
37. Piplani P, Singh P, Sharma A (2013) Synthesis, molecular docking and antiamnesic activity of selected 2-naphthyloxy derivatives. Med Chem 9:371–378
38. Basha SH, Talluri D, Raminni NP (2013) Computational repositioning of ethno medicine elucidated gB-gH-gL complex as novel anti herpes drug target. BMC Complement Altern Med 13:85
39. Hafeez A, Naz A, Naeem S, Bano K, Akhtar N (2013) Computational study on the geometry optimization and excited - state properties of riboflavin by ArgusLab 4.0.1. Pak J Pharm Sci 26:487–493
40. Sardari S, Azadmanesh K, Mahboudi F, Davood A, Vahabpour R, Zabihollahi R et al (2013) Design of small molecules with HIV fusion inhibitory property based on Gp41 interaction assay. Avicenna J Med Biotechnol 5:78–86
41. Song Z, Wang J, Yang B (2014) Spectral studies on the interaction between HSSC and apoCopC. Spectrochim Acta A Mol Biomol Spectrosc 118:454–460
42. Krishnamoorthy M, Balakrishnan R (2014) Docking studies for screening anticancer compounds of Azadirachta indica using Saccharomyces cerevisiae as model system. J Nat Sci Biol Med 5:108–111
43. Sahoo BR, Dubey PK, Goyal S, Bhoi GK, Lenka SK, Maharana J et al (2014) Exploration of the binding modes of buffalo PGRP1 receptor complexed with meso-diaminopimelic acid and lysine-type peptidoglycans by molecular dynamics simulation and free energy calculation. Chem Biol Interact 220:255–268
44. Shaikh RU, Dawane AA, Pawar RP, Gond DS, Meshram RJ et al (2016) Inhibition of Helicobacter pylori and its associate urease by labdane diterpenoids isolated from Andrographis paniculata. Phytother Res 30:412–417
45. Dash R, Uddin MM, Hosen SM, Rahim ZB, Dinar AM, Kabir MS et al (2015) Molecular docking analysis of known flavonoids as duel COX-2 inhibitors in the context of cancer. Bioinformation 11:543–549
46. Jahanban-Esfahlan A, Panahi-Azar V (2016) Interaction of glutathione with bovine serum albumin: spectroscopy and molecular docking. Food Chem 202:426–431
47. Song Z, Yuan W, Zhu R, Wang S, Zhang C, Yang B (2017) Study on the interaction between curcumin and CopC by spectroscopic and docking methods. Int J Biol Macromol 96:192–199
48. Agrahari AK, GPD C (2017) A computational approach to identify a potential alternative drug with its positive impact toward PMP22. J Cell Biochem 118:3730–3743
49. Chaudhary NK, Mishra P (2017) Metal complexes of a novel Schiff Base based on penicillin: characterization, molecular modeling, and antibacterial activity study. Bioinorg Chem Appl 2017:6927675
50. Mohammadi T, Ghayeb Y (2018) Atomic insight into designed carbamate-based derivatives as acetylcholine esterase (AChE) inhibitors: a computational study by multiple molecular docking and molecular dynamics simulation. J Biomol Struct Dyn 36:126–138
51. Berg S, Bergh M, Hellberg S, Högdin K, Lo-Alfredsson Y, Söderman P et al (2012) Discovery of novel potent and highly selective glycogen synthase kinase-3β (GSK3β) inhibitors for Alzheimer's disease: design, synthesis, and characterization of pyrazines. J Med Chem 55:9107–9119
52. Morgan DO (1995) Principles of CDK regulation. Nature 374:131–134
53. Murray AW (1994) Cyclin-dependent kinases: regulators of the cell cycle and more. Chem Biol 1:191–195
54. Kim SH, Schulze-Gahmen U, Brandsen J, de Azevedo Junior WF (1996) Structural basis for chemical inhibition of CDK2. Prog Cell Cycle Res 2:137–145
55. De Azevedo WF Jr, Mueller-Dieckmann HJ, Schulze-Gahmen U, Worland PJ, Sausville E, Kim SH (1996) Structural basis for specificity and potency of a flavonoid inhibitor of human CDK2, a cell cycle kinase. Proc Natl Acad Sci U S A 93:2735–2740
56. Canduri F, de Azevedo WF Jr (2005) Structural basis for interaction of inhibitors with cyclin-dependent kinase 2. Curr Comput Aided Drug Des 1:53–64
57. Krystof V, Cankar P, Frysová I, Slouka J, Kontopidis G, Dzubák P (2006) 4-arylazo-

3,5-diamino-1H-pyrazole CDK inhibitors: SAR study, crystal structure in complex with CDK2, selectivity, and cellular effects. J Med Chem 49:6500–6509

58. Levin NM, Pintro VO, de Ávila MB, de Mattos BB, De Azevedo WF Jr (2017) Understanding the structural basis for inhibition of cyclin-dependent kinases. New pieces in the molecular puzzle. Curr Drug Targets 18:1104–1111
59. de Ávila MB, Xavier MM, Pintro VO, de Azevedo WF (2017) Supervised machine learning techniques to predict binding affinity. A study for cyclin-dependent kinase 2. Biochem Biophys Res Commun 494:305–310
60. Levin NMB, Pintro VO, Bitencourt-Ferreira G, Mattos BB, Silvério AC, de Azevedo WF Jr (2018) Development of CDK-targeted scoring functions for prediction of binding affinity. Biophys Chem 235:1–8
61. Volkart PA, Bitencourt-Ferreira G, Souto AA, de Azevedo WF (2019) Cyclin-dependent kinase 2 in cellular senescence and cancer. A structural and functional review. Curr Drug Targets 20:716–726. https://doi.org/10.2174/1389450120666181204165344
62. De Bondt HL, Rosenblatt J, Jancarik J, Jones HD, Morgan DO, Kim SH (1993) Crystal structure of cyclin-dependent kinase 2. Nature 363:595–602
63. Berman HM, Westbrook J, Feng Z, Gilliland G, Bhat TN, Weissig H et al (2000) The Protein Data Bank. Nucleic Acids Res 28:235–242
64. Berman HM, Battistuz T, Bhat TN, Bluhm WF, Bourne PE, Burkhardt K et al (2002) The Protein Data Bank. Acta Crystallogr D Biol Crystallogr 58:899–907
65. Westbrook J, Feng Z, Chen L, Yang H, Berman HM (2003) The Protein Data Bank and structural genomics. Nucleic Acids Res 31:489–491

Chapter 14

Web Services for Molecular Docking Simulations

Nelson J. F. da Silveira, Felipe Siconha S. Pereira, Thiago C. Elias, and Tiago Henrique

Abstract

Docking process is one of the most significant activities for the analysis of protein–protein or protein–ligand complexes. These tools have become of unique importance when allocated in web services, collaborating scientifically with several areas of knowledge in an interdisciplinary way. Among the several web services dedicated to carrying out molecular docking simulations, we selected the DockThor web service. To illustrate the application of DockThor to protein–ligand docking simulations, we analyzed the docking of a ligand against the structure of epidermal growth factor receptor, an essential molecular marker in cancer research.

Key words Web docking, Web services, Docking affinity, Score function, Complex, Protein–protein, Protein–ligand

1 Introduction

With the termination of the human genome sequencing project, many protein targets for the development of new drugs have been identified [1]. One of the essential tools for the development of new drugs is molecular docking [2]. The creation of web tools for performing molecular docking procedures has become very important for the dissemination of docking for the rational structure-based drug design. The in silico analysis has become a great ally of the experimental methodologies, filtering data for experimentation, allowing the optimization of time and cost for the experiments [3, 4]. Such docking methodologies result in reduced computational cost and improved accuracy in obtaining simulation results.

There are dozens of web services available for protein–ligand docking simulations; in this chapter, we focus our study on the DockThor. It was used for web docking simulation with the complex epidermal growth factor receptor (EGFR)-hydrazone, that is, one molecular marker related to cancer [5].

Walter Filgueira de Azevedo Jr. (ed.), *Docking Screens for Drug Discovery*, Methods in Molecular Biology, vol. 2053, https://doi.org/10.1007/978-1-4939-9752-7_14,

2 Materials

2.1 Web Docking Overview

Currently, the Internet offers a set of web servers for performing molecular docking available for the scientific community. Table 1 lists some of these web services dedicated to docking.

Protein–protein interactions (PPIs) are essential in biological research due to their role in cell signaling, cell regulation, enzyme inhibition, and immune response [15, 18]. These interactions can be analyzed by X-ray crystallography or NMR, but these

Table 1
Docking servers available on web

Web server	Site	Docking Type	Reference	Notes
DockThor	http://dockthor.lncc.br	Rigid-protein/flexible-small ligand	[6]	1
CABS-dock	http://biocomp.chem.uw.edu.pl/CABSdock	Rigid-protein/flexible small peptide	[7]	2
PatchDock	http://bioinfo3d.cs.tau.ac.il/PatchDock/	Rigid-protein/rigid-protein	[8]	3
FireDock	http://bioinfo3d.cs.tau.ac.il/FireDock/	Flexible side chain-protein/flexible side chain-protein refinement	[9]	4
FiberDock	http://bioinfo3d.cs.tau.ac.il/FiberDock/	Flexible protein/flexible protein refinement	[10]	5
SymmDock	http://bioinfo3d.cs.tau.ac.il/SymmDock/	Rigid-protein symmetric complex docking	[8]	6
GRAMM-X	http://vakser.compbio.ku.edu/resources/gramm/grammx	Rigid-protein/rigid-protein	[11]	7
HADDOCK	http://milou.science.uu.nl/services/HADDOCK2.2/haddockserver-easy.html	Protein/protein, protein/DNA, protein/small ligand, all cases flexible	[12]	8
HexServer	http://hexserver.loria.fr	Rigid-protein/rigid-protein	[13]	9
MEDock	http://medock.ee.ncku.edu.tw	Rigid-protein/flexible ligand	[14]	10
RosettaDock	http://rosie.graylab.jhu.edu/docking2/submit	Rigid-protein/rigid-protein	[15]	11
SwissDock	http://www.swissdock.ch	Rigid-protein/flexible-ligand	[16]	12
TarFisDock	http://www.dddc.ac.cn/tarfisdock/	Rigid-protein/flexible-ligand (reverse docking)	[17]	13
ZDOCK	http://zdock.umassmed.edu	Rigid-protein/rigid-protein	[18]	14
ParDOCK	http://www.scfbio-iitd.res.in/dock/pardock.jsp	Rigid-protein/rigid ligand	[19]	15

experimental techniques are expensive, have their limitations, and fewer protein–protein complexes are available [12]. Thus, several protein–protein docking programs have been developed to study protein–protein interactions. Most of them are available as web servers, which facilitates docking simulation and avoid difficulties as installing and updating software [6–19]. Critical assessment of predicted interactions (CAPRI) initiative [20] is an initiative for the advancement of protein–protein docking simulations.

The web interface is in some way similar among protein–protein docking servers. The user uploads PDB structures of proteins to be docked; the largest protein is named "receptor" while another is the "ligand." Alternatively, the PDB code for one or both proteins can also be entered; in this case, PDB structures are automatically downloaded from PDB (http://www.rcsb.org/pdb/home/home.do). Since a protein structure is relatively large, it is hard to find correct protein–protein orientation if the search is on the whole macromolecule structure. Thus, web servers have a way for the users to define protein regions for receptor and ligand. In PatchDock web server, the user can upload a file containing residue number and residue chain. This procedure is carried out one residue per line, for all residues in the receptor that must be in contact with residues in the ligand, a similar file for the ligand can also be uploaded. In the GRAMM-X web server, the user writes in a text the residue number followed by colon and the chain identifier and/or a residue number range for interacting residues for receptor and ligand. In the Hex Server, the user can define an interface residue for both receptor and ligand, the alpha Carbon atom of those residues are located on the intermolecular *z*-axis; in the ZDOCK, the users select interacting residues from a drop-down list for receptor and ligand.

The protein–ligand docking simulation is used to predict the interaction between a protein and a small molecule, generally to search a drug candidate. While protein structure is treated as a rigid body, the ligand is flexible, having freedom in torsional angles. Protein input is usually uploaded in PDB format, ligand input can be in PDB or mol2 format, according to the web server. Some docking web servers allow the user to define protein region in which ligand is expected to be docked; this task is performed by tipping the *x*, *y*, *z* Cartesian center, and the *x*, *y*, *z* Cartesian size for the search box, such procedure is performed in DockThor and SwissDock. In ParDOCK, the definition of the search box is not necessary, since the protein receptor must contain a co-crystallized reference ligand, whose mass center is used to define search box, while MEDock predicts the binding site using a global search (whole receptor structure) exploring maximum entropy property of the Gaussian probability distribution function. TarFisDock is a specialized docking server since it performs a reverse docking, in which the user uploads only ligand structure that is docked against a set of target proteins into its database. Another specialized docking server is CABS-dock, designed to dock small peptides onto a protein structure in blind docking (search done in whole protein structure), users upload receptor file and input small peptide primary sequence in a text box. The peptide structure is automatically constructed in the server.

Molecular docking methodologies are composed by search algorithms and an energy-scoring function for generating and evaluating ligand poses [21]. Search algorithms include Genetic Algorithm (DockThor), Fast Fourier Transformation (FFT) (ZDOCK and GRAMM-X), Spherical polar Fourier (SPF) approach (HexServer), shape complementarity principles (PatchDock), and Monte Carlo-based algorithm (RosettaDock). Servers such as FireDock and FiberDock perform flexible docking, and they can be used for refining docking results provided by other servers. Generally, scoring functions are formed by combinations of terms regarding van der Waals interactions, electrostatic interactions, desolvation effects, and entropy.

3 Methods

3.1 Example for Web Docking

In this section, we show the procedures to perform molecular docking with DockThor Web Server [6]. We selected, as receptor, the structure of human EGFR kinase, a molecular target in lung cancer treatment, complexed with Hydrazone, a dual inhibitor (PDB code: 2RGP) [22]. For the redocking experiment, structures of receptor and ligand were manually separated. After entering the web server page (http://dockthor.lncc.br/index.php?pg=home), the user clicks "Docking" button; a new page will display five tabs corresponding to the steps necessary in docking procedures. In the first tab, "Protein," the user uploads the protein structure in the PDB format. By clicking on the "Prepare" button, the organization of the input file is carried out. It is possible to change protonation state for six residue types (Cys, Lys, Arg, His, Asp, Glu) and reprepare protein; in this experiment, the protonation state remains as default for all residues. Clicking "NEXT" button sends prepared protein to the server and passs to the next step in the "Ligand" tab, where the user uploads ligand structure in PDB format. The ligand is prepared by clicking "Prepare" button, if desired, hydrogen atoms are added checking "Add hydrogens" checkbox. Rotatable ligand bonds are detected automatically, but the user can select among them what will be rotatable in "Rotatable bonds to be flexible during docking" box; it was chosen to add hydrogen atoms and use all found rotatable bonds. Again, clicking "NEXT" button sends prepared ligand to server and passes to next step. In "Cofactors" tab, cofactors files (i.e., metal atoms and waters) can be uploaded and prepared, including adding hydrogen atoms.

Table 2
Results of PDB 2RGP redocking experiment provided by DockThor web service

Run	Model	T. Energy (kcal/mol)	I. Energy (kcal/mol)	RMSD (Å)	Affinity Score (kcal/mol)
16	1	19.865	−43.266	0.985	−10.433
23	7	24.446	−35.670	2.366	−10.195
13	9	28.992	−34.577	3.743	−9.950
12	9	31.783	−31.694	10.026	−9.598
1	11	31.840	−30.544	9.904	−9.421
23	11	33.026	−32.262	6.085	−9.209
16	13	34.011	−25.627	8.786	−8.899
9	12	34.581	−26.338	7.462	−9.197
7	13	34.616	−24.423	11.619	−9.010
18	10	34.721	−27.502	7.249	−8.854

Cofactors are treated as rigid bodies; no cofactors are included in EGFR redocking. In the next tab, the "Docking" step, an e-mail address must be specified for which server will send results link. In the process, the grid center is defined (in this case, it is the center of mass of original ligand co-crystallized in 2RGP PDB structure: *x*: 16.764 Å, *y*: 35.706 Å, *z*: 91.272 Å), as well as grid dimensions (that is taken as default values of 22 Å in each coordinate axis), discretization of the grid energy (also taken as default value of 0.25 Å), and a job label. Genetic algorithm parameters can be changed in the number of evaluations, population size, number of runs, and seed (in this simulation, only the population size parameter was changed to 750; others were kept as default). Finally, clicking "Dock!" performs docking simulation. Table 2 shows the results exhibited in the tab "Results and Analyzes" after finishing the docking calculation.

The column "Run" shows the number of runs obtained by the genetic algorithm in a ranking, and the column "Model" shows the number of the models with better energy score (in this case, ranked by Total Energy). The column "T. Energy" shows the values of the total energy of the complex, in kcal/mol, the column "I. Energy" shows the values of internal energy of the complex, in kcal/mol. The column "RMSD" shows the values of root mean square deviation between a reference pose ligand (i.e., crystallized ligand) and best docking solution, or when not existent, a crystalized ligand, the best docking solution is assumed as a pose reference. The column "Score" shows the values of protein–ligand binding affinity of the complex, in kcal/mol. This affinity score can be correlated with inhibition constant determined by the equation below,

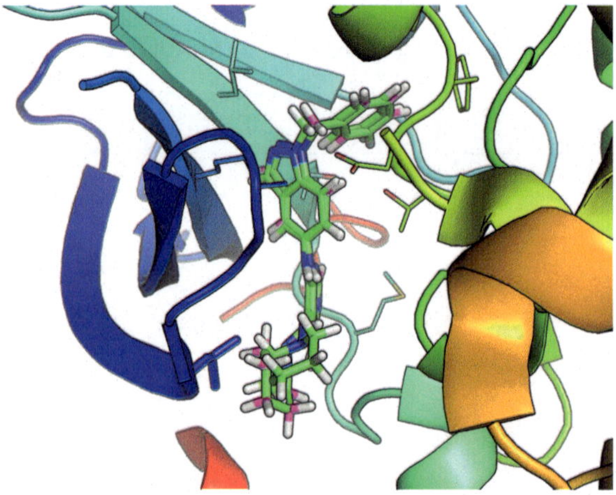

Fig. 1 Visualization of the best docking solution of the complex PDB 2RGP provided by the DockThor web service

$$\Delta G_{\text{bind}} = RT \ln K_{\text{i}}$$

where ΔG_{bind} is the score in kcal/mol, R is the universal gas constant ($R = 1.98$ cal/mol$*$K), T is the temperature ($T = 298$ K), and K_{i} is the inhibition constant of the molecular compound. Figure 1 shows the best ligand pose observed in the complex simulated with DockThor.

3.2 DockThor Profile

The DockThor program was developed by Molecular Modeling of Biological Systems Group (GMMSB), a multidisciplinary research group at National Laboratory for Scientific Computing (LNCC), located in Petrópolis, RJ, Brazil. Several current docking programs exhibit difficulty to treat the pose prediction of large and highly flexible ligands (i.e., ligands with a larger amount of rotatable bonds) [23], so the DockThor was initially developed to perform docking studies of highly flexible ligands and to explore distinct and valuable ligand-binding modes of more reliable way.

The current version of DockThor is freely available since 2013 in a web portal that allows the online execution of steps of file preparation, molecular docking, and analysis of the results, supported by GMMSB/LNCC using the infrastructure provided by Brazilian High-Performance Platform (SINAPAD). The DockThor Portal uses in-house auxiliary programs, all of them developed by GMMSB/LNCC, to automated parametrization and carry out the docking simulation: (1) PdbThorBox [24] and (2) MMFFLigand [25], for automatic parametrization of protein and ligands, respectively, and (3) DTStatistic [26, 27] for automatic clustering and analyses of docking results. The DockThor Portal allows an easy way to variate the protonation states of the amino acid residues, online execution, and visualization of many steps of a docking experiment. The user can also customize the main parameters of the energy grid and the genetic algorithm. The portal provides a ranked set of best energy docking solutions as output and allows the download of them. The results are available from a specific link, sent to the user by e-mail, and can be analyzed by visual inspection on the website using the JSmol tool.

DockThor method employs multiple solutions steady-state genetic algorithm as the search method and evaluates the ligand poses using a scoring function (Eq. 1) based on MMFF94s force field [6, 23, 24]. The binding affinity prediction of the docking solutions is calculated by empirical scoring functions [22], developed by training utilized the dataset PDBbind v2013 [23]. DockThor performs a rigid-receptor/flexible-ligand docking, and explores the conformational and configurational (i.e., translational and rotational) ligand degrees of freedom, while the protein is kept fixed.

$$\text{Score} = E_{\text{torsional}} + E_{\text{vdW}} + E_{\text{eletrostactic}} \quad (1)$$

The DockThor-VS Portal, an established version of the program for virtual screening experiments, scheduled for launch in 2019, will count on several empirical scoring functions, developed by GMMSB/LNCC group using machine-learning techniques, to predict protein–ligand binding affinity. This virtual screening web service will allow researchers to perform large-scale virtual screening experiments in drug design studies.

3.3 Web Interface

The layout of DockThor Portal is shown below. Figure 2 displays the home page of the portal, where it is possible to visualize all the functions of the web portal. A brief description of the program and the web portal is described in the body of the page. The top bar exhibits the buttons (1) Home, (2) Docking, (3) References, (4) About, and (5) Support. The "Home" button shows the initial home page. The "Docking" button directs the user to the molecular docking function, which provides the execution of the pipeline described previously in Subheading 3.1. The "References" button exhibits the articles and works related to the development of the DockThor program. The "About" button shows a brief description of the team responsible for the development and maintenance of the DockThor Portal. The "Support" button displays the options (1) "Help" and (2) "Contact," where "Help" provides tutorial files and "Contact" provides a way to send a message to DockThor team. The current version of DockThor Portal allows the subscription of an e-mail in DockThor e-Newsletters, to receive information about the news, portals released, and versions of the DockThor.

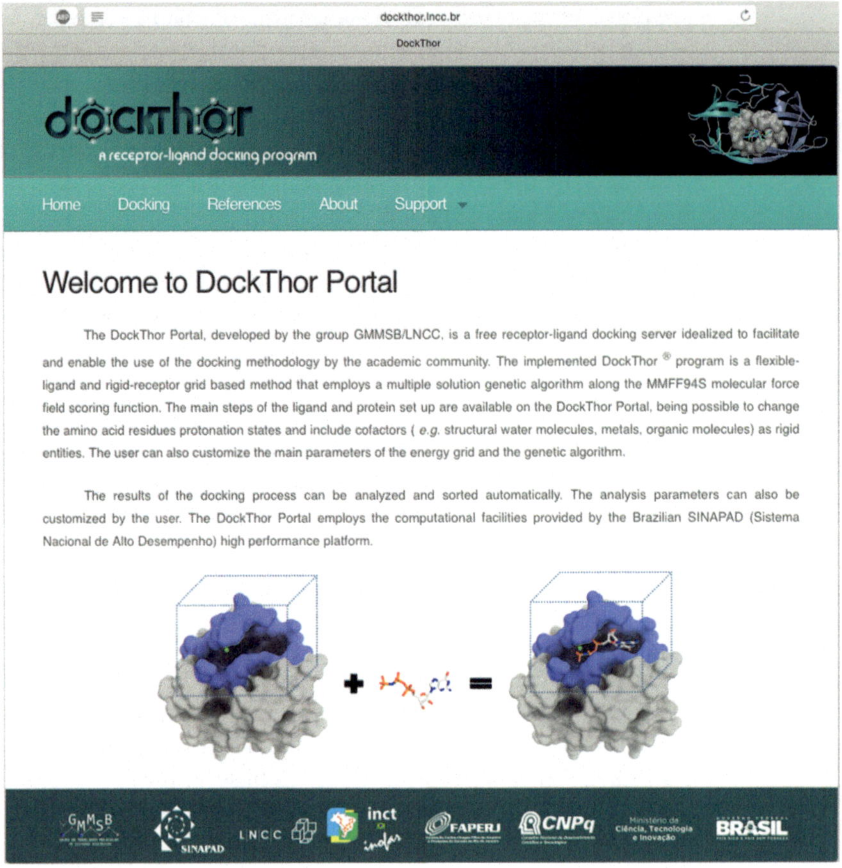

Fig. 2 Home page of DockThor Portal

Acknowledgments

This work was supported by LNCC/MCTIC, SINAPAD, INCT-Inofar, FAPERJ, CNPq, and CAPES.

References

1. Gazdar AF (2009) Activating and resistance mutations of EGFR in non-small-cell lung cancer: role in clinical response to EGFR tyrosine kinase inhibitors. Oncogene 28(Suppl 1):24–31
2. Mukesh B, Rakesh K (2011) Molecular docking: a review. IJRAP 2:1746–1751
3. Vakser IA (2014) Protein-protein docking: from interaction to interactome. Biophys J 107:1785–1793
4. Meng XY, Zhang HX, Mezei M, Cui M (2011) Molecular docking: a powerful approach for structure-based drug discovery. Curr Comput Aided Drug Des 7:146–157
5. Seshacharyulu P, Ponnusamy MP, Haridas D, Jain M, Ganti AK, Batra SK (2012) Targeting the EGFR signaling pathway in cancer therapy. Expert Opin Ther Targets 16:15–31
6. de Magalhães CS, Almeida DM, Barbosa HJC, Dardenne LE (2014) A dynamic niching genetic algorithm strategy for docking of highly flexible ligands. Inform Sci 289:206–224
7. Kurcinski M, Jamroz M, Blaszczyk M, Kolinski A, Kmiecik S (2015) CABS-dock web server for the flexible docking of peptides to proteins without prior knowledge of the binding site. Nucleic Acids Res 43:419–424

8. Schneidman-Duhovny D, Inbar Y, Nussinov R, Wolfson HJ (2005) PatchDock and SymmDock: servers for rigid and symmetric docking. Nucleic Acids Res 33:363–367
9. Mashiach E, Schneidman-Duhovny D, Andrusier N, Nussinov R, Wolfson HJ (2008) FireDock: a web server for fast interaction refinement in molecular docking. Nucleic Acids Res 36:229–232
10. Mashiach E, Nussinov R, Wolfson HJ (2010) FiberDock: a web server for flexible induced-fit backbone refinement in molecular docking. Nucleic Acids Res 38:457–461
11. Tovchigrechko A, Vakser IA (2006) GRAMM-X public web server for protein-protein docking. Nucleic Acids Res 34:310–314
12. Vries SJ, Dijk MY, Bonvin AMJJ (2010) The HADDOCK web server for data-driven biomolecular docking. Nat Protoc 5:883–897
13. Macindoe G, Mavridis L, Venkatraman V, Devignes MD, Ritchie DW (2010) HexServer: an FFT-based protein docking server powered by graphics processors. Nucleic Acids Res 38:445–449
14. Chang DTH, Oyang YJ, Lin JH (2005) MEDock: a web server for efficient prediction of ligand binding sites based on a novel optimization algorithm. Nucleic Acids Res 33:233–238
15. LysKov S, Gray JJ (2008) The RosettaDock server for local protein-protein docking. Nucleic Acids Res 36:233–238
16. Grosdidier A, Zoete V, Michielin O (2011) SwissDock, a protein-small molecule docking web service based on EADock DSS. Nucleic Acids Res 39:270–277
17. Li H, Gao Z, Kang L, Zhang H, Yang K, Yu K et al (2006) TarFisDock: a web server for identifying drug targets with docking approach. Nucleic Acids Res 34:219–224
18. Pierce BG, Wiehe K, Hwang H, Kim BH, Vreven T, Weng Z (2014) ZDOCK server: interactive docking prediction of protein-protein complexes and symmetric multimers. Bioinformatics 30:1771–1773
19. Gupta A, Gandhimathi A, Sharma P, Jayaram B (2007) ParDOCK: an all atom energy based Monte Carlo docking protocol for protein-ligand complexes. Protein Pept Lett 14:632–646
20. Janin J (2002) Welcome to CAPRI: a critical assessment of predicted interactions. Proteins 47:257
21. Guedes IA, de Magalhães CS, Dardenne LE (2014) Receptor–ligand molecular docking. Biophys Rev 6:75–87
22. Xu G, Abad MC, Connolly PJ, Neeper MP, Struble GT, Springer BA et al (2008) 4-Amino-6-arylamino-pyrimidine-5-carbaldehyde hydrazones as potent ErbB-2/EGFR dual kinase inhibitors. Bioorg Med Chem Lett 18:4615–4619
23. Almeida DM (2011) Dockthor: Implementação, Aprimoramento e Validação de um Programa de Docking Receptor-Ligante. MSc Dissertation, Laboratório Nacional de Computação Científica-LNCC, Petrópolis, RJ
24. Halgren TA (1999) MMFF VII. Characterization of MMFF94, MMFF94s, and other widely available force fields for conformational energies and for intermolecular-interaction energies and geometries. J Comput Chem 20:730–748
25. Guedes IA (2016) Development of empirical scoring functions for predicting protein-ligand binding affinity. Doctoral dissertation, Laboratório Nacional de Computação Científica-LNCC, Petrópolis, RJ
26. Li Y, Liu Z, Li J, Han L, Liu J, Zhao Z et al (2014) Comparative assessment of scoring functions on an updated benchmark: 1. Compilation of the test set. J Chem Inf Model 54:1700–1716
27. Dardenne LE (2000) Propriedades Eletrostáticas do Sítio Ativo de Cisteíno Proteinases da Família da Papaína. Doctoral dissertation, Universidade Federal do Rio de Janeiro-UFRJ, Rio de Janeiro, Brasil

Chapter 15

Homology Modeling of Protein Targets with MODELLER

Gabriela Bitencourt-Ferreira and Walter Filgueira de Azevedo Jr.

Abstract

Homology modeling is a computational approach to generate three-dimensional structures of protein targets when experimental data about similar proteins are available. Although experimental methods such as X-ray crystallography and nuclear magnetic resonance spectroscopy successfully solved the structures of nearly 150,000 macromolecules, there is still a gap in our structural knowledge. We can fulfill this gap with computational methodologies. Our goal in this chapter is to explain how to perform homology modeling of protein targets for drug development. We choose as a homology modeling tool the program MODELLER. To illustrate its use, we describe how to model the structure of human cyclin-dependent kinase 3 using MODELLER. We explain the modeling procedure of CDK3 apoenzyme and the structure of this enzyme in complex with roscovitine.

Key words Homology modeling, MODELLER, Cyclin-dependent kinase 3, Drug design, Molecular recognition

1 Introduction

For docking simulations, the primary demand is the availability of the three-dimensional structure of the protein target [1–21]. This structural information can be from X-ray crystallography [22], nuclear magnetic resonance spectroscopy [23], or others techniques such as neutron crystallography, electron micrography (EM), and hybrid methods [24]. X-ray diffraction crystallography is the dominant technique for analysis of protein-ligand complexes.

Considering the structural information available at the Protein Data Bank (PDB) [25–27] and filtering the data to take only protein structures for which ligand-binding affinity information is available, we have over 90% of the structural information originated from X-ray diffraction crystallography [24]. The second most significant technique is nuclear magnetic resonance spectroscopy. All methods combined generated 149,424 structures deposited in the PDB (search carried out on March 1, 2019) (http://www.rcsb.org/pdb/results/results.do?tabtoshow=Current&qrid=6D6E995). Although the success of the experimental

Walter Filgueira de Azevedo Jr. (ed.), *Docking Screens for Drug Discovery*, Methods in Molecular Biology, vol. 2053,
https://doi.org/10.1007/978-1-4939-9752-7_15,

techniques is unquestionable, we are still far from having the structural information for all proteins targets necessary for structure-based drug discovery. Even worst, considering the current data available at the PDB, there is much redundancy in the data stored on it. Many of the deposited structures are for the same protein. For example, considering the available structures of cyclin-dependent kinase 2 (CDK2), we have 436 crystallographic structures of this vital protein target, all obtained through X-ray crystallography (search carried out on March 1, 2019) (http://www.rcsb.org/pdb/results/results.do?tabtoshow=Current&qrid=2DC19CD9). So it is clear that for docking screens for drug discovery purposes, the experimental techniques are not enough to provide all necessary structural information.

To fill this gap of information, we have to make use of computational methodologies. We may divide the computational prediction of the three-dimensional protein structure into two primary techniques: ab initio methods [28–30] and homology modeling approaches [31, 32]. The first technique relies on the fold prediction from physical chemical principles. The second approach uses an experimental structure as a template to build a structural homology model based on the atomic coordinates.

Our focus here is on homology modeling. In this technique, we may use more than one template. There are two major concerns in the modeling of a new protein structure. First is the sequence identity between the template (experimental structure) and the protein to be modeled. If the protein sequence has high sequence identity (>30%) to the template, the homology recognition is fairly straightforward which is typically performed by sequence alignment [33]. The primary computational tool for sequence alignment is the program Basic Local Alignment Search Tool (BLAST) (http://www.ncbi.nlm.nih.gov/blast/) [34] that seeks sequence databases for the best local alignments to the protein sequence. The BLAST tool works well with proteins where the identity is higher than 30%.

Second is the quality of the structural information of the template, as highlighted previously; most of this structural information came from X-ray diffraction crystallography [24], and to select the most reliable templates, we usually consider crystallography resolution, R-factor, R-free [35], and overall stereochemical quality [36] of the templates. Another feature to study is the presence of an inhibitor bound to the crystallographic structure of the template, or any other active ligand bound to the structure.

Thinking about the use of the modeled structure for docking screens for drug discovery, the presence of an inhibitor of any ligand bound to the structure of the template may guide the process of structure-based drug design, where we generate a homology model with the inhibitor already attached to the structure [37]. Furthermore, considering possible conformational changes due to the ligand binding [38], the modeling of a structure, taking the coordinates of a complexed crystallographic structure, may generate a reliable structural model for docking screens.

In this chapter, we describe a tutorial explaining the application of homology modeling to generate the structure of human cyclin-dependent kinase 3. Owing to the ease of use and free availability of the program, we choose the MODELLER software [39]. This program carries out homology modeling based on the satisfaction of spatial restraints present on the template structures and their alignment with the model sequence [39].

2 Biological System

Our objective in this chapter is to describe how to carry out homology modeling of protein targets for drug development. We show how to perform homology modeling of cyclin-dependent kinase 3 (CDK3) (EC 2.7.11.22) with the program MODELLER [39]. We used a closely related serine/threonine protein kinase, CDK2, as a template. In 1993, the research group led by Prof. Sung-Hou Kim (University of California at Berkeley) solved the structure of CDK2 [40] to 2.4 Å. Using the atomic coordinates of the first CDK2 structure, we see that the N-terminal domain of this protein is mainly built by a distorted beta-sheet and a short alpha helix. A helix bundle forms the C-terminal. The two lobes of the CDK2 structure allow the binding of the ATP molecule. Several CDK inhibitors bind to the ATP-binding pocket of CDKs, which includes palbociclib, an FDA-approved drug to treat breast cancer in postmenopause women [41–44]. Palbociclib is a CDK4/6 inhibitor, and structural analysis of the complex between this inhibitor and CDK6 (PDB access code: 5L2I) indicates that it binds to the ATP-binding pocket [45]. Figure 1 shows the intermolecular interactions between Palbociclib and CDK6. There are intermolecular hydrogen bonds involving residues Val 101 and Asp 163. We identify this pattern of intermolecular interactions in several CDK-inhibitor complexes [46–54].

3 Graphical Tutorial

Here, we show how to model the three-dimensional structure of cyclin-dependent kinase (CDK3), using available experimental structures. Because of their role in the cell-cycle progression, CDKs are the protein targets for the development of anticancer drugs. Specifically for CDK3, this enzyme is overexpressed in breast cancer [55], which indicates the potential to use inhibitors of CDK3 to treat this type of malignancy. There are hundreds of CDK structures available in the Protein Data Bank, but not even one for human CDK3. We'll use the program MODELLER [39] to carry out homology modeling of CDK3 structure. We reported the

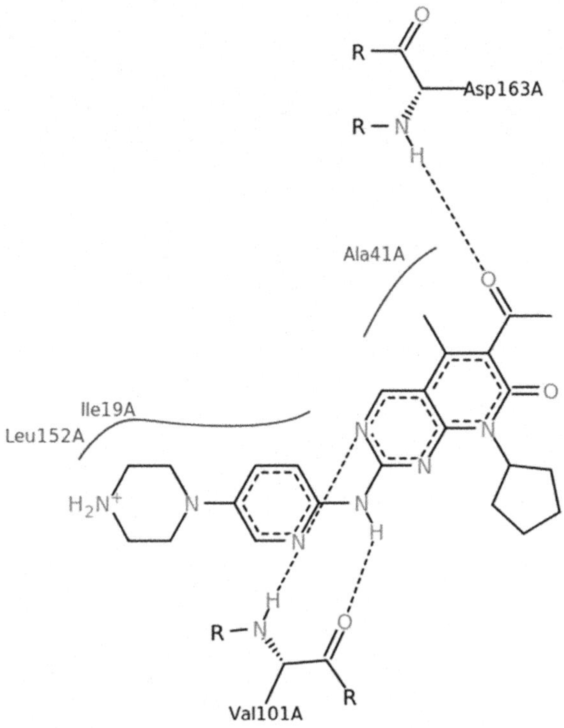

Fig. 1 Intermolecular hydrogen bonds observed for the structure of CDK6 in complex with Palbociclib (PDB access code: 5L2I)

homology modeling and molecular dynamics simulation of human CDK3 in 2009 [56].

For this tutorial, it is necessary to have access to the internet and the latest version of MODELLER installed on a computer. In the flowchart shown in Fig. 2, we can see that the main steps to homology model a protein structure, using structures available in the Protein Data Bank (PDB). The following paragraphs describe the steps to be followed in homology modeling.

First access the Genbank [57] at http://www.ncbi.nlm.nih.gov/genbank/. Then choose Protein tab and type in protein name and click on the Search button. We will get the entries for the keywords. Click on the first entry, which has the sequence for human CDK3 as shown in Fig. 3. We will get additional information about CDK3, and then click on FASTA. Figure 4 shows the amino acid sequence for CDK3. Download this file and copy it to the directory where homology modeling will be carried out.

Next, open the FASTA file with an editor, as vi, for instance, and copy the sequence that will be used to search the PDB (http://www.rcsb.org/pdb/home/home.do). In the PDB, click on the Advanced Search button. Choose Sequence (BLAST/FASTA/PSI-BLAST) option. Then, change the Search Tool to PSI-BLAST. Now we can copy (<Ctrl> C) the sequence in the field Sequence (Fig. 5).

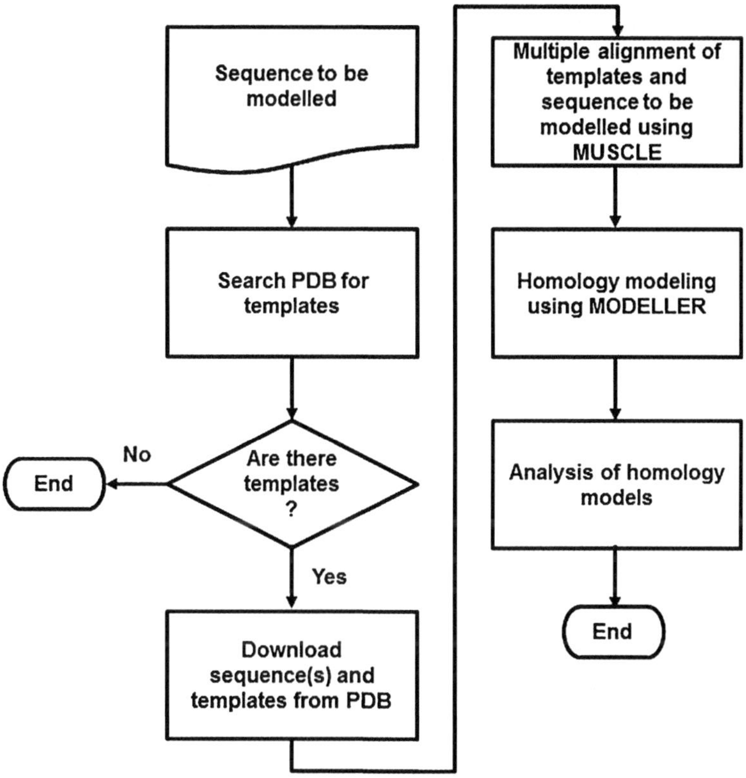

Fig. 2 Schematic flowchart for the modeling process

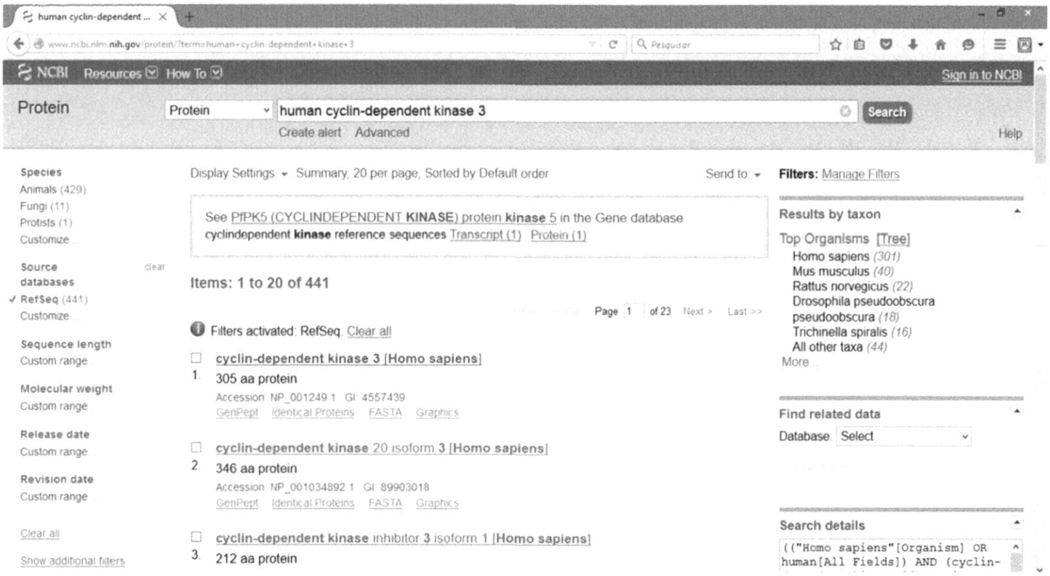

Fig. 3 GenBank website

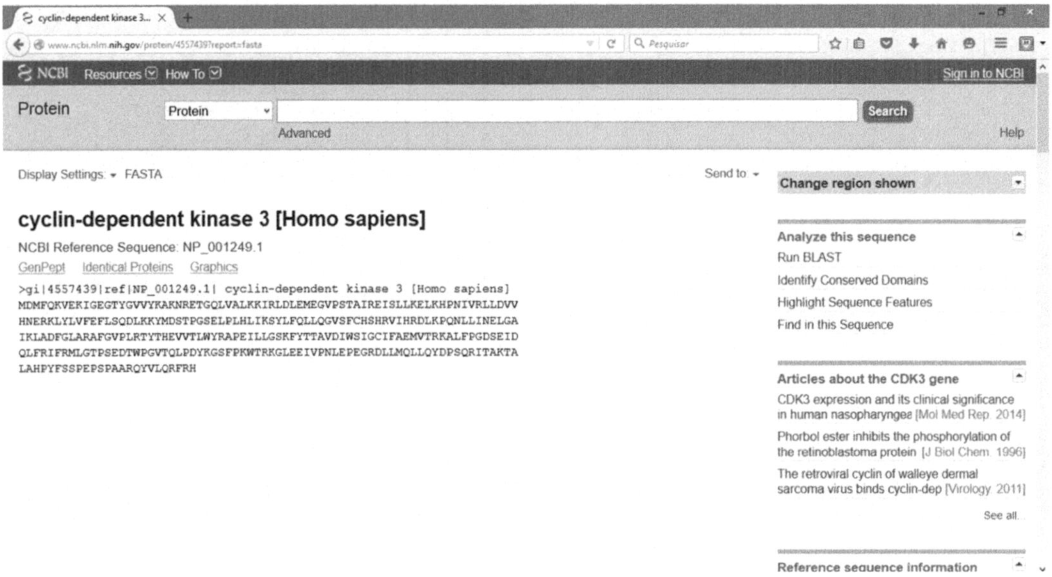

Fig. 4 CDK3 sequence

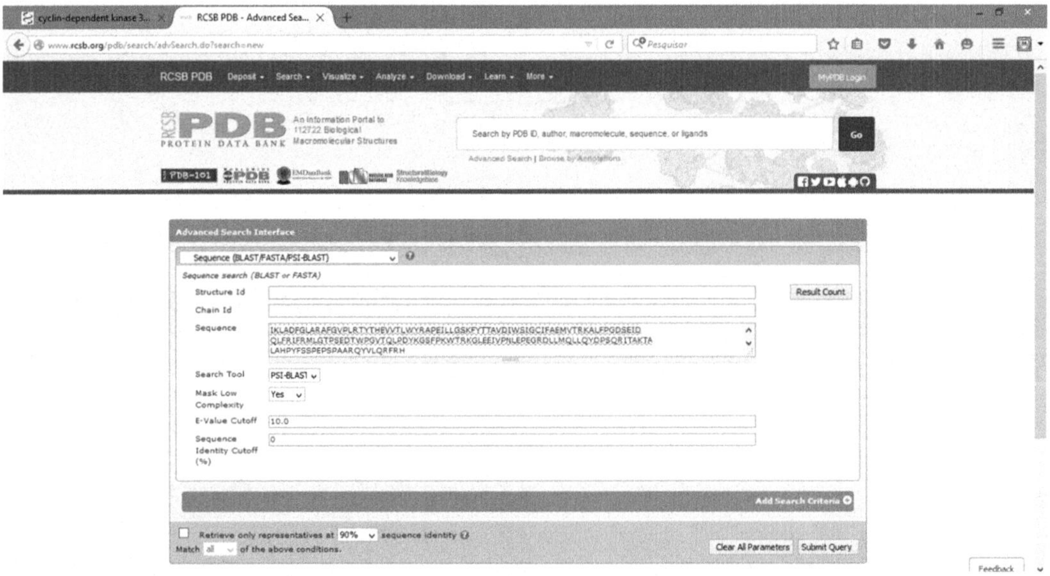

Fig. 5 PDB website

We have the protein sequence now. Then click on Submit Query. The PDB returns all structures that show similarity with the probe sequence. The alignment is shown in Fig. 6. Next, uncheck all structures to pick up only ten structures solved to a resolution better than 2.0 Å. We may choose only one structure if we want or as many templates as we think are necessary. To download PDB and FASTA files, click on *Filter>Download Checked*, as shown in Fig. 7. Then click on Launch Download Application.

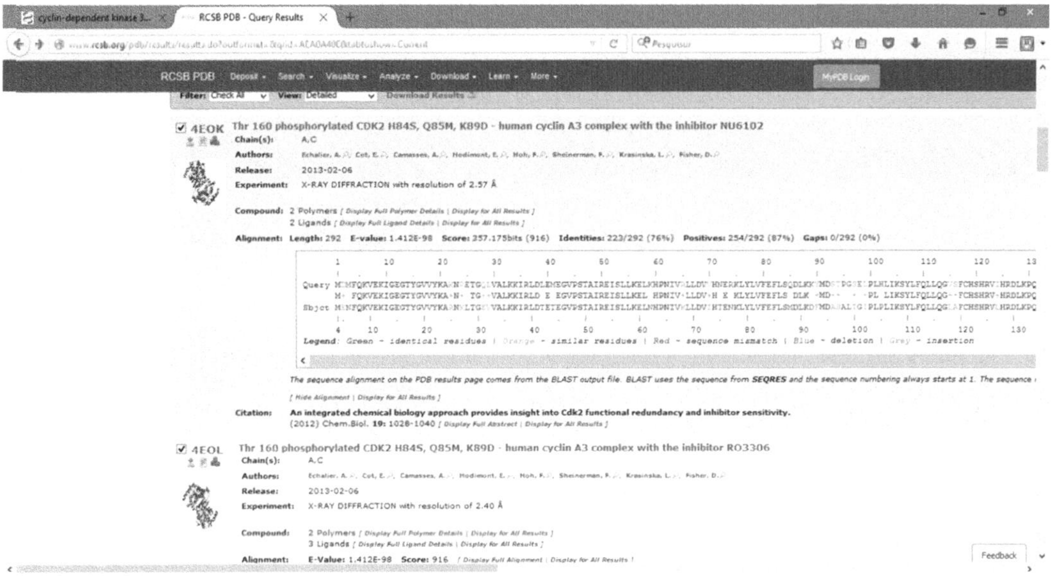

Fig. 6 Sequence alignment of CDK3 and templates

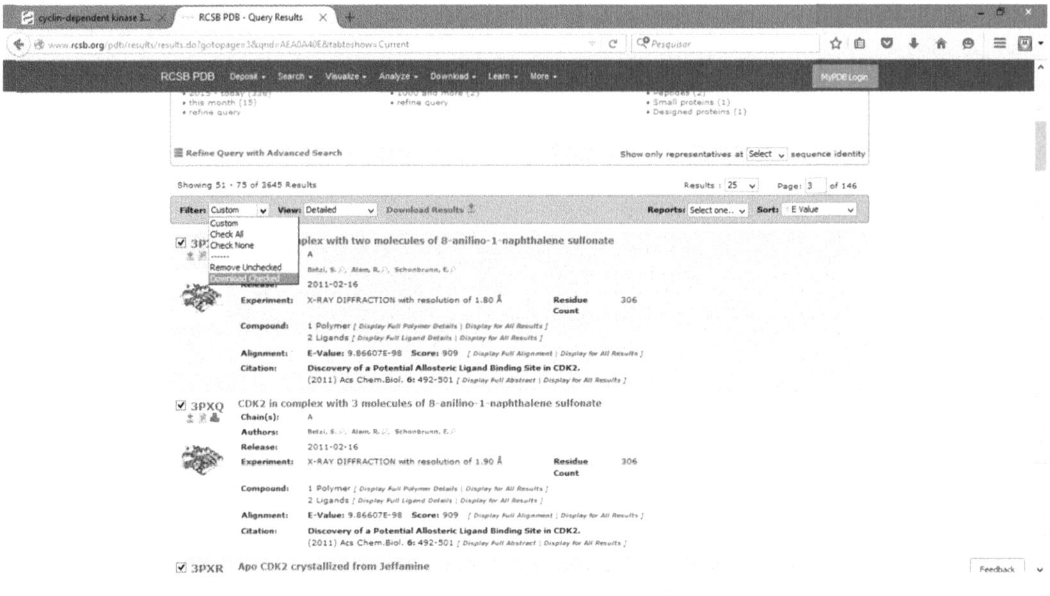

Fig. 7 Sequence alignment of CDK3 and templates

Follow all the steps to download PDB files as separated structures and FASTA as one file.

Later, access MUSCLE [58] at http://www.ebi.ac.uk/Tools/msa/muscle/ to carry out the alignment of the model sequence against the sequences of the templates. Copy (<Ctrl> C) the model sequence and the sequences for all templates obtained from the PDB, as shown in Fig. 8. Then select FASTA as the output format. Next, click on the Submit button (Fig. 9). Then, we get the aligned

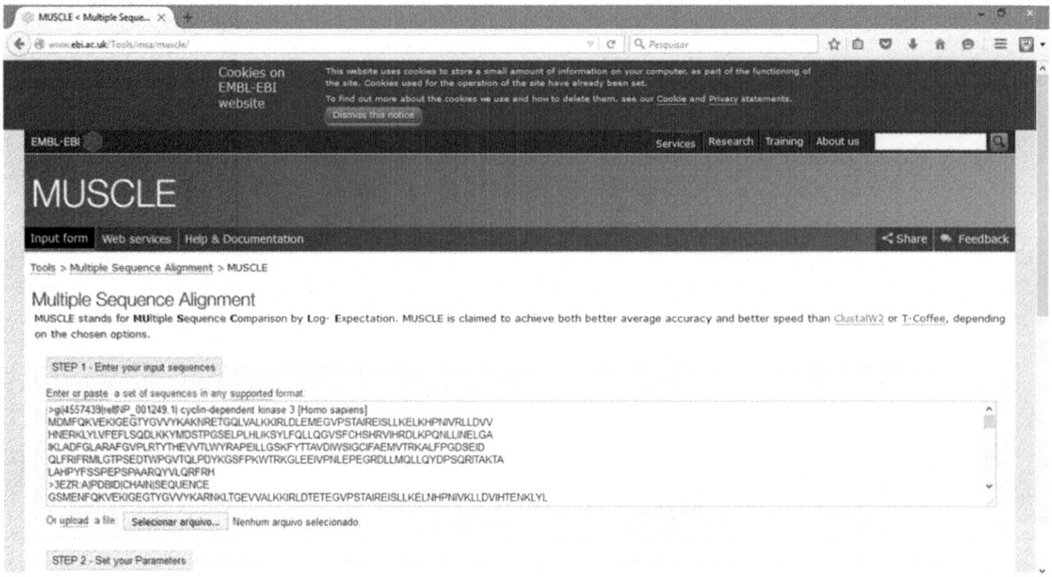

Fig. 8 MUSCLE website

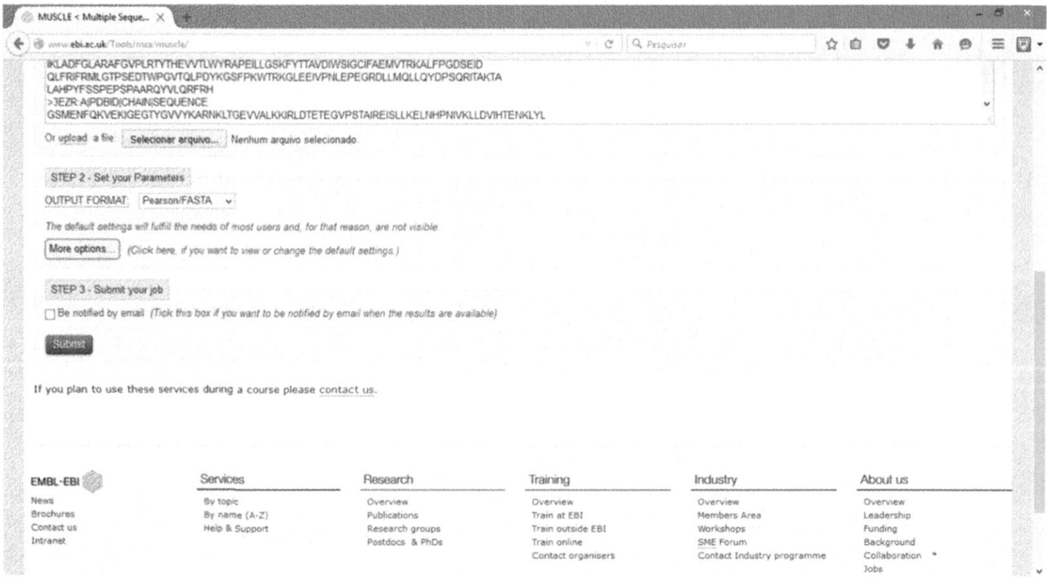

Fig. 9 MUSCLE website

sequences, as shown in Fig. 10. These aligned sequences have to be saved to be used as input to run MODELLER for homology modeling. To run the program MODELLER, there is a need for the PDB files for all templates, the Python input file, and the sequence alignment file (*mult.ali*). We have part of the file *mult.ali* as shown in Fig. 11.

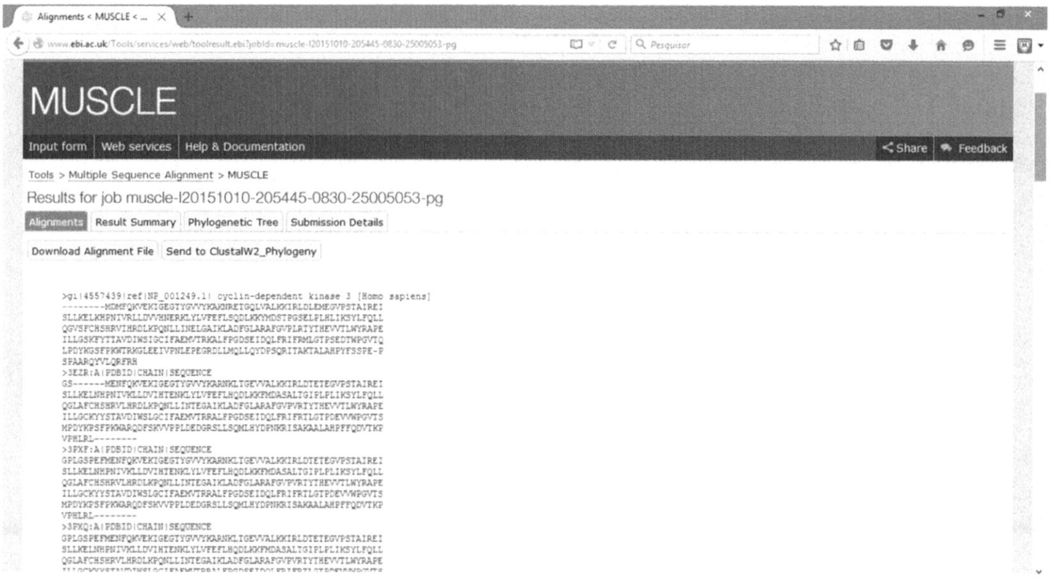

Fig. 10 MUSCLE website

```
>P1;3ezr
structureX:3ezr:1:A:298 :A:CDK2:Homo sapiens: 1.80:0.20
MENFQKVEKIGEGTYGVVYKARNKLTGEVVALKKIR-------VPSTAIREISLLKELNH
PNIVKLLDVIHTENKLYLVFEFLHQDLKKFMDASALTGIPLPLIKSYLFQLLQGLAFCHS
HRVLHRDLKPQNLLINTEGAIKLADF------------------TLWYRAPEILLGCKYY
STAVDIWSLGCIFAEMVTRRALFPGDSEIDQLFRIFRTLGTPDEVVWPGVTSMPDYKPSF
PKWARQDFSKVVPPLDEDGRSLLSQMLHYDPNKRISAKAALAHPFFQDVTKPVPHLRL--
------*

>P1;3pxf
structureX:3pxf:1:A:298 :A:CDK2:Homo sapiens: 1.80:0.20
MENFQKVEKIGEGTYGVVYKARNKLTGEVVALKKIRLDTETEGVPSTAIREISLLKELNH
PNIVKLLDVIHTENKLYLVFEFLHQDLKKFMDASALTGIPLPLIKSYLFQLLQGLAFCHS
HRVLHRDLKPQNLLINTEGAIKLADFGLARAFGVPVRTYTHEVVTLWYRAPEILLGCKYY
STAVDIWSLGCIFAEMVTRRALFPGDSEIDQLFRIFRTLGTPDEVVWPGVTSMPDYKPSF
PKWARQDFSKVVPPLDEDGRSLLSQMLHYDPNKRISAKAALAHPFFQDVTKPVPHLRL--
------*
...... (More sequences)

>P1;HsCDK3
sequence:HsCDK3:::::::0.00: 0.0
MDMFQKVEKIGEGTYGVVYKAKNRETGQLVALKKIRLDLEMEGVPSTAIREISLLKELKH
PNIVRLLDVVHNERKLYLVFEFLSQDLKKYMDSTPGSELPLHLIKSYLFQLLQGVSFCHS
HRVIHRDLKPQNLLINELGAIKLADFGLARAFGVPLRTYTHEVVTLWYRAPEILLGSKFY
TTAVDIWSIGCIFAEMVTRKALFPGDSEIDQLFRIFRMLGTPSEDTWPGVTQLPDYKGSF
PKWTRKGLEEIVPNLEPEGRDLLMQLLQYDPSQRITAKTALAHPYFSS-PEPSPAARQYV
LQRFRH*
```

Fig. 11 Sequence alignment for CDK3 and CDK2 templates

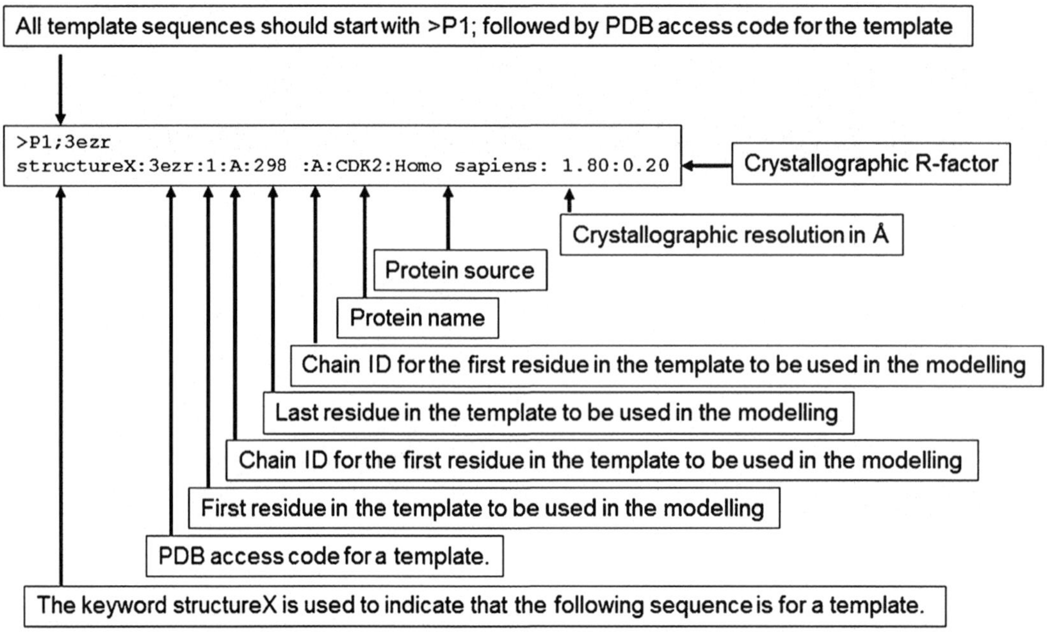

Fig. 12 Keywords used in alignment input for MODELLER

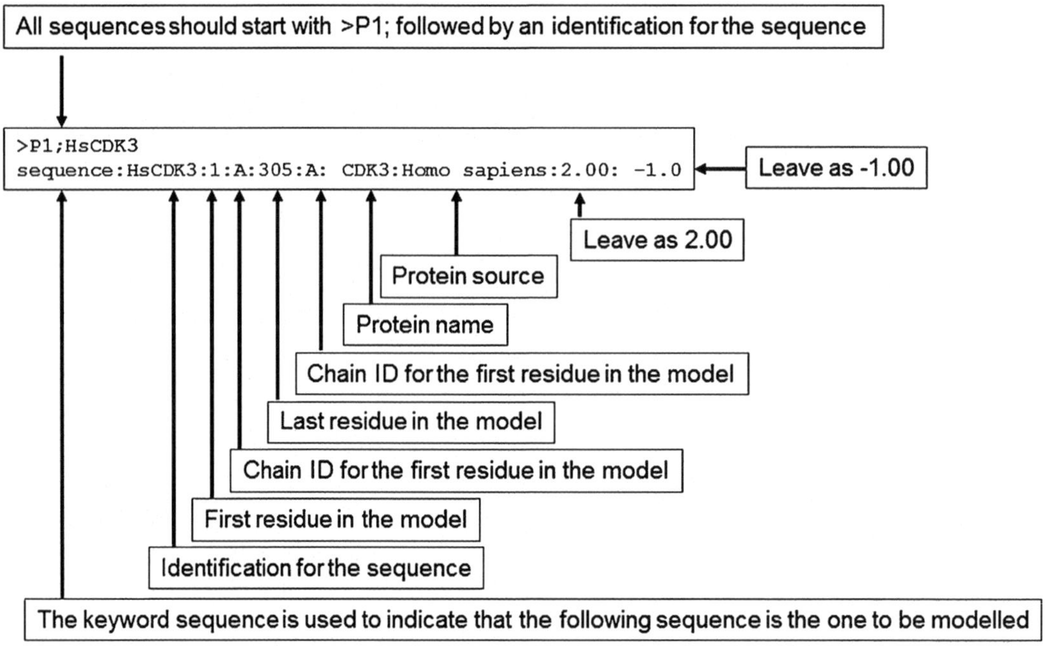

Fig. 13 Keywords used in alignment input for MODELLER

In Fig. 12, there is a description of each field for the header of a template sequence, as shown in the *mult.ali* file. In the following Fig. 13, there is a description of each field for the header of the model sequence, as shown in the *mult.ali* file. We used the file

```
# Comparative modeling with multiple templates for Human CDK3
from modeller import *                  # Load standard Modeller classes
from modeller.automodel import *        # Load the automodel class
log.verbose()                           # request verbose output
env = environ()                         # create a new MODELLER environment to build this model
a = automodel(env,
          alnfile = 'mult.ali',         # alignment filename
          knowns = ('3ezr','3pxf',
                    '3pxq','3pxr',
                    '3pxy','3pxz',
                    '3py0','3ql8',
                    '3qqf','3qqg'),     # PDB access codes of the templates
          sequence = 'HsCDK3')          # code of the target to be modelled
a.starting_model = 1                    # index of the first model
a.ending_model = 100                    # index of the last model
a.make()                                # do the actual comparative modeling
```

Fig. 14 Keywords used in the *model_mult.py* input file for MODELLER

model_mult.py as input to run homology modeling with multiple templates (Fig. 14). In this Python script, we have the explanation of each line after the # symbol.

There are versions of the program MODELLER for Windows, Mac OS X, and Linux. Here the commands to run on Windows have been described. First, click on the Command Prompt. A Command Prompt is a terminal for typing DOS commands in the Command Prompt window. At the Command Prompt, we can execute programs by typing their names.

All files needed to run MODELLER should be in the same directory. In this tutorial, they are in the directory *C:\Users\Walter \Teaching1\Tutorials\HomologyModeling\HsCDK3.*

Type *cd C:\Users\Walter\Teaching1\Tutorials\HomologyModeling\HsCDK3* to go to this directory. Don't forget to press <Enter> after typing the command. The command *cd* means "change directory," it changes from the present directory *C:\User \Walter* to the new directory C:*\Users\Walter\Teaching1\Tutorials \HomologyModeling\HsCDK3*. Type the command *dir* to check all files in the directory. We have ten PDB files (templates), the Python file (*model_mult.py*), and the alignment file (*mult.ali*). We are ready to go.

Type *python model_mult.py > model_mult.log*. This command will run MODELLER using *model_mult.py* as an input file. We will create a log file, named *model_mult.log*, which will be in the same directory and can be used to check the results. Press <Enter> and the command to run MODELLER. Since we asked to generate 100 models, this may take several minutes. There are several ways to evaluate the quality of the models. MODELLER creates a log file (*model_mult.log*) with a table with the MODELLER objective function for each generated model, which we can use to select the best model. We show the structure of the homology model *HsCDK3.B99990064.pdb* in Fig. 15. This structure has the lowest value of the MODELLER objective function among the 100 generated models.

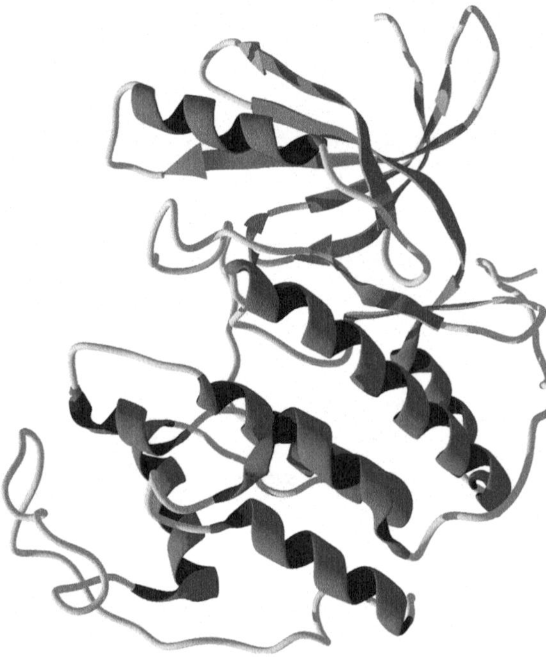

Fig. 15 CDK3 model generated by MODELLER. We used the program Molegro Virtual Docker [60] to generate this figure

The previously described modeling process generated the model of human CDK3 without any ligands bound to the structure (apoenzyme). As we highlighted in the introduction of this chapter, homology modeling may be of interest to have a complex structure involving the protein target and a non-covalent inhibitor. To do so using the program MODELLER, we need only a slight modification on the input files. To illustrate the modeling of the structure of CDK3 in complex with the inhibitor roscovitine, we consider the crystallographic structure 2A4L as a template [59]. The sequence alignment and file preparation are what we have previously described for the CDK3 without any ligands. The novelty here relies on the alignment file. We have to add the structural information about the inhibitor, to do this, we add a point symbol (.) right before the * at the end of the sequence, as shown in Fig. 16. We named this file *align-ligand.ali*. We also need to update the Python script file to add the new alignment file name (*align-ligand.ali*) and to *set env.io.hetatm* to *True* (*env.io.hetatm* = *True*) as shown in Fig. 17. We named this Python script file as *model-ligand.py*. To run the homology modeling, we type *python model-ligand.py* > *model-ligand.log*.

```
>P1;2a41
structureX:2a41:1:A:300 :A:CDK2:Homo sapiens: 2.40:0.173
MENFQKVEKIGEGTYGVVYKARNKLTGEVVALKKI------------AIREISLLKELNH
PNIVKLLDVIHTENKLYLVFEFLHQDLKKFMDASALTGIPLPLIKSYLFQLLQGLAFCHS
HRVLHRDLKPQNLLINTEGAIKLADFGLARAFGVPVRTYTHEVVTLWYRAPEILLGCKYY
STAVDIWSLGCIFAEMVTRRALFPGDSEIDQLFRIFRTLGTPDEVVWPGVTSMPDYKPSF
PKWARQDFSKVVPPLDEDGRSLLSQMLHYDPNKRISAKAALAHPFFQDVTKPVPHLRL--
------.*

>P1;HsCDK3
sequence:HsCDK3:1:A:305:A:CDK3:Homo sapiens:2.00: -1.0
MDMFQKVEKIGEGTYGVVYKAKNRETGQLVALKKIRLDLEMEGVPSTAIREISLLKELKH
PNIVRLLDVVHNERKLYLVFEFLSQDLKKYMDSTPGSELPLHLIKSYLFQLLQGVSFCHS
HRVIHRDLKPQNLLINELGAIKLADFGLARAFGVPLRTYTHEVVTLWYRAPEILLGSKFY
TTAVDIWSIGCIFAEMVTRKALFPGDSEIDQLFRIFRMLGTPSEDTWPGVTQLPDYKGSF
PKWTRKGLEEIVPNLEPEGRDLLMQLLQYDPSQRITAKTALAHPYFSS-PEPSPAARQYV
LQRFRH.*
```

Fig. 16 Keywords used in alignment input for MODELLER

```
# Comparative modeling with ligand transfer from the template
from modeller import *               # Load standard Modeller classes
from modeller.automodel import *     # Load the automodel class

log.verbose()    # request verbose output
env = environ()  # create a new MODELLER environment to build this model in

# Read in HETATM records from template PDBs
env.io.hetatm = True

a = automodel(env,
              alnfile  = 'align-ligand.ali',      # alignment filename
              knowns   = '2a41',                  # codes of the templates
              sequence = 'HsCDK3')                # code of the target
a.starting_model= 1                               # index of the first model
a.ending_model  = 100                             # index of the last model
                                                  # (determines how many models to calculate)
a.make()                                          # do the actual comparative modeling
```

Fig. 17 Keywords used in the *model-ligand.py* input file for MODELLER

4 Availability

All necessary material to run this tutorial is available at http://azevedolab.net/resources/HsCDK3_ready2run.zip.

5 Colophon

We created Figs. 2, 11–14, 16, and 17 using Microsoft PowerPoint 2016. We used the program Molegro Virtual Docker [60] to generate Fig. 15. We captured the screens related to each program described in the text to create Figs. 1 and 3–10. We performed homology modeling described in this chapter using a Desktop PC with 4GB of memory, a 1 TB hard disk, and an Intel® Core® i3-2120 at 3.30 GHz processor running Windows 8.1.

6 Final Remarks

Homology modeling is the computational alternative when we need to have a three-dimensional model for protein without experimental information about its structure. Considering that a structural template is available and satisfies the sequence identity cutoff (sequence identity between template and model >30%), we can carry out modeling quite straightforward. We described, here, a graphical tutorial to generate a model for CDK3 in the apo form (without ligands) and complexed with an inhibitor. For both models, we used the program MODELLER. Homology modeling with MODELLER was able to create models for a wide range of different protein targets for drug discovery, such as transmembrane proteins [61] and enzymes [62–69]. Structural analysis of protein-ligand complex is a vital step in the understanding the essential features responsible ligand-binding affinity [60, 65, 70–105]. The constant development of this software and the strong support of the community interested in homology modeling established MODELLER as an essential tool for computational studies aiming analysis of these complexes' structures.

Acknowledgments

This work was supported by grants from CNPq (Brazil) (308883/2014-4). This study was financed in part by the Coordenação de Aperfeiçoamento de Pessoal de Nivel Superior—Brasil (CAPES)—Finance Code 001. GBF acknowledges support from PUCRS/BPA fellowship. WFA is a senior researcher for CNPq (Brazil) (Process Numbers: 308883/2014-4 and 309029/2018-0).

References

1. Filgueira de Azevedo W Jr, dos Santos GC, dos Santos DM, Olivieri JR, Canduri F, Silva RG et al (2003) Docking and small angle X-ray scattering studies of purine nucleoside phosphorylase. Biochem Biophys Res Commun 309:923–928
2. da Silveira NJ, Arcuri HA, Bonalumi CE, de Souza FP, Mello IM, Rahal P et al (2005) Molecular models of NS3 protease variants of the hepatitis C virus. BMC Struct Biol 5:1
3. Silveira NJ, Uchôa HB, Pereira JH, Canduri F, Basso LA, Palma MS et al (2005) Molecular models of protein targets from *Mycobacterium tuberculosis.* J Mol Model 11:160–166
4. da Silveira NJ, Bonalumi CE, Uchõa HB, Pereira JH, Canduri F, de Azevedo WF (2006) DBMODELING: a database applied to the study of protein targets from genome projects. Cell Biochem Biophys 44:366–374
5. da Silveira NJF, Bonalumi CE, Arcuri HA, de Azevedo WF Jr (2007) Molecular modeling databases: a new way in the search of proteins targets for drug development. Curr Bioinforma 2:1–10
6. Marques MR, Pereira JH, Oliveira JS, Basso LA, de Azevedo WF Jr, Santos DS et al (2007) The inhibition of 5-enolpyruvylshikimate-3-phosphate synthase as a model for development of novel antimicrobials. Curr Drug Targets 8:445–457
7. Breda A, Basso LA, Santos DS, de Azevedo WF Jr (2008) Virtual screening of drugs: score functions, docking, and drug design. Curr Comput Aided Drug Des 4:265–272
8. de Azevedo WF Jr, Dias R (2008) Computational methods for calculation of ligand-binding affinity. Curr Drug Targets 9:1031–1039
9. Dias R, de Azevedo WF Jr (2008) Molecular docking algorithms. Curr Drug Targets 9:1040–1047
10. Arcuri HA, Zafalon GF, Marucci EA, Bonalumi CE, da Silveira NJ, Machado JM et al (2010) SKPDB: a structural database of shikimate pathway enzymes. BMC Bioinformatics 11:12
11. De Azevedo WF Jr (2010) MolDock applied to structure-based virtual screening. Curr Drug Targets 11:327–334
12. Ducati RG, Basso LA, Santos DS, de Azevedo WF Jr (2010) Crystallographic and docking studies of purine nucleoside phosphorylase from *Mycobacterium tuberculosis.* Bioorg Med Chem 18:4769–4774
13. Heberlé G, de Azevedo WF Jr (2011) Bio-inspired algorithms applied to molecular docking simulations. Curr Med Chem 18:1339–1352
14. Rocha BA, Delatorre P, Oliveira TM, Benevides RG, Pires AF, Sousa AA et al (2011) Structural basis for both pro- and anti-inflammatory response induced by mannose-specific legume lectin from *Cymbosema roseum.* Biochimie 93:806–816
15. Vianna CP, de Azevedo WF Jr (2012) Identification of new potential *Mycobacterium tuberculosis* shikimate kinase inhibitors through molecular docking simulations. J Mol Model 18:755–764
16. Moraes FP, de Azevedo WF Jr (2012) Targeting imidazoline site on monoamine oxidase B through molecular docking simulations. J Mol Model 18:3877–3886
17. Azevedo LS, Moraes FP, Xavier MM, Pantoja EO, Villavicencio B, Finck JA et al (2012) Recent Progress of molecular docking simulations applied to development of drugs. Curr Bioinforma 7:352–365
18. Coracini JD, de Azevedo WF Jr (2014) Shikimate kinase, a protein target for drug design. Curr Med Chem 21:592–604
19. de Avila MB, de Azevedo WF (2014) Data Mining of Docking Results. Application to 3-Dehydroquinate Dehydratase. Curr Bioinforma 9:361–379
20. Teles CB, Moreira-Dill LS, Silva Ade A, Facundo VA, de Azevedo WF Jr, da Silva LH et al (2015) A lupane-triterpene isolated from *Combretum leprosum* Mart. fruit extracts that interferes with the intracellular development of *Leishmania* (L.) *amazonensis in vitro.* BMC Complement Altern Med 15:165
21. de Azevedo WF Jr (2016) Opinion paper: targeting multiple cyclin-dependent kinases (CDKs): a new strategy for molecular docking studies. Curr Drug Targets 17:2
22. Canduri F, de Azevedo WF (2008) Protein crystallography in drug discovery. Curr Drug Targets 9:1048–1053
23. Fadel V, Bettendorff P, Herrmann T, de Azevedo WF Jr, Oliveira EB, Yamane T et al (2005) Automated NMR structure determination and disulfide bond identification of the myotoxin crotamine from *Crotalus durissus terrificus.* Toxicon 46:759–767
24. Heck GS, Pintro VO, Pereira RR, de Ávila MB, Levin NMB, de Azevedo WF (2017) Supervised machine learning methods applied

to predict ligand-binding affinity. Curr Med Chem 24:2459–2470
25. Berman HM, Westbrook J, Feng Z, Gilliland G, Bhat TN, Weissig H et al (2000) The Protein Data Bank. Nucleic Acids Res 28:235–242
26. Berman HM, Battistuz T, Bhat TN, Bluhm WF, Bourne PE, Burkhardt K et al (2002) The Protein Data Bank. Acta Crystallogr D Biol Crystallogr 58:899–907
27. Westbrook J, Feng Z, Chen L, Yang H, Berman HM (2003) The Protein Data Bank and structural genomics. Nucleic Acids Res 31 (1):489–491
28. Ingwall RT, Scheraga HA, Lotan N, Berger A, Katchalski E (1968) Conformational studies of poly-L-alanine in water. Biopolymers 6:331–368
29. Lesk AM (1997) CASP2: report on ab initio predictions. Proteins Suppl 1:151–166
30. Zemla A, Venclovas C, Reinhardt A, Fidelis K, Hubbard TJ (1997) Numerical criteria for the evaluation of *ab initio* predictions of protein structure. Proteins Suppl 1:140–150
31. Kolinski A, Rotkiewicz P, Ilkowski B, Skolnick J (1999) A method for the improvement of threading-based protein models. Proteins 37:592–610
32. Rost B, Fariselli P, Casadio R (1996) Topology prediction for helical transmembrane proteins at 86% accuracy. Protein Sci 5:1704–1718
33. Xiang Z (2006) Advances in homology protein structure modeling. Curr Protein Pept Sci 7:217–227
34. Altschul SF, Gish W, Miller W, Myers EW, Lipman DJ (1990) Basic local alignment search tool. J Mol Biol 215:403–410
35. Brünger AT (1992) Free R value: a novel statistical quantity for assessing the accuracy of crystal structures. Nature 355:472–475
36. Ramachandran GN, Ramakrishnan C, Sasisekharan V (1963) Stereochemistry of polypeptide chain configurations. J Mol Biol 7:95–99
37. Fanelli F, De Benedetti PG (2006) Inactive and active states and supramolecular organization of GPCRs: insights from computational modeling. J Comput Aided Mol Des 20:449–461
38. Wierenga RK, Borchert TV, Noble ME (1992) Crystallographic binding studies with triosephosphate isomerases: conformational changes induced by substrate and substrate-analogues. FEBS Lett 307:34–39
39. Sali A, Blundell TL (1993) Comparative protein modelling by satisfaction of spatial restraints. J Mol Biol 234:779–815
40. De Bondt HL, Rosenblatt J, Jancarik J, Jones HD, Morgan DO, Kim SH (1993) Crystal structure of cyclin-dependent kinase 2. Nature 363:595–602
41. Spring LM, Wander SA, Zangardi M, Bardia A (2019) CDK 4/6 inhibitors in breast cancer: current controversies and future directions. Curr Oncol Rep 21:25
42. Roskoski R Jr (2019) Cyclin-dependent protein serine/threonine kinase inhibitors as anticancer drugs. Pharmacol Res 139:471–488
43. Choo JR, Lee SC (2018) CDK4-6 inhibitors in breast cancer: current status and future development. Expert Opin Drug Metab Toxicol 14:1123–1138
44. Zardavas D, Pondé N, Tryfonidis K (2017) CDK4/6 blockade in breast cancer: current experience and future perspectives. Expert Opin Investig Drugs 26:1357–1372
45. Chen P, Lee NV, Hu W, Xu M, Ferre RA, Lam H et al (2016) Spectrum and degree of CDK drug interactions predicts clinical performance. Mol Cancer Ther 15:2273–2281
46. Canduri F, Perez PC, Caceres RA, de Azevedo WF Jr (2008) CDK9 a potential target for drug development. Med Chem 4:210–218
47. Krystof V, Cankar P, Frysová I, Slouka J, Kontopidis G, Dzubák P et al (2006) 4-arylazo-3,5-diamino-1H-pyrazole CDK inhibitors: SAR study, crystal structure in complex with CDK2, selectivity, and cellular effects. J Med Chem 49:6500–6509
48. Leopoldino AM, Canduri F, Cabral H, Junqueira M, de Marqui AB, Apponi LH et al (2006) Expression, purification, and circular dichroism analysis of human CDK9. Protein Expr Purif 47:614–620
49. Canduri F, de Azevedo WF Jr (2005) Structural basis for interaction of inhibitors with cyclin-dependent kinase 2. Curr Comput Aided Drug Des 1:53–64
50. Canduri F, Uchoa HB, de Azevedo WF Jr (2004) Molecular models of cyclin-dependent kinase 1 complexed with inhibitors. Biochem Biophys Res Commun 324:661–666
51. Filgueira de Azevedo W Jr, Gaspar RT, Canduri F, Camera JC Jr, Freitas da Silveira NJ (2002) Molecular model of cyclin-dependent kinase 5 complexed with roscovitine. Biochem Biophys Res Commun 297:1154–1158

52. de Azevedo WF Jr, Canduri F, da Silveira NJ (2002) Structural basis for inhibition of cyclin-dependent kinase 9 by flavopiridol. Biochem Biophys Res Commun 293:566–571
53. De Azevedo WF Jr, Mueller-Dieckmann HJ, Schulze-Gahmen U, Worland PJ, Sausville E, Kim SH (1996) Structural basis for specificity and potency of a flavonoid inhibitor of human CDK2, a cell cycle kinase. Proc Natl Acad Sci U S A 93:2735–2740
54. Kim SH, Schulze-Gahmen U, Brandsen J, de Azevedo Júnior WF (1996) Structural basis for chemical inhibition of CDK2. Prog Cell Cycle Res 2:137–145
55. Cui J, Yang Y, Li H, Leng Y, Qian K, Huang Q et al (2015) MiR-873 regulates era transcriptional activity and tamoxifen resistance via targeting CDK3 in breast cancer cells. Oncogene 34:3895–3907
56. Perez PC, Caceres RA, Canduri F, de Azevedo WF Jr (2009) Molecular modeling and dynamics simulation of human cyclin-dependent kinase 3 complexed with inhibitors. Comput Biol Med 39:130–140
57. Benson DA, Cavanaugh M, Clark K, Karsch-Mizrachi I, Lipman DJ, Ostell J et al (2013) GenBank. Nucleic Acids Res 41:36–42
58. Edgar RC (2004) MUSCLE: multiple sequence alignment with high accuracy and high throughput. Nucleic Acids Res 32:1792–1797
59. De Azevedo WF, Leclerc S, Meijer L, Havlicek L, Strnad M, Kim SH (1997) Inhibition of cyclin-dependent kinases by purine analogues: crystal structure of human cdk2 complexed with roscovitine. Eur J Biochem 243:518–526
60. Thomsen R, Christensen MH (2006) MolDock: a new technique for high-accuracy molecular docking. J Med Chem 49:3315–3321
61. Abdelmonsef AH, Dulapalli R, Dasari T, Padmarao LS, Mukkera T, Vuruputuri U (2016) Identification of novel antagonists for Rab38 protein by homology modeling and virtual screening. Comb Chem High Throughput Screen 19:875–892
62. Filgueira de Azevedo W Jr, Canduri F, Simões de Oliveira J, Basso LA, Palma MS, Pereira JH et al (2002) Molecular model of shikimate kinase from *Mycobacterium tuberculosis*. Biochem Biophys Res Commun 295:142–148
63. Konno K, Hisada M, Fontana R, Lorenzi CC, Naoki H, Itagaki Y et al (2001) Anoplin, a novel antimicrobial peptide from the venom of the solitary wasp Anoplius samariensis. Biochim Biophys Acta 1550:70–80
64. Pereira JH, Canduri F, de Oliveira JS, da Silveira NJ, Basso LA, Palma MS et al (2003) Structural bioinformatics study of EPSP synthase from *Mycobacterium tuberculosis*. Biochem Biophys Res Commun 312:608–614
65. Rádis-Baptista G, Moreno FB, de Lima Nogueira L, Martins AM, de Oliveira Toyama D, Toyama MH et al (2006) Crotacetin, a novel snake venom C-type lectin homolog of convulxin, exhibits an unpredictable antimicrobial activity. Cell Biochem Biophys 44:412–423
66. Freitas PG, Elias TC, Pinto IA, Costa LT, de Carvalho PVSD, Omote DQ et al (2018) Computational approach to the discovery of phytochemical molecules with therapeutic potential targets to the PKCZ protein. Lett Drug Des Discov 15:488–499
67. Uchôa HB, Jorge GE, Freitas Da Silveira NJ, Camera JC Jr, Canduri F, De Azevedo WF Jr (2004) Parmodel: a web server for automated comparative modeling of proteins. Biochem Biophys Res Commun 325:1481–1486
68. Arcuri HA, Borges JC, Fonseca IO, Pereira JH, Neto JR, Basso LA et al (2008) Structural studies of shikimate 5-dehydrogenase from *Mycobacterium tuberculosis*. Proteins 72:720–730
69. Arcuri HA, Apponi LH, Valentini SR, Durigon EL, de Azevedo WF Jr, Fossey MA et al (2008) Expression and purification of human respiratory syncytial virus recombinant fusion protein. Protein Expr Purif 62:146–152
70. de Azevedo WF Jr (2010) Structure-based virtual screening. Curr Drug Targets 11:261–263
71. de Ávila MB, Bitencourt-Ferreira G, de Azevedo WF Jr (2018) Structural basis for inhibition of enoyl-[acyl carrier protein] reductase (InhA) from *Mycobacterium tuberculosis*. Curr Med Chem. https://doi.org/10.2174/0929867326666181203125229
72. Volkart PA, Bitencourt-Ferreira G, Souto AA, de Azevedo WF (2019) Cyclin-dependent kinase 2 in cellular senescence and cancer. A structural and functional review. Curr Drug Targets 20:716–726. https://doi.org/10.2174/1389450120666181204165344
73. Canduri F, Fadel V, Basso LA, Palma MS, Santos DS, de Azevedo WF Jr (2005) New catalytic mechanism for human purine nucleoside phosphorylase. Biochem Biophys Res Commun 327:646–649

74. Canduri F, Teodoro LG, Fadel V, Lorenzi CC, Hial V, Gomes RA et al (2001) Structure of human uropepsin at 2.45 A resolution. Acta Crystallogr D Biol Crystallogr 57:1560–1570
75. de Azevedo WF Jr, Dias R (2008) Experimental approaches to evaluate the thermodynamics of protein-drug interactions. Curr Drug Targets 9:1071–1076
76. Delatorre P, Rocha BA, Souza EP, Oliveira TM, Bezerra GA, Moreno FB et al (2007) Structure of a lectin from *Canavalia gladiata* seeds: new structural insights for old molecules. BMC Struct Biol 7:52
77. de Azevedo WF Jr, Canduri F, dos Santos DM, Pereira JH, Bertacine Dias MV, Silva RG et al (2003) Crystal structure of human PNP complexed with guanine. Biochem Biophys Res Commun 312:767–772
78. Canduri F, Perez PC, Caceres RA, de Azevedo WF Jr (2007) Protein kinases as targets for antiparasitic chemotherapy drugs. Curr Drug Targets 8:389–398
79. Dias MV, Borges JC, Ely F, Pereira JH, Canduri F, Ramos CH et al (2006) Structure of chorismate synthase from *Mycobacterium tuberculosis*. J Struct Biol 154:130–143
80. Dias MV, Ely F, Palma MS, de Azevedo WF Jr, Basso LA, Santos DS (2007) Chorismate synthase: an attractive target for drug development against orphan diseases. Curr Drug Targets 8:437–444
81. Silva RG, Pereira JH, Canduri F, de Azevedo WF Jr, Basso LA, Santos DS (2005) Kinetics and crystal structure of human purine nucleoside phosphorylase in complex with 7-methyl-6-thio-guanosine. Arch Biochem Biophys 442:49–58
82. Timmers LF, Caceres RA, Vivan AL, Gava LM, Dias R, Ducati RG et al (2008) Structural studies of human purine nucleoside phosphorylase: towards a new specific empirical scoring function. Arch Biochem Biophys 479:28–38
83. de Azevedo WF Jr (2011) Molecular dynamics simulations of protein targets identified in *Mycobacterium tuberculosis*. Curr Med Chem 18:1353–1366
84. de Azevedo WF Jr (2011) Protein targets for development of drugs against *Mycobacterium tuberculosis*. Curr Med Chem 18:1255–1257
85. Caceres RA, Saraiva Timmers LF, Dias R, Basso LA, Santos DS, de Azevedo WF Jr (2008) Molecular modeling and dynamics simulations of PNP from *Streptococcus agalactiae*. Bioorg Med Chem 16:4984–4993
86. Dias MV, Faím LM, Vasconcelos IB, de Oliveira JS, Basso LA, Santos DS et al (2007) Effects of the magnesium and chloride ions and shikimate on the structure of shikimate kinase from *Mycobacterium tuberculosis*. Acta Crystallogr Sect F Struct Biol Cryst Commun 63:1–6
87. de Azevedo WF Jr, Ward RJ, Canduri F, Soares A, Giglio JR, Arni RK (1998) Crystal structure of piratoxin-I: a calcium-independent, myotoxic phospholipase A2-homologue from *Bothrops pirajai* venom. Toxicon 36:1395–1406
88. Dias R, Timmers LF, Caceres RA, de Azevedo WF Jr (2008) Evaluation of molecular docking using polynomial empirical scoring functions. Curr Drug Targets 9:1062–1070
89. da Silveira NJ, Uchôa HB, Canduri F, Pereira JH, Camera JC Jr, Basso LA et al (2004) Structural bioinformatics study of PNP from *Schistosoma mansoni*. Biochem Biophys Res Commun 322:100–104
90. de Azevedo WF Jr, Dias R (2008) Evaluation of ligand-binding affinity using polynomial empirical scoring functions. Bioorg Med Chem 16:9378–9382
91. Bezerra GA, Oliveira TM, Moreno FB, de Souza EP, da Rocha BA, Benevides RG et al (2007) Structural analysis of *Canavalia maritima* and *Canavalia gladiata* lectins complexed with different dimannosides: new insights into the understanding of the structure-biological activity relationship in legume lectins. J Struct Biol 160:168–176
92. Canduri F, Fadel V, Dias MV, Basso LA, Palma MS, Santos DS et al (2005) Crystal structure of human PNP complexed with hypoxanthine and sulfate ion. Biochem Biophys Res Commun 326:335–338
93. Timmers LF, Pauli I, Caceres RA, de Azevedo WF Jr (2008) Drug-binding databases. Curr Drug Targets 9:1092–1099
94. Delatorre P, Rocha BA, Gadelha CA, Santi-Gadelha T, Cajazeiras JB, Souza EP et al (2006) Crystal structure of a lectin from *Canavalia maritima* (ConM) in complex with trehalose and maltose reveals relevant mutation in ConA-like lectins. J Struct Biol 154:280–286
95. Nolasco DO, Canduri F, Pereira JH, Cortinóz JR, Palma MS, Oliveira JS et al (2004) Crystallographic structure of PNP from *Mycobacterium tuberculosis* at 1.9A resolution. Biochem Biophys Res Commun 324:789–794
96. Arcuri HA, Canduri F, Pereira JH, da Silveira NJ, Camera Júnior JC, de Oliveira JS et al (2004) Molecular models for shikimate

pathway enzymes of *Xylella fastidiosa*. Biochem Biophys Res Commun 320:979–991

97. Soares MB, Silva CV, Bastos TM, Guimarães ET, Figueira CP, Smirlis D et al (2012) Anti-*Trypanosoma cruzi* activity of nicotinamide. Acta Trop 12:224–229
98. Manhani KK, Arcuri HA, da Silveira NJ, Uchôa HB, de Azevedo WF Jr, Canduri F (2005) Molecular models of protein kinase 6 from *Plasmodium falciparum*. J Mol Model 12:42–48
99. Marques MR, Vaso A, Neto JR, Fossey MA, Oliveira JS, Basso LA et al (2008) Dynamics of glyphosate-induced conformational changes of *Mycobacterium tuberculosis* 5-enolpyruvylshikimate-3-phosphate synthase (EC 2.5.1.19) determined by hydrogen-deuterium exchange and electrospray mass spectrometry. Biochemistry 47:7509–7522
100. Cavada BS, Moreno FB, da Rocha BA, de Azevedo WF Jr, Castellón RE, Goersch GV et al (2006) cDNA cloning and 1.75 A crystal structure determination of PPL2, an endochitinase and N-acetylglucosamine-binding hemagglutinin from *Parkia platycephala* seeds. FEBS J 273:3962–3974
101. Moreno FB, de Oliveira TM, Martil DE, Viçoti MM, Bezerra GA, Abrego JR et al (2008) Identification of a new quaternary association for legume lectins. J Struct Biol 161:133–143
102. Xavier MM, Heck GS, de Avila MB, Levin NM, Pintro VO, Carvalho NL et al (2016) SAnDReS a computational tool for statistical analysis of docking results and development of scoring functions. Comb Chem High Throughput Screen 19:801–812
103. de Ávila MB, Xavier MM, Pintro VO, de Azevedo WF (2017) Supervised machine learning techniques to predict binding affinity. A study for cyclin-dependent kinase 2. Biochem Biophys Res Commun 494:305–310
104. Bitencourt-Ferreira G, de Azevedo WF Jr (2018) Development of a machine-learning model to predict Gibbs free energy of binding for protein-ligand complexes. Biophys Chem 240:63–69
105. Levin NMB, Pintro VO, Bitencourt-Ferreira-G, Mattos BB, Silvério AC, de Azevedo WF Jr (2018) Development of CDK-targeted scoring functions for prediction of binding affinity. Biophys Chem 235:1–8

Chapter 16

Machine Learning to Predict Binding Affinity

Gabriela Bitencourt-Ferreira and Walter Filgueira de Azevedo Jr.

Abstract

Recent progress in the development of scientific libraries with machine-learning techniques paved the way for the implementation of integrated computational tools to predict ligand-binding affinity. The prediction of binding affinity uses the atomic coordinates of protein-ligand complexes. These new computational tools made application of a broad spectrum of machine-learning techniques to study protein-ligand interactions possible. The essential aspect of these machine-learning approaches is to train a new computational model by using technologies such as supervised machine-learning techniques, convolutional neural network, and random forest to mention the most commonly applied methods. In this chapter, we focus on supervised machine-learning techniques and their applications in the development of protein-targeted scoring functions for the prediction of binding affinity. We discuss the development of the program SAnDReS and its application to the creation of machine-learning models to predict inhibition of cyclin-dependent kinase and HIV-1 protease. Moreover, we describe the scoring function space, and how to use it to explain the development of targeted scoring functions.

Key words Machine learning, Regression, Scoring function space, SAnDReS, Binding affinity, Cyclin-dependent kinase, HIV-1 protease

1 Introduction

Studies using machine-learning methods to evaluate biological systems are not new. For example, there is a report of a survey about the application of artificial neural networks to systems biology, as old as 1985 [1]. If we focus our analysis on applications of supervised machine-learning techniques to the evaluation of ligand-binding affinity, we can find reports dating back to 1994 [2, 3]. In recent years, we have witnessed significant progress in the development of machine-learning models for the prediction of protein-ligand binding affinity, for recent reviews see Heck et al., Levin et al., de Azevedo, and Ain et al. [4–8]. This progress is mostly due to the availability of free scientific libraries such as NumPy (http://www.numpy.org/), SciPy (https://scipy.org/), TensorFlow (https://www.tensorflow.org/), and scikit-learn (https://scikit-learn.org/stable/) [9]. All these libraries are

Walter Filgueira de Azevedo Jr. (ed.), *Docking Screens for Drug Discovery*, Methods in Molecular Biology, vol. 2053,
https://doi.org/10.1007/978-1-4939-9752-7_16,

intended to be used with Python programming language (https://www.python.org/). The ease of programming in the Python language and the integration of the libraries mentioned above created a favorable scenario for the development of a new generation of scoring function dedicated to the prediction of protein-ligand binding affinity.

Among the most successful scoring functions, we may highlight the development of machine-learning models to predict binding affinity [10–21]. The basic idea of such computational approaches is to train a novel scoring function by making use of machine-learning techniques such as convolutional neural network [22–24], random forest [25–31], and supervised machine-learning techniques [17], to mention the most commonly used methods.

We may classify these machine-learning approaches for the development of new scoring functions into two major types. The first type, named targeted scoring functions, makes use of energy terms to compose a predictive model and calibrate them to obtain the relative weights of the energy terms for a specific biological system. For instance, we may consider all crystallographic structures of the cyclin-dependent kinase (CDK) for which ligand-binding affinity data are available and then, using supervised machine-learning techniques, generate a novel scoring function targeted to CDK system [15, 18]. Combining structural and ligand-binding affinity data allows us to create a novel scoring function with the strong support of experimental information.

The second type of machine-learning approach to the development of a scoring function considers a broader spectrum of biological systems. For instance, we may take all crystallographic structures solved to high resolution, for which Gibbs free energy (ΔG) experimental data are available. We call this type of machine-learning model a nonspecific scoring function. We have applied such an approach to a dataset of crystallographic structures solved to a resolution higher than 1.5 Å [11], with predictive performance higher than standard scoring functions available in the programs Molegro Virtual Docker [32–34], AutoDock4 [35–38], and AutoDock Vina [39].

These previously mentioned machine-learning models [11, 15, 18] were developed using the program SAnDReS [20]. SAnDReS draws inspiration from several studies focused on protein-ligand complexes that we have been working on in the past decades. These projects began in the 1990s with pioneering studies focused on intermolecular interactions between CDK and inhibitors [40–42]. SAnDReS is a free and open-source general public license (GNU) computational environment for the development of machine-learning models for prediction of ligand-binding affinity.

The program SAnDReS is also a tool for statistical analysis of docking simulations and evaluation of the predictive performance of computational models developed to calculate binding affinity.

We have implemented machine-learning techniques to generate regression models based on experimental binding affinity and scoring functions such as PLANTS and MolDock scores [20]. SAnDReS makes use of the scikit-learn library to implement a broad spectrum of supervised machine-learning techniques for regression, such as Ordinary Least Squares and Ridge Regression. SAnDReS was developed using Python programming language and SciPy, NumPy, Matplotlib, and scikit-learn libraries. With SAnDReS, we can handle data obtained from any protein-ligand docking program; the only requisite is to have protein structures in Protein Data Bank (PDB) format, ligands in Structure Data File (SDF) format, and docking and scoring function data in comma-separated values (CSV) format.

SAnDReS is an acronym for Statistical Analysis of Docking Results and Scoring Functions and has been successfully applied to a wide range of biological systems [3–18, 20, 43–61]. In these studies, SAnDReS predicted binding affinity for protein-ligand complexes with superior performance when compared with traditional scoring functions. SAnDReS also has a user-friendly interface that allows the user to carry out protein-ligand docking simulations without preparing the necessary input files. The latest version of SAnDReS can run MVD, AutoDock4, and AutoDock Vina.

Classical scoring functions are theoretical models to predict binding affinity based on the atomic coordinates of protein-ligand complexes [62–64]. The development of these scoring functions started with the innovative work of Böhm in the early 1990s [65–70]. Scoring functions implemented in docking programs such as AutoDock, AutoDock Vina, and Molegro Virtual Docker employ a computational model that somehow operates analogously to scoring function developed by Böhm. The differences among these scoring functions reside in the energy terms added to the computational model [63], and how they calculate them.

In this chapter, we describe the application of supervised machine-learning techniques to predict ligand-binding affinity. To illustrate the potential of this approach, we explain the development of machine-learning models to predict binding affinity of cyclin-dependent kinases and HIV-1 protease.

2 SAnDReS

The program SAnDReS [20] makes use of supervised machine-learning techniques to generate polynomial equations to predict ligand-binding affinity, which allows improvement of native scoring functions. SAnDReS works through the training of a model making it specific for a biological system (targeted scoring function). Let us consider the HIV-1 Protease system [17]; we could make use of a standard scoring function, such as PLANTS score [71] and fine-

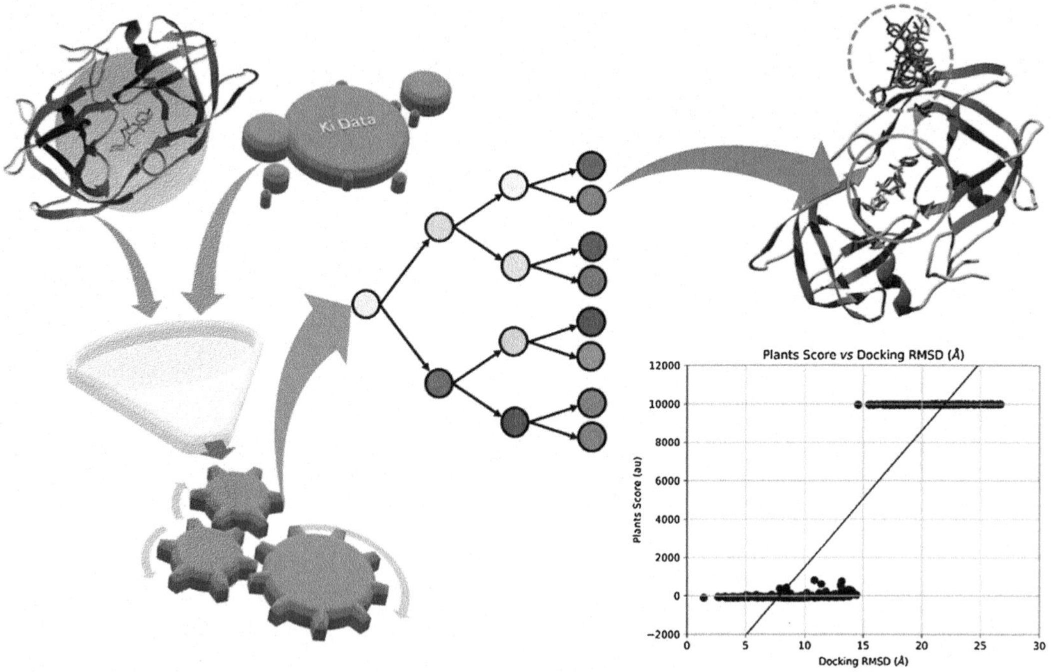

Fig. 1 Schematic illustrating the development of a target-based scoring function to predict the inhibition of HIV-1 Protease [17]

tuning its terms to adjust it to predict inhibition HIV-1 protease [17]. We could say that we are integrating computational systems biology and machine-learning techniques to improve the predictive power of scoring functions, which gives us the flexibility to test different scenarios for a specific biological system.

Figure 1 illustrates the main ideas behind the application of the program SAnDReS for the development of a targeted scoring function. Briefly, we start with the downloading of crystallographic structures of protein target for which ligand-binding data are available. This dataset should have at least 20 different structures; we need to have enough data to have training and test sets. We use the training set to calibrate our scoring function through regression analysis and the test set to evaluate the predictive performance of the scoring function using data not employed for the calibration of the model. The program SAnDReS uses a polynomial equation composed of up to nine explanatory variables. This polynomial empirical scoring function was first described in the development of the program Polscore [72, 73]. Briefly, we consider three energy terms available in the standard scoring functions of docking programs such as Molegro Virtual Docker [32–34], AutoDock4 [35–38], and AutoDock Vina [39]. We take these energy terms as the explanatory variable x_1, x_2, and x_3 and build a polynomial equation as follows:

$$\begin{aligned} f = {} & \gamma_0 + \gamma_1 x_1 + \gamma_2 x_2 + \gamma_3 x_3 \\ & + \gamma_4 x_1 x_2 + \gamma_5 x_1 x_3 + \gamma_6 x_2 x_3 \\ & + \gamma_7 x_1^2 + \gamma_8 x_2^2 + \gamma_9 x_3^2 \end{aligned} \tag{1}$$

where f is the predicted binding affinity, γ_0 the regression constant, the other γs are the relative weights of each explanatory variable of the polynomial equation. Considering that we have nine regression weights for the explanatory variables, the program SAnDReS generates a total of $2^9-1 = 511$ polynomial equations. The predictive performance is determined by statistical analysis using Spearman's rank (ρ) and Pearson (R) correlation coefficients.

Besides the development of machine-learning models based on the polynomial equational with a combination of three explanatory variables, SAnDReS allows the generation of computational models with a higher number of explanatory variables; in this case, without the combination of quadratic or mixed terms of explanatory variables.

3 Supervised Machine-Learning Methods

In the development of a machine-learning model to predict the binding affinity, for instance, the goal is to determine the relative weight (γ_j) of the explanatory variables, to bring the predicted values (f_i) close to the experimental values (y_i). Below we indicate the Eq. 2. In this equation, we have the response variable (f) expressed as a function of the explanatory variables (x_j),

$$f(x_1, \ldots, x_N) = \gamma_0 + \sum_{j=1}^{N} \gamma_j x_j \tag{2}$$

where N indicates the number of explanatory variables and γ_0 represents the regression constant. The explanatory variables could have complex forms, as shown in Eq. 1, where we have mixed and quadratic terms.

Among the supervised machine-learning techniques, the oldest method is the ordinary linear regression method. The first statement of the ordinary linear regression method comes out in the form of an appendix entitled "Sur la Méthode des moindres quarrés" in Legendre's Nouvelles méthodes pour la détermination des orbites des comètes, Paris 1805 [74]. Legendre originally proposed this method in 1805 in a study of orbits of comets. The significant progress in the research of celestial mechanics that occurred during the early years of the nineteenth century was mainly due to the development of the ordinary linear regression method. The basic idea behind the ordinary linear regression method is to minimize the cost function known as the residual sum of squares (RSS). Some authors call this cost function the sum of squared residuals (SSR) [75, 76]. The equation for RSS is as follows:

$$\mathrm{RSS} = \sum_{i=1}^{M} \left(y_i - f(x_1, \ldots, x_N)\right)^2 \tag{3}$$

In the above equation, M is the number of observations, y_i is the experimental value, and f_i is the predicted value. RSS is the sum of the differences between the experimental value (y_i) and the predicted value (f_i). The regression method optimizes the weights (γ_j) in Eq. 2 to minimize the RSS.

We could achieve improvements in the predictive performance of the original ordinary linear regression method by adding terms to the RSS equation. Tikhonov [77] proposed a variation of the ordinary linear regression method in 1963; this method is named Ridge method. In the Ridge method, we add a penalty term to the original expression of RSS (Eq. 3). The penalty term takes a form of a sum of the squared weights (γs), as follows:

$$\mathrm{RSS} = \sum_{i=1}^{M} \left(y_i - f(x_1, \ldots, x_N)\right)^2 + \lambda_2 \sum_{j=1}^{N} \left|\gamma_j\right|^2 \tag{4}$$

In the above equation, $\lambda_2 \neq 0$ is the regularization parameter. The second summation is taken over all regression weights (γs). The Ridge method performs L2 regularization.

Tibshirani developed another variation of the ordinary linear regression method in 1996 [78]. This new regression method is called the least absolute shrinkage and selection operator; also Lasso or LASSO. The Lasso method adds a term involving the sum of the absolute values of the relative weights to the RSS equation, as indicated below,

$$\mathrm{RSS} = \sum_{i=1}^{M} \left(y_i - f(x_1, \ldots, x_N)\right)^2 + \lambda_1 \sum_{j=1}^{N} \left|\gamma_j\right| \tag{5}$$

As observed for Eq. 4, the second summation considers the γs. In Eq. 5, the term $\lambda_1 \neq 0$ indicates a coefficient responsible for controlling the strength of the penalty. The more significant is the value of the penalty; the higher is the shrinkage. We call this additional term added to the original RSS equation as the penalty term.

In Lasso method, the regression carries out the L1 regularization. This method can generate sparse models with fewer coefficients when compared with the ordinary linear regression method. Furthermore, some factors can be zero. When we increase the penalties, the consequences are coefficient values closer to zero. This situation is ideal for producing models with fewer explanatory variables.

In 2005, Zou and Hastie [79] proposed a combination of the Ridge and Lasso methods in one equation as follows:

$$\text{RSS} = \sum_{i=1}^{M} \left(y_i - f(x_1, \ldots, x_N)\right)^2 + \lambda_1 \sum_{j=1}^{N} \left|\gamma_j\right| + \lambda_2 \sum_{j=1}^{N} \left|\gamma_j\right|^2 \quad (6)$$

In the above equation, the terms $\lambda_1 \neq 0$ and $\lambda_2 \neq 0$ are the two regularization parameters.

These supervised machine-learning methods are available in the scikit-learn library [9] and implemented in the program SAnDReS [20].

4 Scoring Functions

To illustrate the potential of the use of supervised machine-learning methods in the improvement of the predictive performance of conventional scoring functions, we will describe the AutoDock4 and MolDock scoring functions. We can use the energy terms found in the scoring functions of these docking programs as explanatory variables in a machine-learning model targeted to a specific protein.

The program AutoDock4 [37, 38] employs a semiempirical free energy force field scoring function to evaluate the binding affinities of protein-ligand complexes. The pairwise energetic terms of the equation of the AutoDock4 scoring function (V) are determined as follows:

$$V = \gamma_{\text{vdw}} \sum_{i,j} \left(\frac{A_{ij}}{r_{ij}^{12}} - \frac{B_{ij}}{r_{ij}^{6}} \right) + \gamma_{\text{HB}} \sum_{i,j} E(t) \left(\frac{C_{ij}}{r_{ij}^{12}} - \frac{D_{ij}}{r_{ij}^{10}} \right) + \gamma_{\text{elec}} \sum_{i,j} \frac{q_i q_j}{\varepsilon(r_{ij}) r_{ij}} + \gamma_{\text{sol}} \sum_{i,j} \left(S_i V_j + S_j V_i\right) e^{\left(-r_{ij}^2/2\sigma^2\right)} + \gamma_{\text{tor}} N_{\text{tor}} \quad (7)$$

In the above equation, the γ's indicate the relative weight of each energy term. The first energy term evaluates the van der Waals potential using the Lennard-Jones approximation [80]. The second term calculates the hydrogen bond potential using a variation of Lennard-Jones based on a 10/12 potential. The third term is the Coulombic electrostatic potential. The fourth term represents the desolvation potential, and the final term considers the number of rotatable bonds in the ligand. In the above equation, summation operates over all pairs of ligand atoms (i) and protein atoms (j) besides all pairs of atoms in the ligand that are apart by three or more bonds.

The docking program Molegro Virtual Docker (MVD) employs the scoring function MolDock Score (V). The MolDock Score is as follows:

$$V = V_{\text{inter}} + V_{\text{intra}} \tag{8}$$

where V_{inter} is the intermolecular energy of the ligand–protein interaction and is determined by the following equation:

$$V_{\text{inter}} = \sum_{i \in \text{ligand}}^{\text{M1}} \sum_{j \in \text{receptor}}^{\text{M2}} \left[V_{\text{PLP}}(r_{ij}) + 332.0 \frac{q_i q_j}{4 r_{ij}} \right] \tag{9}$$

In the above equation, the limits M1 and M2 refer to the quantities of atoms of the ligand and receptor. The component V_{PLP} indicates the piecewise linear potential [32] and r_{ij} is the interatomic distance. The last term in the equation shows the Coulombic electrostatic potential, q_i being the electric charges for the ligand and q_j the receptor charge.

The component V_{intra} indicates the intramolecular energy, as follows:

$$V_{\text{intra}} = \sum_{i \in \text{ligand}}^{\text{M1}} \sum_{j \in \text{ligand}}^{\text{M1}} V_{\text{PLP}}(r_{ij}) + \sum_{\text{flexible bonds}} A[1 - \cos(m \cdot \theta - \theta_0)] + V_{\text{clash}} \tag{10}$$

In the above equation, the M1 and r_{ij} terms have the same meaning as the Eq. 9, in this equation, the double summation is between all non-hydrogen atoms in the ligand M1. The second part is a torsional energy term, determined by torsional angles present in the ligand. The component θ is the torsional angle of the bond and the terms m, θ_0, and A have been previously described elsewhere [32]. Moreover, the V_{class} term is a penalty term of 1000, if the intra-atomic distance is less than 2.0 Å.

5 Statistical Analysis

To evaluate the predictive performance of the machine-learning models, we employ two correlation coefficients, the squared correlation coefficient (R^2) and the Spearman's rank correlation coefficient (ρ) [81]. We calculate the coefficient R^2 by the following equation:

$$R^2 = 1 - \frac{\text{RSS}}{\text{TSS}} \tag{11}$$

The residual sum of squares (RSS) is determined by Eqs. 3–6, depending on the machine-learning method. We calculate the total sum of squares (TSS) as follows:

$$\text{TSS} = \sum_{i=1}^{N} (y_i - \langle y \rangle)^2 \tag{12}$$

The variables y_i are the experimental observations, $<y>$ is the mean value for y, and N the number of observations. We define the Spearman's rank correlation coefficient (ρ) by the following expression:

$$\rho = 1 - \frac{6\sum_{i=1}^{N} d_i^2}{N(N^2 - 1)} \tag{13}$$

In the above equation, the term d_i indicates the difference in the ranks for a given observation [20].

In the analysis of the predictive performance of machine-learning models, it is common to evaluate the root mean squared error (RMSE) defined as follows:

$$\text{RMSE} = \sqrt{\frac{1}{N}\sum_{i=1}^{N}\left(y_i - \langle y\rangle\right)^2} \tag{14}$$

As highlighted for the terms of Eq. 13, the variables y_i are the experimental data, $<y>$ is the mean value for y, and N the number of observations. RMSE is a quadratic scoring rule that also evaluated the average intensity of the error between the predicted and the experimental values.

6 CDK2 Dataset

Here we discuss the application of the machine-learning methods to predict binding affinity for CDK2. This enzyme has been intensively studied as a target for the development of anticancer drugs [40, 41, 82–85]. The first crystallographic structure of human CDK2 was determined in 1993 by Prof. Sung-Hou Kim and collaborators [86]. Structural analysis of the CDK2 showed a typical bilobal architecture of serine/threonine protein kinases (EC 2.7.11.1). Analysis of the CDK2 indicates that the N-terminal domain is mostly built by a distorted beta-sheet and a short alpha helix. A helix bundle forms the C-terminal. The two lobes of the CDK2 structure permit the binding of the ATP molecule [87], as we can see in Fig. 2.

Let's consider the development of a scoring function to predict binding affinity for CDK2. We used the program SAnDReS to develop this scoring function targeted to CDK2. We created a dataset of CDK2 for which crystallographic and inhibition constant (K_i) data are available. We identified a total of 27 structures satisfying both criteria. Table 1 shows the PDB access codes and the ligand data for each structure.

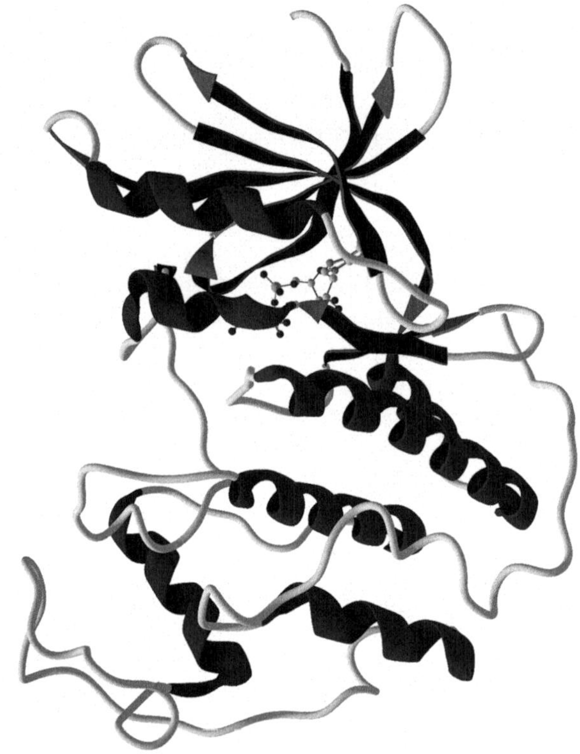

Fig. 2 Crystallographic structure of human CDK2 in complex with ATP. This figure was generated using Molegro Virtual Docker (MVD) [32]. PDB access code: 1HCK [87]

Table 1
List of the structures used to build machine-learning models for human CDK2 dataset

PDB	Ligand Code	Ligand Chain	Ligand Number	K_i (nM)	Test Set
1E1V	CMG	A	401	8400	1
1E1X	NW1	A	401	1300	1
1H1S	4SP	A	1298	6	0
1JSV	U55	A	400	2000	0
1PXN	CK6	A	500	195	1
1PXO	CK7	A	500	2	0
1PXP	CK8	A	500	220	0
1PYE	PM1	A	700	386	0
3DDQ	RRC	A	299	250	0

(continued)

Table 1
(continued)

PDB	Ligand Code	Ligand Chain	Ligand Number	K_i (nM)	Test Set
2CLX	F18	A	1299	13,300	0
2EXM	ZIP	A	400	78,000	0
2FVD	LIA	A	299	3	0
2XMY	CDK	A	500	0.11	0
2XNB	Y8L	A	1299	149	1
3LFN	A27	A	299	3160	0
3LFS	A07	A	299	2500	1
3MY5	RFZ	A	300	65,000	0
4ACM	7YG	A	1302	210	0
4BCK	T3E	A	1298	4	0
4BCM	T7Z	A	1297	123	0
4BCN	T9N	A	1299	12	0
4BCO	T6Q	A	1299	131	1
4BCP	T3C	A	1299	568	0
4BCQ	TJF	A	1296	147	0
4EOP	1RO	A	301	890	0
4NJ3	2KD	A	301	140	0

We indicated the structures used as test set with "1" in the respective column

7 HIV-1 Protease Dataset

In this chapter, we also examine the development of a machine learning model for the prediction of the inhibition of HIV-1 protease (Enzyme Classification, (EC) 3.4.23.16). This enzyme is an essential target for the development of drugs to treat infection by the type 1 human immunodeficiency virus (HIV-1), for reviews see [88, 89]. The HIV-1 protease is a member of the aspartyl protease family, and its activity is necessary for the breaking of a chemical bond in the Gag and Gag-Pol polyprotein precursors during HIV-1 infection. Different from other members of the aspartyl protease family [90], the HIV-1 protease shows a dimeric quaternary structure [91, 92]. Its quaternary structure has two identical symmetrical subunits (each 99 residues long) [92]. Each HIV-1 protease monomer shows three domains: a flap domain (residues 33-62), a

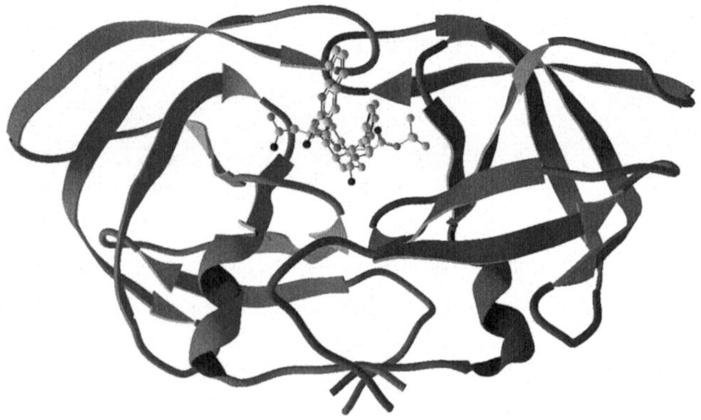

Fig. 3 Crystallographic structure of HIV-1 protease in complex with FDA-approved drug saquinavir. This figure was generated using Molegro Virtual Docker (MVD) [32]. PDB access code: 3D1Y [93]

core domain (10-32 and 63-85), and a terminal domain (1-4 and 96-99). Figure 3 brings the dimeric structure of HIV-1 protease with the inhibitor saquinavir bound in the cleft between the chains [93]. This HIV-1 protease inhibitor (brand name: Invirase) was developed by F. Hoffmann-La Roche Ltd. (Basel, Switzerland). The inhibitor saquinavir was the first FDA-approved HIV-1 protease inhibitor employed for the treatment of HIV-1 infection [94].

From the machine-learning standpoint, HIV-1 proteases comprise an appealing protein target for a combined analysis of three-dimensional data and ligand-binding affinity information. A recent study of the structures of HIV-1 protease available in the protein data bank [95] indicated that there are over 500 crystallographic structures for HIV-1 protease, a search carried out on February 2, 2019. Since PDB permits to filter data for inhibition constant (K_i), we can link crystallographic structures with affinity information and build up a dataset with structures for which inhibition data are known. This abundance of functional and crystallographic information opens the possibility for the development of a machine-learning model to predict ligand-binding affinity for this target protein.

In a recent publication [17], we described the use of the program SAnDReS to develop a targeted scoring function for HIV protease. We built a dataset of HIV-1 protease, for which crystallographic structures and inhibition constant (K_i) data are available. There are 70 structures in this dataset. Table 2 shows the PDB access codes and the ligand data for each structure. We describe the details about the predictive performance of this machine-learning model in Subheading 8.

Table 2
List of the structures used to build machine-learning models for HIV-1 protease dataset

PDB	Ligand Code	Ligand Chain	Ligand Number	K_i (nM)	Test Set
1A8G	2Z4	A	100	7.4	0
1AJV	NMB	A	501	20.05	0
1AJX	AH1	A	500	12.2	1
1BWB	146	B	641	1.911	1
1D4H	BEH	B	501	0.1	1
1D4I	BEG	A	501	1.4	0
1D4J	MSC	B	501	4.4	0
1D4K	PI8	A	201	0.6	1
1D4L	PI9	A	201	1.7	0
1D4Y	TPV	A	501	0.008	0
1EBW	BEI	A	501	0.9	0
1EBY	BEB	B	501	0.2	0
1EBZ	BEC	B	501	0.4	0
1EC0	BED	A	501	3.2	1
1EC1	BEE	A	501	1.2	0
1EC2	BEJ	B	501	0.15	1
1EC3	MS3	A	501	0.92	1
1G35	AHF	B	501	7.3	0
1HIH	C20	B	101	9	0
1HPO	UNI	B	100	0.666667	0
1HVH	Q82	B	265	11	0
1HXW	RIT	B	301	0.015	1
1IIQ	0ZR	A	201	355.333	0
1MTR	PI6	B	101	4	0
1ODW	0E8	A	201	100	0
1ODY	LP1	A	201	8	1
1PRO	A88	A	301	0.005	0
1TCX	IM1	A	400	112	0
1VIK	BAY	B	201	0.3	0
1W5V	BE3	A	1100	7.1	0

(continued)

Table 2
(continued)

PDB	Ligand Code	Ligand Chain	Ligand Number	K_i (nM)	Test Set
1W5W	BE4	A	1100	1.6	0
1W5X	BE5	A	501	4	0
1W5Y	BE6	A	1100	3.3	1
1XL2	189	A	1001	1500	1
1XL5	190	B	1001	45	0
1ZJ7	0ZT	A	201	57.3833	0
1ZSF	0ZS	B	201	0.12	0
1ZTZ	CB5	B	1002	66	0
2AID	THK	A	201	15,000	0
2AVM	2NC	B	300	2000	1
2AVS	MK1	B	902	113.013	0
2BPV	1IN	B	902	21.2	0
2BPY	3IN	B	902	39.8	0
2BQV	A1A	A	1100	9	1
2CEJ	1AH	B	1200	2.4	0
2CEM	2AH	B	1200	12	0
2CEN	4AH	B	1200	5	1
2HS1	017	A	201	3.3	0
2PYN	1UN	A	1001	4.5	1
2RKG	AB1	B	501	8.2	1
2UPJ	U02	A	100	41	0
2UXZ	HI1	A	1100	3.3	0
2UY0	HV1	B	1200	120	0
2WKZ	5AH	B	1200	1.7	0
3AID	ARQ	A	401	137	0
3D1Y	ROC	A	201	32.26	0
3MXD	K53	A	200	1.47	0
3MXE	K54	A	200	0.097	1
3OXX	DR7	A	100	0.2845	0
3QAA	G04	A	401	0.0029	1
3QIP	NVP	A	561	18,200	0

(continued)

Table 2
(continued)

PDB	Ligand Code	Ligand Chain	Ligand Number	K_i (nM)	Test Set
3UPJ	U03	A	100	560	0
4CP7	9 MW	A	1101	7.8	0
4FE6	0TQ	A	200	0.2	1
4HE9	G52	A	401	3.5	0
4U8W	G10	A	201	0.0058	0
4UPJ	U04	A	100	160	0
5UPJ	UIN	B	100	75	1
6UPJ	NIU	A	100	480	0
7UPJ	INU	A	100	3.15	0

We indicated the structures used as test set with "1" in the respective column

8 Development of Scoring Functions for CDK2

We carried out all ligand-binding evaluations using the crystallographic positions of the ligand and the protein. The charges were assigned using the Partial Equalization of Orbital Electronegativity (PEOE) algorithm [96] available in the program AutoDockTools4 tools [38] for the binding affinity evaluation using AutoDock 4. For the MVD, we used the default values of charges of the MolDock scoring function.

The Polscore methodology implemented in the program SAnDReS [20] makes it possible to test different scoring schemes, using polynomial equations where their terms are taken from the original scoring functions generated by the molecular docking programs. Here, we consider a polynomial equation involving the energy terms available in the program AutoDock4 [37, 38]. We generated 511 polynomial equations with the program SAnDReS; the highest correlation among them was observed for the polynomial scoring function number 504 (Polscore#514). Table 3 brings the predictive performance of the scoring functions (Free Energy Score [AutoDock4], MolDock Score [MVD], Ligand Efficiency Scores 1 and 3 [MVD], and PolScore#514 [SAnDReS]).

The values of ρ range from −0.057 to 0.629, the highest correlation obtained for the PolScore504. This polynomial equation was obtained through a regression analysis using the elastic net method available in the program SAnDReS. This predictive model uses as explanatory variables the energy terms found in the AutoDock4 scoring function (vdW+Hbond+desolv Energy [T_1], final

Table 3
Predictive performance for the structures (CDK2 dataset) in the training set

Scoring Function	P	*p*-value (ρ)	RMSE	R2	*p*-value (R2)
Free energy score (AutoDock4)	0.242	0.2915	6049.29	0.046	0.3503
MolDock score (MVD)	0.226	0.3246	112.038	0.073	0.2374
Ligand efficiency 1 score (MVD)	−0.057	0.8057	2.78186	0	0.936
Ligand efficiency 3 score (MVD)	0.229	0.319	3.52467	0.067	0.2561
Polscore#514 (SAnDReS)	0.629	0.002274	1.1453	0.382	0.002839

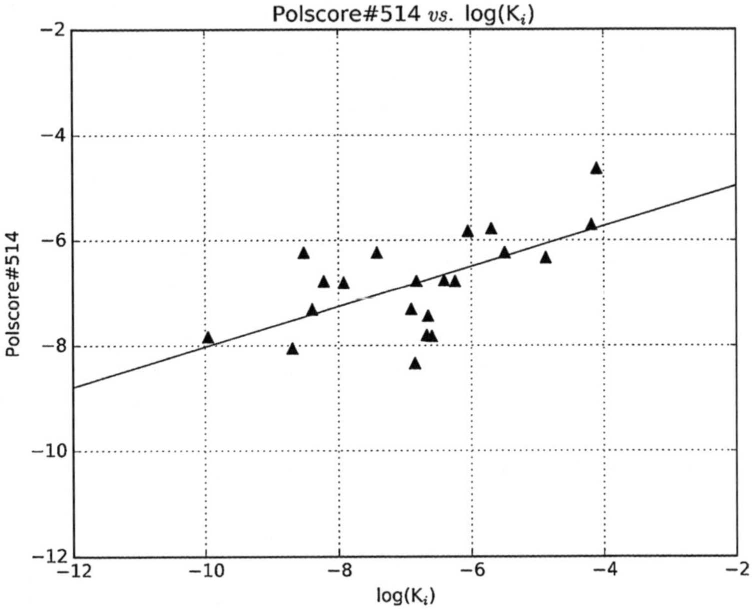

Fig. 4 Scatter plot for experimental and predicted binding affinities. We used the program SAnDReS to generate this plot

total internal energy [T_2], torsional free energy [T_3]). This polynomial equation (Polscore#514) has the following expression,

$$\text{PBA} = -3.061068 - 0.000159T_1 - 0.018819T_2 - 1.785568T_3$$

where PBA means predicted binding affinity (PBA = log [K_i]). Figure 4 shows the scattering plot for the PBA (Polscore#504) and the experimental binding affinity (log [K_i]).

To further validate the predictive performance of the Polscore#504, we calculated the binding affinity using structures of the test set, not used to obtain the relative weights of the polynomial equation. Table 4 brings the statistical analysis of the predictive

Table 4
Predictive performance for the structures (CDK2 dataset) in the test set

Scoring Function	*P*	*p*-value (ρ)	RMSE	R2	*p*-value (R2)
Free energy score (AutoDock4)	0.143	0.7872	843.736	0.124	0.4929
MolDock score (MVD)	0.771	0.0724	103.542	0.731	0.03004
Ligand efficiency 1 score (MVD)	−0.6	0.208	1.75141	0.131	0.4801
Ligand efficiency 3 score (MVD)	−0.314	0.5441	2.16765	0.115	0.5107
Polscore#514 (SAnDReS)	0.771	0.0724	0.797785	0.335	0.2291

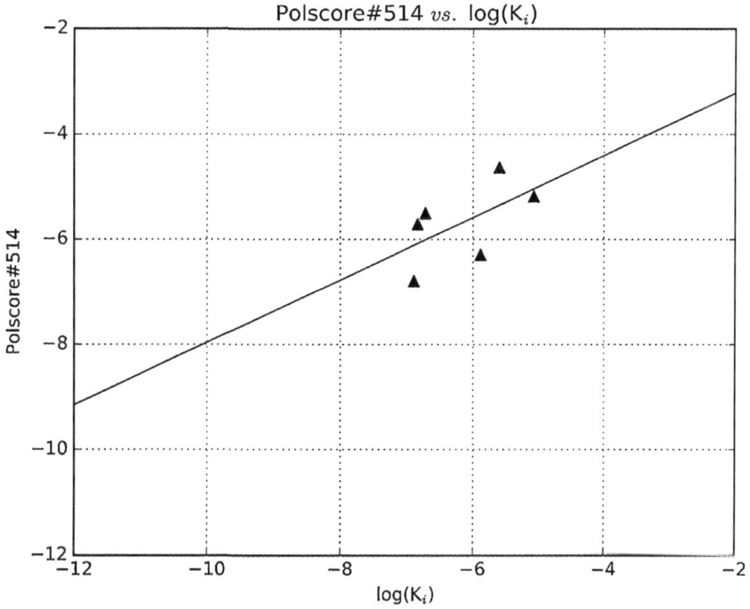

Fig. 5 Scatter plot for experimental and predicted binding affinities. We used the program SAnDReS to generate this plot

performance for the test set. The ρ ranges from −0.6 to 0.771, the highest correlations obtained for the MolDock scoring function and Polscore#504. Analysis of the RMSE values indicated that Polscore#504 has the lowest value, which suggests that this machine-learning model has superior performance when compared with the native scoring functions available in the programs MVD and AutoDock4. Figure 5 brings the scatter plot for the PBA (Polscore#504) and the experimental binding affinity (log(K_i) for the test set.

As we can see for the CDK2 system, the application of the machine-learning technique generated a model with superior predictive power when compared with standard scoring functions available in the programs AutoDock4 and MVD.

9 Development of Scoring Functions for HIV-1 Protease

In our previously published study, we employed the crystallographic position of ligands for the structures in HIV-1 protease dataset and applied machine-learning techniques using as explanatory variables the scoring functions and energy terms available in the program MVD [32–34] to predict binding affinity.

We show the statistical analysis (ρ for training (a) and test (b) sets) of the predictive performance of the MVD scoring functions and the best machine-learning model in Table 5. The polynomial scoring number 504 presents the most significant correlation (ρ). As we can see in Table 5, the predictive performance of the polynomial scoring function is superior to MVD scoring functions. Below we have polynomial equation 504 (Polscore#504), with coefficients determined by regression analysis,

$$\begin{aligned}\text{PBA} = & -5.685144 + 0.011990T_1 + 0.004743T_2 \\ & + 0.001676T_3 + 0.000024T_1T_2 + 0.000106T_1T_3 \\ & + 000040T_2T_3\end{aligned}$$

where T_1 is the PLANTS score function, T_2 is the interaction energy term of the MolDock scoring function, and T_3 is the ligand efficiency 3 score. These all scoring functions were determined with the program MVD [32–34] and combined as a polynomial equation with hybrid terms with the program SAnDReS [20]. We obtained the above-described model using ordinary linear regression available in the scikit-learn library [9].

The highest regression coefficient in the machine-learning model (Polscore#504) is the PLANTS Score. Moreover, among three hybrid terms of the machine-learning model, two explanatory variables (T_1T_2 and T_1T_3) have the contribution of PLANTS Score. A previous study indicated that this scoring function is frequently superior to the other scores at estimating binding affinity [97], which also observed in the HIV-1 protease dataset.

Table 5
Predictive performance for the structures HIV-1 protease dataset

Scoring Function	ρ(a)	*p*-value(a)	ρ(b)	*p*-value(b)
MolDock score (MVD)	0.218	$1.247.10^{-1}$	0.086	$7.193.10^{-1}$
Ligand efficiency 1 score (MVD)	0.187	$1.886.10^{-1}$	0.256	$2.750.10^{-1}$
Ligand efficiency 3 score (MVD)	0.045	$7.559.10^{-1}$	−0.140	$5.563.10^{-1}$
Polscore#504 (SAnDReS)	0.525	$7.707.10^{-5}$	0.368	$1.106.10^{-1}$

10 Availability

Program SAnDReS is available at https://github.com/azevedolab/sandres.

11 Colophon

We employed the program MVD [32] to generate Figs. 1–3. We created Figs. 4 and 5 using the program SAnDReS [20]. We performed the modeling reported on this chapter using a Desktop PC with 4GB of memory, a 1 TB hard disk, and an Intel® Core® i3-2120 at 3.30 GHz processor running Windows 8.1.

12 Final Remarks

The development of scoring functions to predict binding for protein-ligand complexes based on the atomic coordinates is a challenge from the computational point of view [4]. The use of standard scoring functions has successfully been used in the selection of docking poses.

On the other hand, application of docking scoring functions to predict binding affinity doesn't present reliable results [73]. In this chapter, we demonstrated recent successes in the development of targeted-scoring functions through machine-learning techniques implemented in the program SAnDReS [33]. These studies [13–18] indicated that the application of supervised machine-learning techniques to create scoring functions calibrated for a specific protein-ligand system of interest has superior predictive performance when compared with traditional scoring functions.

Acknowledgments

This work was supported by grants from CNPq (Brazil) (308883/2014-4). This study was financed in part by the Coordenação de Aperfeiçoamento de Pessoal de Nivel Superior—Brasil (CAPES)—Finance Code 001. GB-F acknowledges support from PUCRS/BPA fellowship. WFA is a senior researcher for CNPq (Brazil) (Process Numbers: 308883/2014-4 and 309029/2018-0).

References

1. Nanard M, Nanard J (1985) A user-friendly biological workstation. Biochimie 67:429–432
2. Hirst JD, King RD, Sternberg MJ (1994) Quantitative structure-activity relationships by neural networks and inductive logic programming. I. The inhibition of dihydrofolate reductase by pyrimidines. J Comput Aided Mol Des 8:405–420
3. Hirst JD, King RD, Sternberg MJ (1994) Quantitative structure-activity relationships by

neural networks and inductive logic programming. II. The inhibition of dihydrofolate reductase by triazines. J Comput Aided Mol Des 8:421–432
4. Heck GS, Pintro VO, Pereira RR, de Ávila MB, Levin NMB, de Azevedo WF (2017) Supervised machine learning methods applied to predict ligand-binding affinity. Curr Med Chem 24:2459–2470
5. Levin NM, Pintro VO, de Ávila MB, de Mattos BB, De Azevedo WF Jr (2017) Understanding the structural basis for inhibition of cyclin-dependent kinases. New pieces in the molecular puzzle. Curr Drug Targets 18:1104–1111
6. de Azevedo WF Jr (2016) Opinion paper: targeting multiple cyclin-dependent kinases (CDKs): a new strategy for molecular docking studies. Curr Drug Targets 17:2
7. Ain QU, Aleksandrova A, Roessler FD, Ballester PJ (2015) Machine-learning scoring functions to improve structure-based binding affinity prediction and virtual screening. Wiley Interdiscip Rev Comput Mol Sci 5:405–424
8. Xue LC, Dobbs D, Bonvin AM, Honavar V (2015) Computational prediction of protein interfaces: a review of data driven methods. FEBS Lett 589:3516–3526
9. Pedregosa F, Varoquaux G, Gramfort A, Michel V, Thirion B, Grisel O et al (2011) Scikit-learn: machine learning in python. J Mach Learn Res 12:2825–2830
10. Li H, Peng J, Leung Y, Leung KS, Wong MH, Lu G et al (2018) The impact of protein structure and sequence similarity on the accuracy of machine-learning scoring functions for binding affinity prediction. Biomolecules 8:12
11. Bitencourt-Ferreira G, de Azevedo WF Jr (2018) Development of a machine-learning model to predict Gibbs free energy of binding for protein-ligand complexes. Biophys Chem 240:63–69
12. Jiménez J, Škalič M, Martínez-Rosell G, De Fabritiis G (2018) KDEEP: protein-ligand absolute binding affinity prediction via 3D-convolutional neural networks. J Chem Inf Model 58:287–296
13. de Ávila MB, de Azevedo WF Jr (2018) Development of machine learning models to predict inhibition of 3-dehydroquinate dehydratase. Chem Biol Drug Des 92:1468–1474
14. Amaral MEA, Nery LR, Leite CE, de Azevedo Junior WF, Campos MM (2018) Pre-clinical effects of metformin and aspirin on the cell lines of different breast cancer subtypes. Invest New Drugs 36:782–796
15. Levin NMB, Pintro VO, Bitencourt-Ferreira G, Mattos BB, Silvério AC, de Azevedo WF Jr (2018) Development of CDK-targeted scoring functions for prediction of binding affinity. Biophys Chem 235:1–8
16. Freitas PG, Elias TC, Pinto IA, Costa LT, de Carvalho PVSD, Omote DQ et al (2018) Computational approach to the discovery of phytochemical molecules with therapeutic potential targets to the PKCZ protein. Lett Drug Des Discov 15:488–499
17. Pintro VO, Azevedo WF (2017) Optimized virtual screening workflow. Towards target-based polynomial scoring functions for HIV-1 protease. Comb Chem High Throughput Screen 20:820–827
18. de Ávila MB, Xavier MM, Pintro VO, de Azevedo WF (2017) Supervised machine learning techniques to predict binding affinity. A study for cyclin-dependent kinase 2. Biochem Biophys Res Commun 494:305–310
19. Zhang L, Ai HX, Li SM, Qi MY, Zhao J, Zhao Q et al (2017) Virtual screening approach to identifying influenza virus neuraminidase inhibitors using molecular docking combined with machine-learning-based scoring function. Oncotarget 8:83142–83154
20. Xavier MM, Heck GS, de Avila MB, Levin NM, Pintro VO, Carvalho NL et al (2016) SAnDReS a computational tool for statistical analysis of docking results and development of scoring functions. Comb Chem High Throughput Screen 19:801–812
21. Wójcikowski M, Ballester PJ, Siedlecki P (2017) Performance of machine-learning scoring functions in structure-based virtual screening. Sci Rep 7:46710
22. Sunseri J, King JE, Francoeur PG, Koes DR (2019) Convolutional neural network scoring and minimization in the D3R 2017 community challenge. J Comput Aided Mol Des 33 (1):19–34. https://doi.org/10.1007/s10822-018-0133-y
23. Ragoza M, Hochuli J, Idrobo E, Sunseri J, Koes DR (2017) Protein-ligand scoring with convolutional neural networks. J Chem Inf Model 57:942–957
24. Hochuli J, Helbling A, Skaist T, Ragoza M, Koes DR (2018) Visualizing convolutional neural network protein-ligand scoring. J Mol Graph Model 84:96–108
25. Afifi K, Al-Sadek AF (2018) Improving classical scoring functions using random forest: the non-additivity of free energy terms' contributions in binding. Chem Biol Drug Des 92:1429–1434
26. Wang C, Zhang Y (2017) Improving scoring-docking-screening powers of protein-ligand

scoring functions using random forest. J Comput Chem 38:169–177

27. Li H, Leung KS, Wong MH, Ballester PJ (2015) Low-quality structural and interaction data improves binding affinity prediction via random forest. Molecules 20:10947–10962
28. Khamis MA, Gomaa W, Ahmed WF (2015) Machine learning in computational docking. Artif Intell Med 63:135–152
29. Li H, Leung KS, Wong MH, Ballester PJ (2015) Improving AutoDock Vina using random forest: the growing accuracy of binding affinity prediction by the effective exploitation of larger data sets. Mol Inform 34:115–126
30. Zilian D, Sotriffer CA (2013) SFCscore(RF): a random forest-based scoring function for improved affinity prediction of protein-ligand complexes. J Chem Inf Model 53:1923–1933
31. Ballester PJ, Mitchell JB (2010) A machine learning approach to predicting protein-ligand binding affinity with applications to molecular docking. Bioinformatics 26:1169–1175
32. Thomsen R, Christensen MH (2006) MolDock: a new technique for high-accuracy molecular docking. J Med Chem 49:3315–3321
33. Heberlé G, de Azevedo WF Jr (2011) Bio-inspired algorithms applied to molecular docking simulations. Curr Med Chem 18:1339–1352
34. De Azevedo WF Jr (2010) MolDock applied to structure-based virtual screening. Curr Drug Targets 11:327–334
35. Goodsell DS, Olson AJ (1990) Automated docking of substrates to proteins by simulated annealing. Proteins 8:195–202
36. Morris GM, Goodsell DS, Huey R, Olson AJ (1996) Distributed automated docking of flexible ligands to proteins: parallel applications of AutoDock 2.4. J Comput Aided Mol Des 10:293–304
37. Morris GM, Goodsell DS, Halliday RS, Huey R, Hart WE, Belew RK et al (1998) Automated docking using a Lamarckian genetic algorithm and empirical binding free energy function. J Comput Chem 19:1639–1662
38. Morris GM, Huey R, Lindstrom W, Sanner MF, Belew RK, Goodsell DS et al (2009) AutoDock4 and AutoDockTools4: automated docking with selective receptor flexibility. J Comput Chem 30:2785–2791
39. Trott O, Olson AJ (2010) AutoDock Vina: improving the speed and accuracy of docking with a new scoring function, efficient optimization, and multithreading. J Comput Chem 31:455–461
40. Kim SH, Schulze-Gahmen U, Brandsen J, de Azevedo Júnior WF (1996) Structural basis for chemical inhibition of CDK2. Prog Cell Cycle Res 2:137–145
41. De Azevedo WF Jr, Mueller-Dieckmann HJ, Schulze-Gahmen U, Worland PJ, Sausville E, Kim SH (1996) Structural basis for specificity and potency of a flavonoid inhibitor of human CDK2, a cell cycle kinase. Proc Natl Acad Sci U S A 93:2735–2740
42. De Azevedo WF, Leclerc S, Meijer L, Havlicek L, Strnad M, Kim SH (1997) Inhibition of cyclin-dependent kinases by purine analogues: crystal structure of human cdk2 complexed with roscovitine. Eur J Biochem 243:518–526
43. de Ávila MB, Bitencourt-Ferreira G, de Azevedo WF Jr (2018) Structural basis for inhibition of enoyl-[acyl carrier protein] reductase (InhA) from *Mycobacterium tuberculosis*. Curr Med Chem. https://doi.org/10.2174/0929867326666181203125229
44. Volkart PA, Bitencourt-Ferreira G, Souto AA, de Azevedo WF (2019) Cyclin-dependent kinase 2 in cellular senescence and cancer. A structural and functional review. Curr Drug Targets 20(7):716–726. https://doi.org/10.2174/1389450120666181204165344
45. Russo S, De Azevedo WF (2018) Advances in the understanding of the cannabinoid receptor 1 - focusing on the inverse agonists interactions. Curr Med Chem. https://doi.org/10.2174/0929867325666180417165247
46. Pinto-Junior VR, Osterne VJ, Santiago MQ, Correia JL, Pereira-Junior FN, Leal RB et al (2017) Structural studies of a vasorelaxant lectin from *Dioclea reflexa* Hook seeds: Crystal structure, molecular docking and dynamics. Int J Biol Macromol 98:12–23
47. Abbasi WA, Asif A, Ben-Hur A, Minhas FUAA (2018) Learning protein binding affinity using privileged information. BMC Bioinformatics 19:425
48. Kumari M, Tiwari N, Chandra S, Subbarao N (2018) Comparative analysis of machine learning based QSAR models and molecular docking studies to screen potential anti-tubercular inhibitors against InhA of *Mycobacterium tuberculosis*. Int J Comput Biol Drug Des 11:3
49. Masand VH, El-Sayed NNE, Bambole MU, Patil VR, Thakur SD (2019) Multiple quantitative structure-activity relationships (QSARs) analysis for orally active trypanocidal

N-myristoyltransferase inhibitors. J Mol Struct 1175:481–487

50. Maltarollo VG, Kronenberger T, Windshugel B, Wrenger C, Trossini GHG, Honorio KM (2018) Advances and challenges in drug design of PPARδ ligands. Curr Drug Targets 19:144–154
51. Lemos A, Melo R, Preto AJ, Almeida JG, Moreira IS, Dias Soeiro Cordeiro MN (2018) *In silico* studies targeting G-protein coupled receptors for drug research against Parkinson's disease. Curr Neuropharmacol 16:786–848
52. Ribeiro FF, Mendonca Junior FJB, Ghasemi JB, Ishiki HM, Scotti MT, Scotti L (2018) Docking of natural products against neurodegenerative diseases: general concepts. Comb Chem High Throughput Screen 21:152–160
53. Aleksandrov A, Myllykallio H (2019) Advances and challenges in drug design against tuberculosis: application of *in silico* approaches. Expert Opin Drug Discov 14:35–46
54. Safarizadeh H, Garkani-Nejad Z (2019) Investigation of MI-2 analogues as MALT1 inhibitors to treat of diffuse large B-Cell 0lymphoma through combined molecular dynamics simulation, molecular docking and QSAR techniques and design of new inhibitors. J Mol Struct 1180:708–722
55. Joy M, Elrashedy AA, Mathew B, Pillay AS, Mathews A, Dev S et al (2018) Discovery of new class of methoxy carrying isoxazole derivatives as COX-II inhibitors: Investigation of a detailed molecular dynamics study. J Mol Struct 1157:19–28
56. Leal RB, Pinto-Junior VR, Osterne VJS, Wolin IAV, Nascimento APM, Neco AHB et al (2018) Crystal structure of DlyL, a mannose-specific lectin from *Dioclea lasiophylla* Mart. Ex Benth seeds that display cytotoxic effects against C6 glioma cells. Int J Biol Macromol 114:64–76
57. Cavada BS, Araripe DA, Silva IB, Pinto-Junior VR, Osterne VJS, Neco AHB et al (2016) Structural studies and nociceptive activity of a native lectin from *Platypodium elegans* seeds (nPELa). Int J Biol Macromol 107:236–246
58. Usman MSM, Bharbhuiya TK, Mondal S, Rani S, Kyal C, Kumari R (2018) Combined protein and ligand based physicochemical aspects of molecular recognition for the discovery of CDK9 inhibitor. Gene Rep 13:212–219
59. Neco AHB, Pinto-Junior VR, Araripe DA, Santiago MQ, Osterne VJS, Lossio CF et al (2018) Structural analysis, molecular docking and molecular dynamics of an edematogenic lectin from *Centrolobium microchaete* seeds. Int J Biol Macromol 117:124–133
60. Nowaczyk A, Fijałkowski Ł, Zaręba P, Sałat K (2018) Docking and pharmacodynamic studies on hGAT1 inhibition activity in the presence of selected neuronal and astrocytic inhibitors. Part I. J Mol Graph Model 85:171–181
61. Tong J, Lei S, Qin S, Wang Y (2018) QSAR studies of TIBO derivatives as HIV-1 reverse transcriptase inhibitors using HQSAR, CoMFA and CoMSIA. J Mol Struct 1168:56–64
62. Azevedo LS, Moraes FP, Xavier MM, Pantoja EO, Villavicencio B, Finck JA et al (2012) Recent progress of molecular docking simulations applied to development of drugs. Curr Bioinform 7:352–365
63. Dias R, de Azevedo WF Jr (2008) Molecular docking algorithms. Curr Drug Targets 9:1040–1047
64. Breda A, Basso LA, Santos DS, de Azevedo WF Jr (2008) Virtual screening of drugs: score functions, docking, and drug design. Curr Comput Aided Drug Des 4:265–272
65. Böhm HJ (1993) A novel computational tool for automated structure-based drug design. J Mol Recognit 6:131–137
66. Böhm HJ (1994) The development of a simple empirical scoring function to estimate the binding constant for a protein-ligand complex of known three-dimensional structure. J Comput Aided Mol Des 8:243–256
67. Böhm HJ (1996) Towards the automatic design of synthetically accessible protein ligands: peptides, amides and peptidomimetics. J Comput Aided Mol Des 10:265–272
68. Stahl M, Böhm HJ (1998) Development of filter functions for protein-ligand docking. J Mol Graph Model 16:121–132
69. Klebe G, Böhm HJ (1997) Energetic and entropic factors determining binding affinity in protein-ligand complexes. J Recept Signal Transduct Res 17:459–473
70. Böhm HJ, Banner DW, Weber L (1999) Combinatorial docking and combinatorial chemistry: design of potent non-peptide thrombin inhibitors. J Comput Aided Mol Des 13:51–56
71. Korb O, Stützle T, Exner TE (2009) Empirical scoring functions for advanced protein-ligand docking with PLANTS. J Chem Inf Model 49:84–96
72. Dias R, Timmers LF, Caceres RA, de Azevedo WF Jr (2008) Evaluation of molecular docking using polynomial empirical scoring functions. Curr Drug Targets 9:1062–1070
73. de Azevedo WF Jr, Dias R (2008) Evaluation of ligand-binding affinity using polynomial empirical scoring functions. Bioorg Med Chem 16:9378–9382

74. Legendre AM (1805) Nouvelle méthodes pour la détermination des orbites des comètes. Courcier, Paris
75. Bell J (2015) Machine learning. Hands-on for developers and technical professionals. Wiley, Indianapolis, IN
76. Bruce P, Bruce A (2017) Practical statistics for data scientists. 50 essential concepts. O'Reilly Media, Sebastopol
77. Tikhonov AN (1963) On the regularization of ill-posed problems. Dokl Akad Nauk SSSR 153:49–52
78. Tibshirani R (1996) Regression shrinkage and selection via the lasso. J R Stat Soc Series B Stat Methodol 58:267–288
79. Zou H, Hastie T (2005) Regularization and variable selection via the elastic net. J R Stat Soc Series B Stat Methodol 67:301–320
80. Lennard-Jones JE (1931) Cohesion. Proc Phys Soc 43:461–482
81. Zar JH (1972) Significance testing of the Spearman rank correlation coefficient. J Am Stat Assoc 67:578–580
82. Morgan DO (1995) Principles of CDK regulation. Nature 374:131–134
83. Murray AW (1994) Cyclin-dependent kinases: regulators of the cell cycle and more. Chem Biol 1:191–195
84. Canduri F, de Azevedo WF Jr (2005) Structural basis for interaction of inhibitors with cyclin-dependent kinase 2. Curr Comput Aided Drug Des 1:53–64
85. Krystof V, Cankar P, Frysová I, Slouka J, Kontopidis G, Dzubák P et al (2006) 4-arylazo-3,5-diamino-1H-pyrazole CDK inhibitors: SAR study, crystal structure in complex with CDK2, selectivity, and cellular effects. J Med Chem 49:6500–6509
86. De Bondt HL, Rosenblatt J, Jancarik J, Jones HD, Morgan DO, Kim SH (1996) Crystal structure of cyclin-dependent kinase 2. Nature 363:595–602
87. Schulze-Gahmen U, De Bondt HL, Kim SH (1996) High-resolution crystal structures of human cyclin-dependent kinase 2 with and without ATP: bound waters and natural ligand as guides for inhibitor design. J Med Chem 39:4540–4546
88. Pang X, Liu Z, Zhai G (2014) Advances in non-peptidomimetic HIV protease inhibitors. Curr Med Chem 21:1997–2011
89. Berti F, Frecer V, Miertus S (2014) Inhibitors of HIV-protease from computational design. A history of theory and synthesis still to be fully appreciated. Curr Pharm Des 20:3398–3411
90. Canduri F, Teodoro LG, Fadel V, Lorenzi CC, Hial V, Gomes RA et al (2001) Structure of human uropepsin at 2.45 A resolution. Acta Crystallogr D Biol Crystallogr 57:1560–1570
91. Miller M, Jaskólski M, Rao JK, Leis J, Wlodawer A (1989) Crystal structure of a retroviral protease proves relationship to aspartic protease family. Nature 337:576–579
92. Navia MA, Fitzgerald PM, McKeever BM, Leu CT, Heimbach JC, Herber WK et al (1989) Three-dimensional structure of aspartyl protease from human immunodeficiency virus HIV-1. Nature 337:615–620
93. Liu F, Kovalevsky AY, Tie Y, Ghosh AK, Harrison RW, Weber IT (2008) Effect of flap mutations on structure of HIV-1 protease and inhibition by saquinavir and darunavir. J Mol Biol 381:102–115
94. Lv Z, Chu Y, Wang Y (2015) HIV protease inhibitors: a review of molecular selectivity and toxicity. HIV AIDS (Auckl) 7:95–104
95. Berman HM, Westbrook J, Feng Z, Gilliland G, Bhat TN, Weissig H et al (2000) The Protein Data Bank. Nucleic Acids Res 28:235–242
96. Gasteiger J, Marsili M (1980) Iterative partial equalization of orbital electronegativity—a rapid access to atomic charges. Tetrahedron 36:3219–3228
97. Korb O, Stutzle T, Exner TE (2009) Empirical scoring functions for advanced protein-ligand docking with PLANTS. J Chem Inf Model 49:84–96

Chapter 17

Exploring the Scoring Function Space

Gabriela Bitencourt-Ferreira and Walter Filgueira de Azevedo Jr.

Abstract

In the analysis of protein-ligand interactions, two abstractions have been widely employed to build a systematic approach to analyze these complexes: protein and chemical spaces. The pioneering idea of the protein space dates back to 1970, and the chemical space is newer, later 1990s. With the progress of computational methodologies to create machine-learning models to predict the ligand-binding affinity, clearly there is a need for novel approaches to the problem of protein-ligand interactions. New abstractions are required to guide the conceptual analysis of the molecular recognition problem. Using a systems approach, we proposed to address protein-ligand scoring functions using the modern idea of the scoring function space. In this chapter, we describe the fundamental concept behind the scoring function space and how it has been applied to develop the new generation of targeted-scoring functions.

Key words Scoring function, Scoring function space, Protein space, Chemical space, Machine learning, SAnDReS, Binding affinity

1 Introduction

Studies using machine-learning methodologies to create a novel scoring function demonstrated the superior predictive performance of these approaches when compared with standard scoring functions [1–14]. Most of the times, these studies revealed some structural features related to the success of the machine-learning models. Nevertheless, a general description of the reasons for the superior predictive performance of machine-learning models was lacking. Recently, we have proposed an elegant mathematical abstraction to establish a relationship between the chemical space and the protein space [14]. This bridge between these two spaces is named scoring function space. In this chapter, we describe the fundamental concepts behind the scoring function space. We also explain how we can use this novel concept to build robust machine-learning models to predict ligand-binding affinity based on the atomic coordinates of protein-ligand complexes.

In our explanation of the scoring function space, we need to review the significant features of the protein and chemical spaces.

Walter Filgueira de Azevedo Jr. (ed.), *Docking Screens for Drug Discovery*, Methods in Molecular Biology, vol. 2053, https://doi.org/10.1007/978-1-4939-9752-7_17, © Springer Science+Business Media, LLC, part of Springer Nature 2019

The first description of the protein space came out in 1970 [15]. The brief description of the concept had focused on the evolutionary relationships in protein sequences from the close related organisms. As the number protein three-dimensional structures grew in the next decades, the idea of protein space gained a structural view, with the description of the protein structure space, as depicted by Hou et al. 2005 [16]. Briefly, we could visualize the set of all possible protein folds as a finite protein space, where elements of this set with a similar overall structure are close in the schematic representation of this space. Considering kinase protein family, all members of this class of protein could be represented in a three-dimensional space where one axis could be the percentage of alpha helices in the structure, the second axis would represent the percentage of beta-sheet in the protein, and the third axis indicates the portion of the alpha/beta structure in the protein. Such mathematical representation of the protein space facilitate the overall analysis of protein folds and provides a systematic view of how to address elements of this space taking into account the proximity of a component to others of the same class. Figure 1 shows a simple scheme to represent a few elements of the protein space.

The concept of chemical space deals with small molecules that exist [17–22]. To build the chemical space, we may consider all

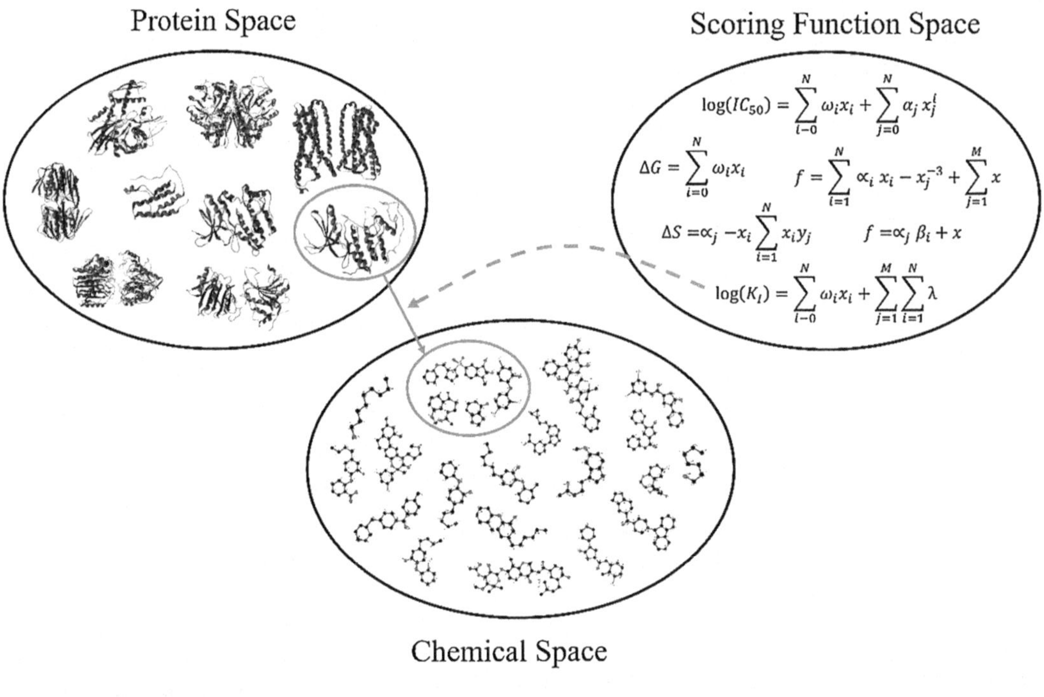

Fig. 1 Representation of the relationships involving protein space, chemical space, and scoring function space. A view of the scoring function space as a way to develop a computational model to predict the ligand-binding affinity. Structures of proteins available with the following PDB access codes: 2OW4, 2OVU, 2IDZ, 2GSJ, 2G85, 2A4I, 1ZTB, 1Z99, 1WE2, 1M73, 1FLH, and 1FHJ

viable molecules and chemical compounds which obey a given set of rules and limits on the number of rings, molecular weight, and the type of atoms. The prediction of the number of elements of the chemical space needs careful analysis of the type of small molecules we will consider to build the chemical space. Several authors believe that the chemical space is composed of Carbon, Hydrogen, Oxygen, Nitrogen, and Sulfur. Moreover, we may consider only molecules with up to 30 non-hydrogen atoms and molecular weight below 500 Da, and a maximum number of rings of four. With these conditions, we have approximately 10^{63} elements in the chemical space [17]. Next, in this chapter, we describe the relationship involving the chemical and protein spaces, and now we could access this relation using the novel concept of the scoring function space.

2 Scoring Functions Space

To establish a mathematical abstraction to describe the functioning of scoring functions, we make use of the scoring function space [14]. In this approach, we see protein-ligand interaction as a result of the relation between the protein space [14, 15] and the chemical space [17–22], and we propose to represent these sets as a unique complex system, where the application of computational methodologies may contribute to generating models to predict protein-ligand binding affinities. Such approaches have the potential to create novel semi-empirical force fields to predict binding affinity with superior predictive power when compared with standard methodologies.

We proposed to use the abstraction of a mathematical space composed of infinite computational models to predict ligand-binding affinity. We named this space as the scoring function space. By the use of supervised machine-learning techniques, it is possible to explore this scoring function space and build a computational model targeted to a specific biological system. For instance, we created targeted scoring functions for coagulation factor Xa [1], cyclin-dependent kinases [2, 8, 12], HIV-1 protease [10], estrogen receptor [7], cannabinoid receptor 1 [13], and 3-dehydroquinate dehydratase [6]. We have also developed a scoring function to predict Gibbs free energy of binding for protein-ligand complexes [4]. We developed the program SAnDReS to generate computational models to predict ligand-binding affinity. SAnDReS is an integrated computational tool to explore the scoring function space.

To understand the fundamental concepts behind the scoring function space, let's first consider the protein space composed of protein structures. This protein space can be represented by the protein structure space, as depicted by Hou et al. 2005 [16]. We take this limited protein space as a starting point to the application of the concept of scoring function space. Figure 1 captures the main ideas necessary to understand the scoring function space and its relationships with protein and chemical spaces. If we pick an element of the protein space, for instance, the cyclin-dependent kinase family, we may identify all ligands that bind to this protein.

Now, let's consider the chemical space, which is formed by small molecules that may bind or not to an element of the protein space. If we take into account a subspace of the chemical space composed of structures that attach to cyclin-dependent kinase family, it is easy to imagine an association involving the cyclin-dependent kinase and this subspace of the chemical space. We represent this relationship as an arrow from the protein space to the chemical space (Fig. 1).

Finally, we consider a mathematical space composed of infinite scoring functions; each element of this space is a mathematical function that uses the atomic coordinates of protein-ligand complexes to predict the binding affinity. In Fig. 1, we have an idealization of the scoring function space.

Moving forward, we propose that there exists at least one scoring function capable of predicting the ligand-binding affinity of the elements of the chemical space for a component of the protein space. We indicate this relationship in Fig. 1 as an arrow from the scoring function space to the arrow indicating the relation between CDK and the chemical space.

So, the basic idea is quite simple: we intend to identify an element of the scoring function space (computational model) that predicts the binding affinity of a component of the protein space (target protein) for all members of the subspace of the chemical space composed of ligands that bind to this target protein.

Under the light of the scoring function space, we may say that the development of machine-learning models for CDK2 and HIV-1 protease was achieved through the exploration of the scoring function space, where SAnDReS found the adequate model to predict binding affinity specific for each enzyme. Such an innovative approach to the analysis of the development of computational models to predict binding provides a robust mathematical framework to develop new predictive models.

3 SAnDReS

The program SAnDReS [1] makes use of supervised machine-learning techniques to generate polynomial equations to predict ligand-binding affinity, which allows improvement of native scoring functions. SAnDReS works through the training of a model, making it specific for a biological system (targeted scoring function).

The program SAnDReS applies a polynomial equation with up to nine explanatory variables. We described this equation in the development of the program Polscore [23, 24]. In the program SAnDReS, we consider three energy terms available in docking programs such as programs Molegro Virtual Docker [25–27], AutoDock4 [28–31], and AutoDock Vina [32]. We use these energy terms as explanatory variables. The regression polynomial equation is as follows:

$$\begin{aligned} \text{PBA} = {} & \alpha_0 + \alpha_1 x_1 + \alpha_2 x_2 + \alpha_3 x_3 \\ & + \alpha_4 x_1 x_2 + \alpha_5 x_1 x_3 + \alpha_6 x_2 x_3 \\ & + \alpha_7 x_1^2 + \alpha_8 x_2^2 + \alpha_9 x_3^2 \end{aligned} \tag{1}$$

where the response variable PBA is the predicted binding affinity, α_0 is the regression constant, the other αs are the relative weights of each explanatory variable in the computational model. Since we have nine weights for the explanatory variables, the program SAnDReS creates a total of $2^9-1 = 511$ computational models. We could think that we are exploring the scoring function space, searching for an adequate model where the predictive performance is assessed by statistical analysis using Spearman's rank (ρ) and Pearson (R) correlation coefficients [33].

4 Availability

Program SAnDReS is available at https://github.com/azevedolab/sandres.

5 Colophon

We employed the program MVD [25–27] to generate Fig. 1.

6 Final Remarks

The development of scoring functions to predict ligand-binding affinity lacked a formal basis for integrating a systems approach to the machine-learning techniques applied to calibrate the weights of novel computational models to predict binding affinity. With the application of the concepts behind the abstraction of scoring function space, we started to establish the basis for a systematic view of the development of computational models to predict binding affinity. Taken together, we may say that we live in a new age of the application of computational methods for drug discovery, where serendipity is gradually overcome by the systems approach to the design of drugs in silico.

Acknowledgments

This work was supported by grants from CNPq (Brazil) (308883/2014-4). This study was financed in part by the Coordenação de Aperfeiçoamento de Pessoal de Nivel Superior—Brasil (CAPES)—Finance Code 001. GB-F acknowledges support from PUCRS/BPA fellowship. WFA is a senior researcher for CNPq (Brazil) (Process Numbers: 308883/2014-4 and 309029/2018-0).

References

1. Xavier MM, Heck GS, de Avila MB, Levin NM, Pintro VO, Carvalho NL et al (2016) SAnDReS a computational tool for statistical analysis of docking results and development of scoring functions. Comb Chem High Throughput Screen 19:801–812
2. de Ávila MB, Xavier MM, Pintro VO, de Azevedo WF (2017) Supervised machine learning techniques to predict binding affinity. A study for cyclin-dependent kinase 2. Biochem Biophys Res Commun 494:305–310
3. Azevedo LS, Moraes FP, Xavier MM, Pantoja EO, Villavicencio B, Finck JA et al (2012) Recent progress of molecular docking simulations applied to development of drugs. Curr Bioinforma 7:352–365
4. Bitencourt-Ferreira G, de Azevedo WF Jr (2018) Development of a machine-learning model to predict Gibbs free energy of binding for protein-ligand complexes. Biophys Chem 240:63–69
5. Jiménez J, Škalič M, Martínez-Rosell G, De Fabritiis G (2018) KDEEP: protein-ligand absolute binding affinity prediction via 3D-convolutional neural networks. J Chem Inf Model 58:287–296
6. de Ávila MB, de Azevedo WF Jr (2018) Development of machine learning models to predict inhibition of 3-dehydroquinate dehydratase. Chem Biol Drug Des 92:1468–1474
7. Amaral MEA, Nery LR, Leite CE, de Azevedo Junior WF, Campos MM (2018) Pre-clinical effects of metformin and aspirin on the cell lines of different breast cancer subtypes. Investig New Drugs 36:782–796
8. Levin NMB, Pintro VO, Bitencourt-Ferreira G, Mattos BB, Silvério AC, de Azevedo WF Jr (2018) Development of CDK-targeted scoring functions for prediction of binding affinity. Biophys Chem 235:1–8
9. Freitas PG, Elias TC, Pinto IA, Costa LT, de Carvalho PVSD, Omote DQ et al (2018) Computational approach to the discovery of phytochemical molecules with therapeutic potential targets to the PKCZ protein. Lett Drug Des Discov 15:488–499
10. Pintro VO, Azevedo WF (2017) Optimized virtual screening workflow. Towards target-based polynomial scoring functions for HIV-1 protease. Comb Chem High Throughput Screen 20:820–827
11. de Ávila MB, Bitencourt-Ferreira G, de Azevedo WF Jr (2019) Structural basis for inhibition of enoyl-[acyl carrier protein] reductase (InhA) from *Mycobacterium tuberculosis*. Curr Med Chem. https://doi.org/10.2174/0929867326666181203125229
12. Volkart PA, Bitencourt-Ferreira G, Souto AA, de Azevedo WF (2019) Cyclin-dependent kinase 2 in cellular senescence and cancer. A structural and functional review. Curr Drug Targets 20(7):716–726. https://doi.org/10.2174/1389450120666181204165344
13. Russo S, De Azevedo WF (2019) Advances in the understanding of the cannabinoid receptor 1 - focusing on the inverse agonists interactions. Curr Med Chem. https://doi.org/10.2174/0929867325666180417165247
14. Heck GS, Pintro VO, Pereira RR, de Ávila MB, Levin NMB, de Azevedo WF (2017) Supervised machine learning methods applied to predict ligand-binding affinity. Curr Med Chem 24:2459–2470
15. Smith JM (1970) Natural selection and the concept of a protein space. Nature 225:563–564
16. Hou J, Jun SR, Zhang C, Kim SH (2005) Global mapping of the protein structure space and application in structure-based inference of protein function. Proc Natl Acad Sci U S A 102:3651–3656
17. Bohacek RS, McMartin C, Guida WC (1996) The art and practice of structure-based drug design: a molecular modeling perspective. Med Res Rev 16:3–50
18. Dobson CM (2004) Chemical space and biology. Nature 432:824–828

19. Kirkpatrick P, Ellis C (2004) Chemical space. Nature 432:823
20. Lipinski C, Hopkins A (2004) Navigating chemical space for biology and medicine. Nature 432:855–861
21. Shoichet BK (2004) Virtual screening of chemical libraries. Nature 432:862–865
22. Stockwell BR (2004) Exploring biology with small organic molecules. Nature 432:846–854
23. Dias R, Timmers LF, Caceres RA, de Azevedo WF Jr (2008) Evaluation of molecular docking using polynomial empirical scoring functions. Curr Drug Targets 9:1062–1070
24. de Azevedo WF Jr, Dias R (2008) Evaluation of ligand-binding affinity using polynomial empirical scoring functions. Bioorg Med Chem 16:9378–9382
25. Thomsen R, Christensen MH (2006) MolDock: a new technique for high-accuracy molecular docking. J Med Chem 49:3315–3321
26. Heberlé G, de Azevedo WF Jr (2011) Bio-inspired algorithms applied to molecular docking simulations. Curr Med Chem 18:1339–1352
27. De Azevedo WF Jr (2010) MolDock applied to structure-based virtual screening. Curr Drug Targets 11:327–334
28. Goodsell DS, Olson AJ (1990) Automated docking of substrates to proteins by simulated annealing. Proteins 8:195–202
29. Morris GM, Goodsell DS, Huey R, Olson AJ (1996) Distributed automated docking of flexible ligands to proteins: Parallel applications of AutoDock 2.4. J Comput Aided Mol Des 10:293–304
30. Morris GM, Goodsell DS, Halliday RS, Huey R, Hart WE, Belew RK et al (1998) Automated docking using a Lamarckian genetic algorithm and empirical binding free energy function. J Comput Chem 19:1639–1662
31. Morris GM, Huey R, Lindstrom W, Sanner MF, Belew RK, Goodsell DS et al (2009) AutoDock4 and AutoDockTools4: automated docking with selective receptor flexibility. J Comput Chem 30:2785–2791
32. Trott O, Olson AJ (2010) AutoDock Vina: improving the speed and accuracy of docking with a new scoring function, efficient optimization, and multithreading. J Comput Chem 31:455–461
33. Zar JH (1972) Significance testing of the Spearman rank correlation coefficient. J Am Stat Assoc 67:578–580

INDEX

Walter Filgueira de Azevedo Jr. (ed.), *Docking Screens for Drug Discovery*, Methods in Molecular Biology, vol. 2053, https://doi.org/10.1007/978-1-4939-9752-7, © Springer Science+Business Media, LLC, part of Springer Nature 2019

N

O

P

Q

R

S

T

Printed by Printforce, the Netherlands